Cognitive Cloud Computing

This book serves as a comprehensive guide, covering the fundamentals of cloud computing, advanced concepts, and practical applications.

- Discusses cyber-physical systems, cloud integration for environmental management, and cloud security intelligence systems for sustainable operations.
- Presents case studies based on computational intelligence-based optimization for sustainable operations and cloud integration.
- Emphasizes how cloud computing revolutionizes traditional processes, enabling direct input of assembly details into a computation model, streamlining development cycles, and reducing costs.
- Bridges the gap between theory and practice by offering guidelines on problem encoding and implementation strategies, empowering readers to apply their acquired knowledge to solve complex industrial problems.
- Covers wireless security in the cloud era, mechatronics and cloud integration, and cloud-enabled manufacturing for eco-friendly operations.

The text is primarily written for senior undergraduates, graduate students, and academic researchers in electrical engineering, electronics and communications engineering, computer engineering, and information technology.

Computational and Intelligent Systems

Series Editor: Ramesh Chandra Poonia
CHRIST (Deemed to be University), Bangalore, Karnataka, India

About the Series:
In today's world, the systems that integrate intelligence into machine-based applications are known as intelligent systems. In order to simplify the man–machine interaction, intelligent systems play an important role. The books under the proposed series will explain the fundamentals of intelligent systems, review the computational techniques and also offer step-by-step solutions of the practical problems. Aimed at senior undergraduate students, graduate students, academic researchers and professionals, the proposed series will focus on broad topics including artificial intelligence, deep learning, Image processing, cyber physical systems, wireless security, mechatronics, cognitive computing, and industry 4.0.

Application of Soft Computing Techniques in Mechanical Engineering
Amar Patnaik, Vikas Kukshal, Pankaj Agarwal, Ankush Sharma and Mahavir Choudhary

Computational Intelligence based Optimization of Manufacturing Process for Sustainable Materials
Deepak Sinwar, Kamalakanta Muduli, Vijaypal Singh Dhaka and Vijander Singh

Advanced Mathematical Techniques in Computational and Intelligent Systems
Sandeep Singh, Aliakbar Montazer Haghighi and Sandeep Dalal

Natural Language Processing and Information Retrieval
Principles and Applications
Edited By Muskan Garg, Sandeep Kumar, Abdul Khader Jilani Saudagar

Soft Robotics for Medical and Healthcare Applications
Edited by Shaik Himam Saheb, Tharakeshwar Appala and Mohammad S Khan

Smart Cyber Physical Systems: Innovations and Real-World Implications
Edited by Ramesh Chandra Poonia, Kamal Upreti and Mohammad S Khan

Cognitive Cloud Computing: Building Intelligent Systems for Tomorrow
Edited by Jossy George, Kamal Upreti, Ramesh Chandra Poonia, Ankit Gautam and Danish Nadeem

For more information on this series page, please visit our website link: https://www.routledge.com/Computational-and-Intelligent-Systems/book-series/CRCCIS

Cognitive Cloud Computing

Building Intelligent Systems for Tomorrow

Edited by
Jossy George, Kamal Upreti,
Ramesh Chandra Poonia,
Ankit Gautam, and Danish Nadeem

CRC Press
Taylor & Francis Group
Boca Raton London New York

CRC Press is an imprint of the
Taylor & Francis Group, an **informa** business

First edition published 2026
by CRC Press
2385 NW Executive Center Drive, Suite 320, Boca Raton FL 33431

and by CRC Press
4 Park Square, Milton Park, Abingdon, Oxon, OX14 4RN
CRC Press is an imprint of Taylor & Francis Group, LLC

ISBN: 978-1-032-94165-3 (hbk)
ISBN: 978-1-032-94170-7 (pbk)
ISBN: 978-1-003-56925-1 (ebk)

DOI: 10.1201/9781003569251

Typeset in Sabon
by Deanta Global Publishing Services, Chennai, India

Contents

Preface

The rapid fusion of artificial intelligence with cloud infrastructure has given rise to a transformative technological paradigm: cognitive cloud computing. At the intersection of cognitive computing and cloud technologies, this field enables systems to learn, reason, and adapt in real time, paving the way for intelligent automation, sustainable digital ecosystems, and resilient infrastructure across industries.

This book, *Cognitive Cloud Computing: Building Intelligent Systems for Tomorrow*, delves into the emerging landscape shaped by this convergence. It investigates how cognitive intelligence enhances the capabilities of cloud environments, driving progress in domains such as healthcare, education, defense, manufacturing, cybersecurity, and sustainable operations. This volume is designed to bridge foundational theory with hands-on innovation, offering insights for researchers, professionals, developers, and students invested in the future of intelligent cloud ecosystems.

Through its multidimensional approach, the book captures a wide spectrum of applications and challenges, illustrating the powerful synergy between cognition and cloud infrastructure. From AI-powered air traffic systems and secure cloud networks to sustainable cloud operations and therapeutic cognitive robotics, each chapter contributes to a broader understanding of how cognitive cloud computing is redefining the digital frontier.

A synopsis of the chapters is as follows:

Chapter 1 explores cognitive robotics in social, educational, and therapeutic contexts, demonstrating how cloud-enabled intelligence can enhance human–machine interaction in emotionally and cognitively responsive environments.

Chapter 2 investigates load balancing strategies in cognitive environments, offering a sustainable approach to cloud computing by optimizing resource allocation and minimizing energy consumption.

Chapter 3 presents a vision for the next generation of cloud intelligence, illustrating how cognitive computing can be a cornerstone for sustainable and adaptive digital infrastructures.

Chapter 4 introduces AI-driven air traffic control systems, showcasing how cognitive cloud architectures are revolutionizing aviation management in an era of rising complexity and sustainability demands.

Chapter 5 delves into quantum-inspired algorithms, highlighting their role in advancing cognitive computing capabilities and addressing computational challenges in cloud-based problem-solving.

Chapter 6 provides a practical framework for implementing smart cyber-physical systems (SCPS) in industrial and urban contexts, emphasizing real-world integration and system resilience.

Chapter 7 discusses the integration of SCPS with wearable technologies, illustrating a novel approach to remote patient monitoring and health analytics powered by cloud cognition.

Chapter 8 examines how cloud security Intelligence strengthens operational resilience, particularly in mission-critical systems where data integrity and threat detection are paramount.

Chapter 9 analyzes the convergence of intelligent transportation and SCPS, offering insights into mobility, logistics, and infrastructure design in smart cities.

Chapter 10 presents machine learning approaches to malware detection, addressing the growing threat landscape and the need for dynamic, adaptive defense mechanisms in cloud environments.

Chapter 11 investigates sustainability in manufacturing through eco-conscious operations and data-driven strategies, shedding light on key performance factors in industrial contexts.

Chapter 12 proposes AI-driven defenses against Sybil and DDoS attacks, underlining the importance of cognitive intelligence in safeguarding distributed cloud networks.

Chapter 13 compares machine learning models for detecting web trackers, emphasizing privacy, data ethics, and computational effectiveness in a cloud-first digital world.

Chapter 14 identifies emerging trends in smart cyber-physical systems, offering a forward-looking perspective on the evolving landscape of cloud-cyber integration.

Chapter 15 discusses computational intelligence in optimizing cloud operations, with a focus on environmentally sustainable computing models.

Chapter 16 reflects on the evolution of cyber-physical systems, tracing their trajectory in the context of cognitive cloud convergence and the future of intelligent infrastructures.

Through a blend of theoretical frameworks, experimental studies, and real-world applications, this book aims to illuminate the boundless potential of cognitive cloud computing. Each chapter contributes to a broader conversation about innovation, sustainability, and digital transformation in the age of intelligent systems.

We hope this volume inspires new research directions and fosters meaningful dialogue across disciplines, catalyzing advancements that shape a more connected, intelligent, and sustainable future.

Editors Biographies

Fr. Jossy George holds a dual degree in computer science and human resources and completed his FDPM from IIM, Ahmedabad. Awarded a doctorate in computer science from CHRIST University, Bengaluru, he specialized in image processing with a thesis titled "Development of Efficient Biometric Recognition Algorithms Based on Fingerprint and Face." An active member of various international associations, his primary research focus is on developing algorithms to enhance accuracy in biometrics.

His extensive research portfolio includes numerous major and minor projects, notably in the realm of technology in higher education. He has authored multiple research papers published in both national and international journals. Beyond his research endeavors, he is a prolific author, having written two motivational books on management, namely, *Boost Your Ability to Make Great Decisions* and *Leadership Par Excellence*. Additionally, he coauthored the books *Data Ethics and Challenges* and *Introduction to Principles of Data Science*.

Recognized for his outstanding contributions, he received the Visionary Leader of 2019 award for his exemplary role in education, skill development, and research. He was also honored with the Certificate of Academic Excellence (2021) for his dedicated contributions to Education, presented by H.D. Deve Gowda, former prime minister of India.

Currently, he serves on several esteemed officiating bodies in the field of computer science and holds numerous patents and copyrights. Since 2002, he has been associated with CHRIST University (formerly Christ College), Bengaluru, India, and other affiliated institutions, assuming various roles. Presently, he serves as the director and CHRIST (Deemed to be University), BGR Campus Bengaluru, Karnataka, India.

Kamal Upreti is currently working as an associate professor in the Department of Computer Science, CHRIST (Deemed to be University), Delhi NCR, Ghaziabad, India. He completed his BTech (Hons.) from UPTU, MTech (Gold Medalist), PGDM (Executive) from IMT Ghaziabad, and PhD from the Department of Computer Science & Engineering. He has completed a postdoc at the National Taipei University of Business, Taiwan, funded by MHRD. He has published 50-plus patents, 32-plus magazine issues, and 135-plus research papers in various reputed journals and international conferences. His areas of interest include modern physics, data analytics, cybersecurity, machine learning, health care, embedded systems, and cloud computing. He has published more than 45 authored and edited books in reputed publications. He has many years of experience in corporate and teaching experience in engineering colleges. He worked with HCL, NECHCL, *Hindustan Times*, Dehradun Institute of Technology, and Delhi Institute of Advanced Studies, with more than 15 years of enriching experience in research, academics, and corporate. He also worked at NECHCL in Japan, where he had a project called "Hydrastore" funded by a collaboration between HCL and NECHCL. Upreti worked on a government project, "Integrated Power Development Scheme (IPDS)," launched by the Ministry of Power, Government of India, to strengthen the subtransmission and distribution network in urban areas. He recently completed work in collaboration with GB PANT & AIIMS Delhi on the funded project of the ICMR scheme on cardiovascular disease prediction strokes using machine learning techniques from 2017 to 2020, which has a fund of 80 lakhs. He got 5 lakhs in funding from DST SERB for conducting the International Conference, ICSCPS-2024. Recently, he got a 10 lakhs fund from AICTE–Inter-Institutional Biomedical Innovations and Entrepreneurship Program (AICTE-IBIP) for 2024–2026. He has attended as a session chairperson at national and international conferences and was a keynote speaker on various platforms such as skill-based training, corporate training, guest faculty, and faculty development programs. He was awarded best teacher, researcher, extra academic performer, and gold medalist in the MTech program.

Ramesh Chandra Poonia completed a postdoctoral fellowship at the Cyber-Physical Systems Laboratory (CPS Lab), Department of ICT and Natural Sciences, Norwegian University of Science and Technology (NTNU), Alesund, Norway. He also completed a part-time post-doctoral fellowship on the International Collaborative Project for a prediction model for pandemic disease using machine learning at Oakland University, USA, and Imam University, Saudi Arabia. He received his PhD in computer science from Apaji Institute of Mathematics & Applied Computer Technology, Banasthali University, Banasthali, India, in July 2013. He has also completed a master of technology (MTech) in data science and engineering, Work Integrated Learning Programme (WILP) from Birla Institute of Technology and Science, Pilani (BITS Pilani), India. He works as a professor at the Department of Computer Science, CHRIST (Deemed to be University), Banglore, India. His research interests are sustainable technologies, cyber-physical systems, computational intelligence, and network protocol evaluation. He is chief editor of *TARU Journal of Sustainable*

Technologies and Computing (*TJSTC*) and associate editor of the *Journal of Sustainable Computing: Informatics and Systems*. He also serves on the editorial boards of a few international journals. He is the principal author and coauthor of 6 books and an editor of more than 25 special issues of journals, books, and conference proceedings. He has also authored/coauthored over 75 research publications in peer-reviewed reputed journals, book chapters, and conference proceedings. Moreover, he is a regular reviewer for international journals and conferences. He also serves as conference chair of the conference series, International Conference on Sustainable Computing in Science, Technology and Management (SUSCOM). He was endorsed with the prestigious Faculty Appreciation Award 2013 for commendable services.

Ankit Gautam, currently a project lead at BUSINESSNEXT with over eight years of industrial experience, excels in cloud-based solutions and full-stack artificial intelligence and data platforms for the banking, financial services, and insurance (BFSI) sector. Holding an MBA from K J Somaiya Institute of Management and a BTech in electrical and electronics engineering, he combines technical expertise with managerial acumen. Having previously contributed significantly to Adani, Bajaj, and Fino Payment Bank, his diverse experience showcases adaptability across industries. He specializes in steering projects toward efficiency and security through cloud computing. His expertise in full-stack development underscores his capability to handle end-to-end solutions, particularly in the BFSI domain, where he navigates complexities related to security and compliance. He emerges as a dynamic professional, seamlessly integrating technology and business strategies to address contemporary challenges in the ever-evolving corporate landscape.

Danish Nadeem is a senior data scientist and technical consultant working in the Netherlands. Along with his education in computer science and artificial intelligence from Germany, he has over 18 years of industry and research experience on the applications of cutting-edge AI-based and machine learning technologies while working within EU-funded framework and Horizon2020 projects. Working in industries such as education, health, pharmaceuticals, media and entertainment, finance, and logistics, he has also contributed about 13 research publications in various international conferences and journals. He spent several years working in R&D at the German Research Center for Artificial Intelligence before moving to the Netherlands to further develop his career by taking up leading roles, advocating innovative information and communication technologies. With years of experience living and working in Europe, he has been actively involved in promoting strategies for tech startups, innovation, and entrepreneurship, as well as mentoring young professionals for careers in academic research and industry.

Contributors

Gaurav Agarwal
Department of Computer Science and Engineering
Invertis University
Bareilly, Uttar Pradesh, India

Rupanshi Agarwal
Department of Computer Science and Engineering
Invertis University
Bareilly, Uttar Pradesh, India

Faraz Ahmad
Department of Economics
CHRIST University
Delhi NCR, India

Bosco Paul Alapatt
Department of Sciences
CHRIST University
Delhi NCR, India

Daksh Alfred
Department of Social Sciences
CHRIST University
Delhi NCR, India

Prabha Arya
School of Management
G D Goenka University
Sohna Gurugram, Haryana, India

Arpita Banerjee
Department of Computer Science
St. Xavier's College
Jaipur Rajasthan, India

C. Banerjee
Amity Institute of Information Technology
Amity University, Jaipur
Jaipur, Rajasthan, India

Neeta Bhide
MGM University
Aurangabad, India

Shivani Chaudhry
Department of Social Sciences
CHRIST University
Delhi NCR, India

Salineeta Chaudhuri
Department of Social Sciences
CHRIST University
Delhi NCR, India

Sapna Dadwal
Department of Management Studies
DPG Institute of Technology and Management
Gurugram, India

Safia Farooqui
Centre for Online Learning
Dr. D. Y. Patil Vidyapeeth, Pune
Sant Tukaram Nagar, Pimpri, Pune, India

Jossy P. George
School of Sciences
CHRIST University
BGR Campus Bengaluru, Karnataka , India

Rituraj Jain
Department of Information Technology
Marwadi University
Rajkot, India

Kushal Johari
Department of Computer Applications
Invertis University
Bareilly, Uttar Pradesh, India

Maria Johnson
School of Business and Management
CHRIST University
Bengaluru, India

Dennis Jose
School of Business and Management
CHRIST University
Bengaluru, India

Nausheen Khilji
JIET Universe
Jodhpur, Rajasthan, India

Amit Kumar
Department of Social Sciences
CHRIST University
Delhi NCR, India

Prachi Kumar
Department of Economics
CHRIST University
Delhi NCR, India

Vishakha Kuwar
Centre for Online Learning
Dr. D. Y. Patil Vidyapeeth, Pune
Sant Tukaram Nagar, Pimpri, Pune, India

Shubham Mahajan
Amity School of Engineering and Technology (ASET)
Amity University
Gurugram, Panchgaon, Haryana, India

Calvin Mathew
School of Sciences
CHRIST University
Delhi NCR, India

Pragya Lilly Mathew
Department of Economics
CHRIST University
Delhi NCR, India

Pratham Nijhawan
T&A Consulting Global Pvt. Ltd.
Gurgugram, India

Arvind Panwar
School of Computer Science and Engineering
Galgotias University
Greater Noida, India

Ramesh Chandra Poonia
School of Sciences
CHRIST University
Delhi NCR, India

Venkateswararao Pulipati
Department of Computer Science and Engineering
Koneru Lakshmaiah Education Foundation
Bowrampet, Hyderabad, Telangana, India

GV Radhakrishnan
Kalinga School of Management
Kalinga Institute of Industrial Technology
Bhubaneswar, India

Akash Sanghi
Department of Computer Applications
Invertis University
Bareilly, Uttar Pradesh, India

Uma Shankar
Faculty of Management and Social Sciences
Qaiwan International University
Sulaymaniyah, Kurdistan, Iraq

Anupam Sharma
Department of Artificial Intelligence and Machine Learning
Raj Kumar Goel Institute of Technology
Ghaziabad, India

Arnav Sharma
School of Sciences
CHRIST University
Delhi NCR, India

Ashish Sharma
Department of Technology
JIET Universe
Jodhpur, Rajasthan, India

Ashish Kumar Sharma
Department of Computer Science and Engineering
Invertis University
Bareilly, Uttar Pradesh, India

Durgansh Sharma
Department of MBA School of Business Management
CHRIST University
Bengaluru, India

Amit Kumar Singh
Amity Institute of Information Technology
Amity University
Rajasthan, Jaipur, India

Nidhi Singh
School of Management
G D Goenka University
Sohna Gurugram, Haryana, India

Vandana Sonwaney
Symbiosis Institute of Operations Management
Symbiosis International University
Pune, India

Urvashi Sugandh
Department of Artificial Intelligence and Machine Learning
Raj Kumar Goel Institute of Technology
Ghaziabad, India

Sumegh Tharewal
School of Advanced Computing
DBS Global University
Dehradun, Uttarakhand, India

Akhilesh Tiwari
Department of Business and Management
CHRIST University
Delhi NCR, India

Kamal Upreti
School of Sciences
CHRIST University
Delhi NCR, India

Shitiz Upreti
Manav Rachna International Institute of Research & Studies (MRIIRS)
Faridabad, Haryana, India

Sunita Yadav
Department of Artificial Intelligence and Machine Learning
Raj Kumar Goel Institute of Technology
Ghaziabad, India

Chapter 1

Cognitive robotics for social, educational, and therapeutic applications

Vishakha Kuwar, Vandana Sonwaney, Neeta Bhide, Shitiz Upreti, Gaurav Agarwal, and Akash Sanghi

1.1 INTRODUCTION

Cognitive robotics is a field where robots are designed to think and learn like humans, combining artificial intelligence, robotics, and cognitive science. These robots can see, understand, and make decisions in useful areas like helping people with disabilities, teaching students, or providing companionship. Cognitive robotics represents a frontier in artificial intelligence development with human-like cognitive capabilities [1]. These systems integrate perception, reasoning, decision-making, and learning and perform complex tasks in dynamic environments. Cognitive robots have gained attention in various fields, particularly in social, educational, and therapeutic settings. Social robots facilitate human–robot interactions and companionship. In education, these robots serve as personalized tutors and learning assistants. In therapeutic applications, they assist individuals with disabilities and support mental health interventions. This chapter explores the current state, applications, challenges, and future potential of cognitive robotics [2].

Recent reports highlight progress in equipping robots with commonsense knowledge and help them to understand everyday tasks like finding objects. Brain–computer interfaces are also advancing, letting people control robots with their thoughts, especially for rehabilitation [3]. The market is growing fast and is expected to reach USD 26.04 billion by 2032, driven by AI and automation needs. These robots are used in social settings to reduce loneliness, personalize learning education, and support cognitive training therapy. However, challenges include high costs, data privacy concerns, and ethical issues like potential dehumanization. Future research aims to make them more affordable and address these concerns. Figure 1.1 illustrates the progression from the foundations of cognitive robotics and includes AI, robotics engineering, and cognitive science. The core capabilities, key enabling technologies, and diverse applications in social, educational, and therapeutic robotics are represented.

Cognitive robotics integrates advanced technologies, enabling the robots to perceive, learn, reason, and interact intelligently. Perception and sensory

DOI: 10.1201/9781003569251-1

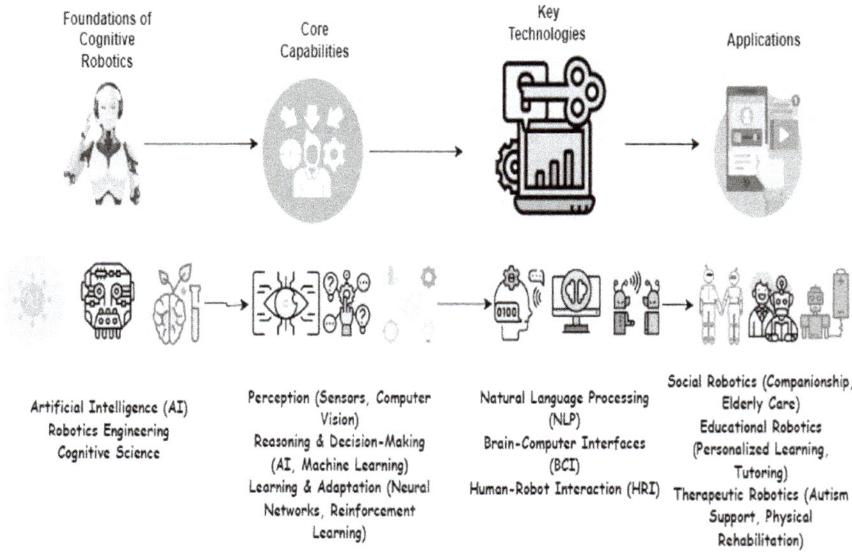

Figure 1.1 Overview of cognitive robotics: Foundations, capabilities, technologies, and applications

processing allow robots to interpret their surroundings using sensors like cameras, LiDAR, and tactile devices [4]. The computer vision, natural language processing (NLP), and auditory processing help them to understand visual, spoken, and environmental cues. Learning and adaptation enable robots to improve timely performance using machine learning (ML) and reinforcement learning, and can allow them to modify their behavior based on past experiences. Reasoning and decision-making involve logic-based AI and probabilistic models, which help robots make informed choices, with planning algorithms to navigate complex tasks. Human–robot interaction (HRI) focuses on communication through speech, gestures, and facial expressions. Robots can work seamlessly with humans in their social intelligence development [5]. Lastly, embodied cognition examines how a robot's physical form influences its intelligence and the relationship between movement and perception in cognitive development. Together, these aspects create intelligent, adaptable robots capable of operating in real-world environments.

1.2 LITERATURE REVIEW

Cognitive robotics has grown from AI to today's machine learning models, letting robots adapt and learn like humans. Social robots like Pepper (SoftBank Robotics) and NAO (Aldebaran Robotics) help in care homes for the reduction of loneliness [6]. In schools, robots tutor kids, boosting learning [7]. In therapy, robots like KASPAR aid autism treatment, and Lokomat helps with physical rehab [8]. But high costs, ethics, and privacy worries remain hurdles. The field of cognitive robotics has witnessed significant advancements. Early research on AI systems leverages machine learning and neural networks [9]. Studies have demonstrated the effectiveness of cognitive robots in autism therapy, elderly care, and special education. Pepper and NAO have also been employed in classrooms and healthcare facilities [10]. These robots show promising results in engagement enhancement and therapeutic outcomes [11]. There are some challenges such as ethical concerns, user acceptance, and technological limitations. This section provides a review of key studies and developments in cognitive robotics across social, educational, and therapeutic applications [12].

1.2.1 Background and importance

Cognitive robotics is an interdisciplinary field that merges artificial intelligence (AI), robotics, and cognitive science for the creation of robotic systems capable of human-like perception, reasoning, and decision-making. These intelligent robots are designed to interact with an adaptive environment. Cognitive processes and makes particularly valuable in social, educational, and therapeutic contexts. The field's importance is to underscore technological solutions to address social isolation, educational disparities, and healthcare challenges, especially for the aging population and conditions like autism spectrum disorder [13].

A Scopus literature analysis presents insightful trends in cognitive robotics and related fields. Figure 1.2 shows the United States leads in research output, followed by Italy, the United Kingdom, and Germany, and highlights these nations as key contributors to advancements in robotics. Figure 1.3 identifies A. Cangelosi as the most prolific author, with other notable contributors including R.J. Duro, M. Cong, and S.H. Scott, indicating strong individual contributions to the field.

Figure 1.4 reveals that computer science (31.6%) and engineering (19.6%) dominate the field, followed by mathematics, medicine, and neuroscience, showcasing the multidisciplinary nature of cognitive robotics research. Psychology, social sciences, and biochemistry also contribute, indicating the integration of human cognition and biological studies in robotics. Figure 1.5 highlights that most publications are journal articles (42.0%) and conference papers (38.1%), underlining the significance of both theoretical and applied research in the field. Reviews, book chapters, and conference reviews also play a role in disseminating knowledge. Overall, these visualizations illustrate the global, interdisciplinary, and evolving nature of cognitive robotics research, with strong contributions from leading nations, researchers, and diverse subject areas.

Figure 1.6 illustrates the trends in research publications and citations over time (2015–2025). The bars represent the number of published documents each year and while the black dotted line indicates the corresponding

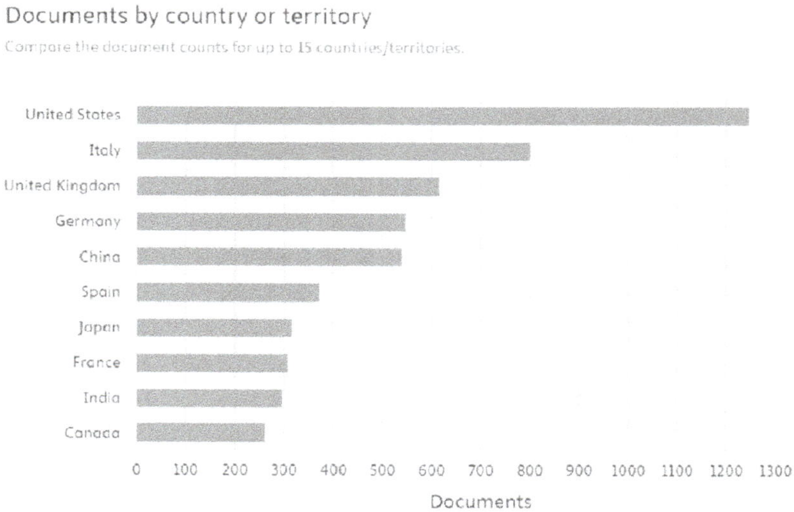

Figure 1.2 Documents by country or territory

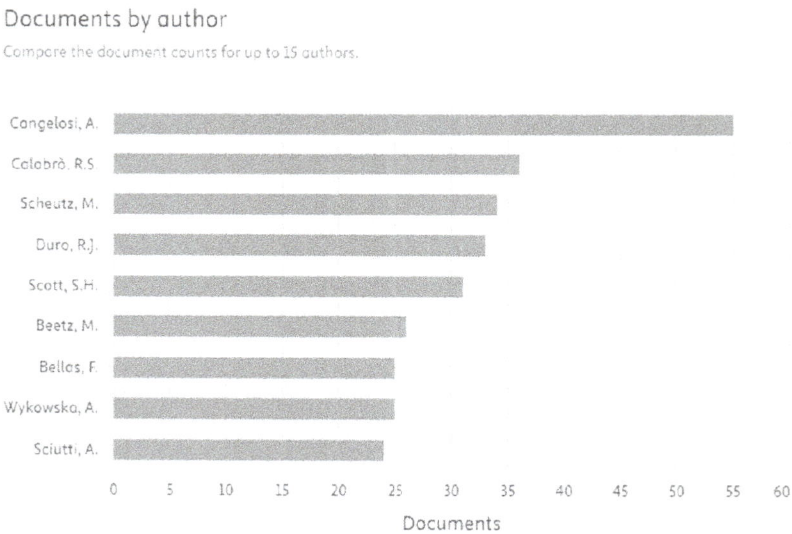

Figure 1.3 Documents by author

citation count. From 2015 to 2022, there was a steady increase in both documents and citations, reflecting a growing interest in the field. The publication rate remains relatively stable between 2017 and 2022, with a slight increase over the years. In 2023 and 2024, citations peaked significantly, suggesting that earlier publications gained more recognition and impact. In 2025, there was a sharp decline in both document count and citations.

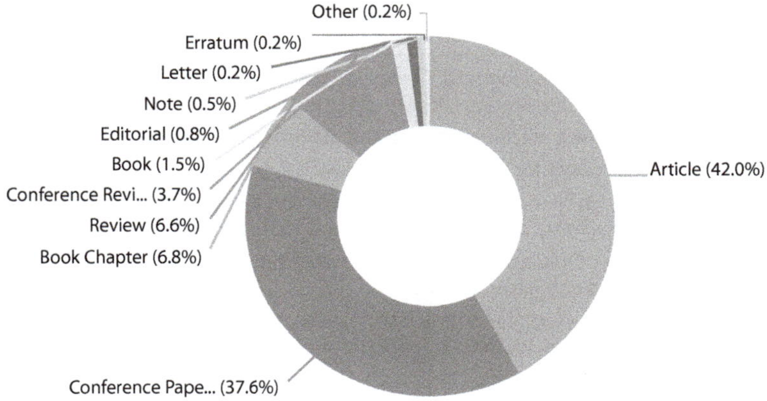

Figure 1.4 Document by type

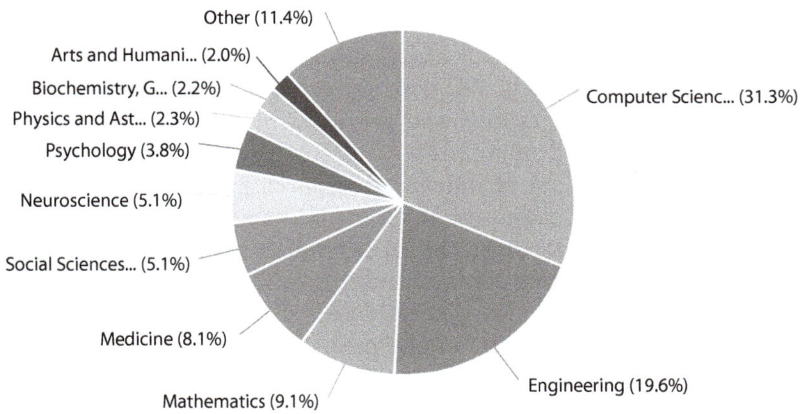

Figure 1.5 Documents by subject area

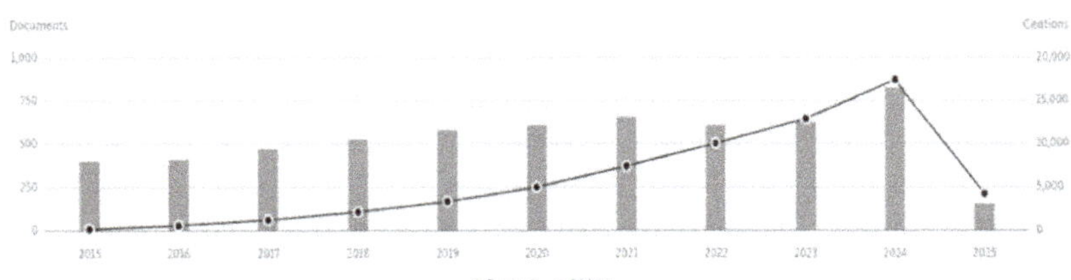

Figure 1.6 Trend in research publications and citations over time (2015–2025)

Table 1.1 Influential research papers with citation counts

Reference	*Document title*	*Citation count*
[15]	"Review of deep learning: Concepts, CNN architectures, challenges, applications, future directions"	4185
[16]	"Probabilistic machine learning and artificial intelligence"	453
[10]	"Social robots for education: A review"	52
[17]	"Survey on human–robot collaboration in industrial settings: Safety, intuitive interfaces and applications"	40
[18]	"Review of assistive strategies in powered lower-limb orthoses and exoskeletons"	39
[19]	"Understanding of machine learning with deep learning: Architectures, workflow, applications and future directions"	25
[20]	"Decoding motor imagery from the posterior parietal cortex of a tetraplegic human"	68
[21]	"Scoping review on the use of socially assistive robot technology in elderly care"	17

This drop may be due to incomplete data for the year or a temporary reduction in research output. Overall, the graph highlights the growing significance of research in this field and with an increasing number of documents contributing to higher citations and particularly peaking in 2024 before experiencing a notable drop in 2025.

Table 1.1 presents a list of influential research papers, along with their citation counts, and highlights the key contributions to artificial intelligence, robotics, and machine learning [14]. Alzubaidi et al. [15] have the highest citation count (4185) with a comprehensive review of deep learning, including CNN architectures, challenges, and future directions. Ghahramani [16] follows with 453 citations, discussing probabilistic machine learning and artificial intelligence. Belpaeme et al. [10] and Villani et al. [17] contribute to human-robot interaction and particularly in education and industrial collaboration, with 52 and 40 citations, respectively. Further, Yan et al. [18] explore assistive robotics and human–robot collaboration, though with relatively lower citation counts (39). Taye [19] discusses deep learning architectures and applications, accumulating 25 citations. Aflalo et al.'s [20] work on motor imagery decoding has received 68 citations. Lastly, Abdi et al. [21] review socially assistive robots for elderly care, with 17 citations. This data underscores the varying impact of research topics, with deep learning and probabilistic AI gaining the most traction.

1.2.2 Historical development and current state

Cognitive robotics has its roots in the early days of AI, with pioneers like Alan Turing proposing the concept of machines that could think and learn like humans [22]. Initially, efforts were focused on rule-based systems that attempted to replicate human reasoning through predefined logic and algorithms on human problem-solving [23]. However, these systems were limited in their ability with real-world complexities due to their inflexibility and lack of adaptability.

The shift toward dynamic and adaptive approaches was catalyzed for the advent of machine learning and neural networks in the late 1990s and early 2000s [24]. Modern cognitive robots integrate advanced technologies such as NLP for human speech [25] and computer vision, visual information integration [26], and reinforcement learning through trial and error [27]. These technologies have enabled robots to better understand and interact with their surroundings and users. In social robotics, cognitive robots are designed and engaged with humans in socially acceptable ways. It recognizes and responds to human emotions, gestures, and speech. Robots like Pepper and NAO [7] have been deployed in various settings from customer service to companionship for the elderly. Studies have shown that robots can reduce loneliness and overall well-being among seniors [28]. It was found that interacting with the seal-like robot Paro led to low loneliness and good mood among elderly residents in a care home.

In educational contexts, cognitive robots are explored as personalized learning assistants. Robots like NAO have been used to teach programming concepts to primary school children [29]. Robo Tutor has been utilized to help students with math problems [30]. These robots can adapt their teaching methods based on individual student performance and learning styles, and learning experience. Chou et al. [31] reported that students using robotic tutors showed high motivation and better academic performance compared to traditional settings. Therapeutic applications of cognitive robotics are particularly promising for individuals with physical or cognitive impairments. Robots like KASPAR have been shown to help children with autism develop social skills through interactive play, and robotic exoskeletons like Lokomat assist in physical rehabilitation for patients with neurological disorders [32]. In mental health, virtual agents and robots are being used to deliver cognitive behavioral therapy and support patients with conditions like depression and anxiety [33].

The VOS viewer network visualization map (Figure 1.7) illustrates the interdisciplinary nature of cognitive robotics research and highlights the key areas and their interconnections. The cluster on the right focuses on human–robot interaction, robotics, and cognitive architectures, and emphasizes advancements in robot programming, social robots [10], and neural networks for intelligent systems. Another cluster relates to human performance and rehabilitation, and shows the strong links between robotics and healthcare, particularly in motor performance, stroke recovery, and assistive technologies. One cluster connects neuroscience, machine learning, and robotics, showcasing research on brain mapping, sensory feedback, and Neurorobotics for cognitive and perceptual enhancements. Another cluster highlights aging and cognitive impairment with studies on robotics for elderly care and cognitive decline management. Finally, one cluster

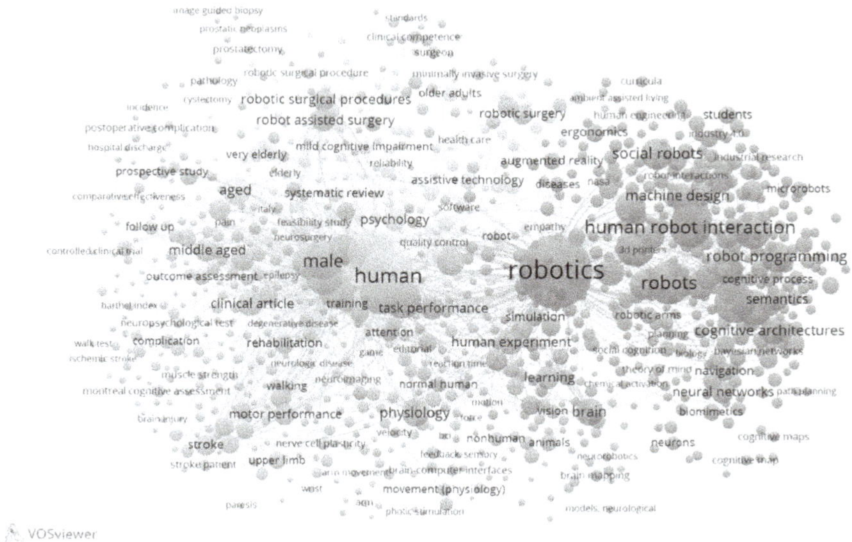

Figure 1.7 VOS viewer analysis of interdisciplinary nature of cognitive robotics

represents robotic surgery and medical applications and demonstrates the role of robotics in minimally invasive surgery and clinical advancements. The visualization underscores how cognitive robotics integrates artificial intelligence, neuroscience, and healthcare, and paves the way for intelligent and adaptive robotic systems in various domains.

Despite these advancements, cognitive robotics faces several challenges. High development costs limit accessibility, especially in underfunded settings [34]. Ethical concerns arise regarding the potential for robots to replace human interaction in sensitive areas like therapy and companionship and potentially leading to dehumanization [35, 36]. Data privacy is another concern, as these robots collect sensitive user information, raising issues about security [37] and continuous human supervision is often required to ensure safety and efficacy, particularly in therapeutic settings [38].

1.3 ANALYSIS AND INTERPRETATION

The effectiveness of cognitive robotics in various studies and real-world implementations has been analyzed. Metrics such as user engagement, learning outcomes, and therapeutic improvements were examined. Findings suggest that cognitive robots can significantly enhance social interactions, personalized learning experiences, and consistent therapeutic support. The integration of natural language processing, computer vision, and reinforcement learning has enabled robots to adapt to user needs. There are limitations such as high costs, data privacy concerns, and continuous human supervision that remain challenges that need to be addressed.

1.3.1 Effectiveness and limitations

The effectiveness of cognitive robotics evaluated from various studies and real-world implementations were analyzed. The study focuses on the metrics such as user engagement, learning outcomes, and therapeutic improvements.

1.3.1.1 Social applications

Research suggests cognitive robots enhance user engagement and provide emotional support. For example, Wada and Shibata [28] found that older adults experienced reduced loneliness and improved mood. A systematic review by Li and Pu [38] examined the use of robots for older adults and found they can improve social connection, though more high-quality studies are needed. Robinson, MacDonald, and Broadbent [12] reported that the social robot Pepper positively impacted patients' emotional states in hospitals, with higher satisfaction scores.

Cognitive robotics for social applications focuses on developing intelligent robots that can interact with humans in meaningful ways. These robots integrate advanced perception, learning, and decision-making capabilities to understand and respond to human emotions, speech, and gestures. Using technologies such as machine learning, natural language processing, and computer vision, these robots can engage in conversations, assist with daily tasks, and provide companionship, particularly for elderly individuals or those with special needs. Socially assistive robots [11] are widely applied in healthcare, education, and customer service, enhancing human–robot collaboration and improving quality of life. By incorporating embodied cognition and adaptive learning, these robots can refine their interactions over time, making them more intuitive and effective in diverse environments. However, challenges such as ethical considerations, privacy concerns, and the need for seamless integration into human-centric spaces remain key areas of research and development. As cognitive robotics advances, its potential to revolutionize social interactions and support human well-being continues to grow.

1.3.1.2 Educational applications

It seems likely that cognitive robots improve learning outcomes, with studies showing higher motivation and performance. Wang et al. [39] found that students using robotic tutors had better academic results, supported by Belpaeme et al. [40] who reviewed social robots for education, noting positive impacts in language learning and mathematics. Nye, Graesser, and Hu [30] found that students using Robo Tutor for math showed significant improvements in understanding and retention compared to controls. Cognitive robotics for educational applications leverages artificial intelligence and robotics to enhance learning experiences and make education more interactive and engaging. These robots use advanced perception, natural language processing, and adaptive learning algorithms to interact with students, personalize instruction, and provide real-time feedback. They can support teachers in repetitive tasks; tutor students in subjects like mathematics, science, and language learning; and foster critical thinking skills through problem-solving activities. Social robots [10] in classrooms help improve student engagement, especially for children with special needs, creating an interactive and inclusive learning environment. Additionally, cognitive robotics enables hands-on learning in STEM education and allows the students to develop programming, engineering, and AI-related skills. Despite their challenges, such as cost and ethical considerations, an effective human-robot collaboration remain key areas of focus. As technology advances, cognitive robots are expected to make the future of education learning more personalized, accessible, and immersive.

1.3.1.3 Therapeutic applications

Technological enablers include NLP for communication, computer vision for visual interpretation, and reinforcement learning for behavior optimization. These allow robots to adapt, but limitations persist. High costs, data privacy concerns, and required human supervision are significant barriers. An unexpected detail is how these technologies also address social sustainability, improving worker safety through automation, less commonly discussed but evident in their integration. Cognitive robotics for therapeutic applications harnesses artificial intelligence and robotics to support physical, cognitive, and emotional rehabilitation. These robots are designed to assist patients recovering from neurological disorders, physical injuries, or mental health challenges by providing interactive and adaptive therapy [41].

Equipped with advanced perception, machine learning, and human–robot interaction capabilities, they can guide patients through rehabilitation exercises, monitor progress, and adjust therapy. Socially assistive robots play a crucial role in mental health therapy by offering companionship, anxiety reduction, and social engagement specifically for individuals with autism, dementia, or depression. In physical rehabilitation, robotic exoskeletons and assistive devices enhance motor recovery, supporting movement and neural pathways [2]. Additionally, these robots improve accessibility to therapy, making treatment more consistent and available to patients in remote areas. Apart from the growing potential, challenges such as ethical concerns, user acceptance, and seamless integration into healthcare systems remain critical considerations. Cognitive robotics evolves its application in therapeutic settings and continues to enhance patient care, recovery outcomes, and overall well-being.

1.4 CHALLENGES AND ETHICAL CONSIDERATIONS

Cognitive robotics face significant challenges as described in the following sections.

1.4.1 Data privacy and security

The security of data collected by cognitive robots is crucial and there are concerns about breaches and misuse, especially given their use in sensitive areas like healthcare [3]. In the context of cognitive robotics, data privacy and security are critical concerns and particularly in sensitive applications such as healthcare, education, and social interactions. Cognitive robots rely on vast amounts of personal and behavioral data for their learning, decision-making, and human interaction capabilities. The dependence raises risks, data breaches, unauthorized access, and potential misuse of sensitive information. In healthcare, robots assist in patient monitoring, rehabilitation, and diagnostic and strict adherence to data protection regulations like HIPAA and GDPR [42]. It ensures that robust encryption, secure storage, and controlled data access is essential to maintain compliance and trust. Additionally, ethical considerations arise in informed consent, data ownership, and transparency in data usage. Cognitive robotics continues and implements strong cybersecurity frameworks, anonymization techniques, and real-time threat detection systems. It is imperative to safeguard user privacy, enabling the benefits of intelligent robotic systems [43].

1.4.2 High initial costs

A cognitive robot's development and deployment can be expensive with high initial investment costs, limiting accessibility in underfunded settings such as small healthcare facilities, schools, and research institutions [2]. Cognitive robots are built using advanced hardware, high-resolution sensors, actuators, processors, and sophisticated ML and AI algorithms. Research development cost and continuous software updates further add to the financial burden. In education and healthcare sectors in which cognitive robots have immense potential for personalized learning and patient care, there is limited funding for adoption [44]. The infrastructure upgrades and skilled personnel needed increases the overall investment. The long-term benefits such as efficiency, automation, and decision-making can offset costs such as initial affordability, which remains a barrier. Efforts should focus on cost-effective designs, open-source AI frameworks, and financial support mechanisms like government grants or public–private partnerships.

1.4.3 Ethical issues

Ethical considerations include the dehumanization potential with robots replacing human interaction in therapy and companionship. AI algorithms and autonomy bias issues must be addressed in responsible deployment [1]. Ethical issues in cognitive robotics are a major concern as well as applications therapy, companionship, and human interaction. The key issue is the potential for dehumanization, robots' reliance on caregivers, and emotional support in human-to-human interactions. Sensitive environments such as elder care, mental health therapy, and education play a vital role in wellbeing. AI algorithm bias poses a risk. Cognitive robots learn from data and reflect societal prejudices, leading to the potential for unfair treatment or exclusion of certain groups. Transparency and fairness in AI decision-making are essential for bias mitigation. The ethical dilemma revolves around the autonomy of cognitive robots in decision-making and high-stakes areas like healthcare and security. Clear guidelines and ethical frameworks are established to prevent unintended consequences. Cognitive robots serve as assistive tools rather than replacements for human roles. The concerns require interdisciplinary collaboration, ongoing ethical reviews, and regulatory policies to ensure the responsible and equitable deployment of cognitive robotics.

1.4.4 Complexity and maintenance

Cognitive robots are complex systems that require ongoing maintenance and updates. Thirty-two of the fifty-two publications in the 2024 review did not publish evaluation data. The gap indicates the need for standardized methods [1]. The complexity and maintenance of cognitive robots present significant challenges. These systems require continuous updates, repairs, and improvements to function effectively. Cognitive robots integrate advanced artificial intelligence, machine learning, and sensor technologies, making upkeep resource-intensive. Regular software updates are necessary for performance, to address security vulnerabilities, and new learning adaptation models. The hardware components such as sensors, actuators, and processors may require frequent calibration or replacement and add to the long-term costs and technical demands [4]. The gap regarding the absence of evaluation data in most of the literature suggests inconsistencies in the effectiveness and reliability of cognitive robots for industry-wide standards. Without standardized maintenance protocols and frameworks, evaluation of long-term sustainability and scalability of cognitive robots remain uncertain. The challenges require collaboration among researchers, developers, and policymakers for best practices for maintenance, performance evaluation, and cost-effective solutions, in order for cognitive robots to remain functional and accessible in real-world applications.

1.5 FUTURE PROSPECTS AND RECOMMENDATIONS

Future prospects of cognitive robotics are promising with continued advancements in AI and robotics leading to more sophistication and accessibility. Research should focus on cost-effective solutions to enhance accessibility, particularly in developing regions. It can leverage cloud-based robotics and open-source platforms. The user-centered design ensures intuitive robots, and a user-friendly focus on stakeholder-centered design meets diverse needs. Ethical frameworks establish the guidelines for data security, AI bias, and autonomy, and ensure responsible deployment and address the ethical concerns [45]. Interdisciplinary collaboration fosters collaboration between engineers, psychologists, and therapists for cognitive robotics integration into social, educational, and therapeutic domains effectively [46].

1.6 DISCUSSION

The potential of cognitive robotics in transforming social, educational, and therapeutic domains is immense. Social robots can serve as companions for individuals facing social isolation. In education, adaptive robots can personalize learning based on student progress. In therapy, robots can assist individuals with cognitive impairments and provide structured interventions. However, ethical considerations regarding autonomy, bias, and data security must be addressed. Future research should focus on developing cost-effective, scalable, and user-friendly cognitive robotic solutions.

1.7 POTENTIAL AND ETHICAL CONSIDERATIONS

The potential of cognitive robotics to transform social, educational, and therapeutic domains is immense. In social settings, they can alleviate isolation, providing companionship for the elderly or disabled [28]. In education, personalized learning can reduce disparities, with robots like NAO enhancing collaboration. In therapy, consistent interventions can augment treatments, with KASPAR aiding autism and Lokomat supporting rehab.

However, ethical considerations are crucial. Autonomy ensures users control interactions and avoid manipulation. Bias in AI algorithms, potentially discriminatory, must be addressed through diverse datasets. Data security is vital, protecting user information from breaches. The risk of human replacement, especially in therapy, could harm mental health, and requires robots to complement, not substitute, human contact. Future research should focus on cost-effectiveness, reducing development expenses, user-centered design for intuitive interfaces, and interdisciplinary collaboration between engineers, psychologists, and therapists. Establishing ethical frameworks will ensure responsible deployment, addressing current limitations.

1.8 RESULTS AND CONCLUSION

The analysis highlights the growing impact of cognitive robotics in enhancing human life. Results indicate that these robots can improve social engagement, learning outcomes, and therapeutic effectiveness. Challenges exist in continuous advancements in AI and robotics, but will further refine these systems to be more accessible and effective. The future of cognitive robotics lies in interdisciplinary collaboration, ethical AI development, and user-centered design. Addressing the current limitations in cognitive robotics can revolutionize multiple domains and foster a more inclusive and technologically advanced society.

1.9 SYNTHESIS AND FUTURE VISION

This analysis highlights the growing impact of cognitive robotics on human life through improved interactions, learning, and therapy. Results indicate loneliness reduction, learning outcomes, and therapeutic efficacy. Continuous advancements in AI and robotics systems can be refined to make them more accessible and effective. The future lies in interdisciplinary collaboration, ethical AI development, and user-centered design, and fostering a more inclusive society by addressing current limitations. To organize findings, Table 1.2 summarizes key aspects.

Table 1.2 underscores the multifaceted nature of cognitive robotics, balancing benefits and challenges across domains. By addressing these, cognitive robotics can revolutionize multiple fields and ensure technological advances align with human needs. For researchers and practitioners, the

Table 1.2 Finding theme with respect to benefits and challenge

Domain	*Key benefit*	*Key challenge*	*Example robot*
Social	Reduces loneliness, enhances mood	High cost, privacy concerns	Paro (IEEE Xplore)
Educational	Improves learning outcomes, motivation	Data security, supervision needs	NAO (Aldebaran Robotics)
Therapeutic	Aids skill development, rehab progress	Ethical risks, human replacement concerns	KASPAR (Applied Bionics)

analysis emphasizes integrating advanced technologies and user needs, and ensuring investments in AI align with ethical goals. Companies should conduct cost–benefit analyses to identify viable robotic initiatives and leverage government incentives where available. Addressing resistance to change through training and communication strategies can facilitate smoother transitions, as suggested in the literature.

1.10 CONCLUSION

Cognitive robotics is revolutionizing social interactions in which robots perceive, learn, and respond to human behaviors in a dynamic and adaptive manner. The integration of artificial intelligence, machine learning, and sensor technologies allows robots to engage in meaningful communication; assist in healthcare, education, and social services; and enhance human–robot collaboration. As cognitive robotics continues to advance, its impact on various industries will expand and offer solutions that improve accessibility, efficiency, and personalized experiences. However, challenges such as ethical considerations, data privacy and the need for seamless human–robot integration remain key areas for future research.

Moving forward, the development of cognitive robotics must balance technological advancements with ethical and social implications. Responsible AI design, decision-making bias, and human trust in robotic systems will be critical to widespread adoption. Collaborative efforts among researchers, policymakers, and industries will shape the trajectory of cognitive robotics and will guide it toward socially beneficial applications. As these systems evolve, they hold the potential to transform human–machine interactions and foster a future where intelligent robots serve as valuable companions, caregivers, and partners in various aspects of daily life.

REFERENCES

1. Cangelosi, A., & Asada, M. (eds.) (2022). *Cognitive Robotics*. MIT Press.
2. Liu, S., Wang, L., & Gao, R. X. (2024). Cognitive neuroscience and robotics: advancements and future research directions. *Robotics and Computer-Integrated Manufacturing*, *85*, 102610.
3. Baniqued, P. D. E., Stanyer, E. C., Awais, M., Alazmani, A., Jackson, A. E., Mon-Williams, M. A., ... & Holt, R. J. (2021). Brain–computer interface robotics for hand rehabilitation after stroke: a systematic review. *Journal of Neuroengineering and Rehabilitation*, *18*, 1–25.
4. Hernandez, M. E., Sowers, R., Thompson, N., Krishnan, G., & Hsiao-Wecksler, E. T. (2022). Robotics in virtual reality. *Virtual Reality: Recent Advancements, Applications and Challenges*, *1*, 99.

5. Junior, W. T. L., Welter, R. A., Ferreira, W. P., Souza, R. F., & Eduardo, T. (2025). Human robot interaction (HRI): an artificial cognitive autonomy approach to enhance decision-making. *Cognitive Systems Research*, *91*, 101336.
6. Shibata, T., Wada, K., Ikeda, Y., & Sabanovic, S. (2009). Cross-cultural studies on subjective evaluation of a seal robot. *Advanced Robotics*, *23*(4), 443–458.
7. Wu, X., & Li, R. (2024). Effects of robot-assisted language learning on English-as-a-foreign-language skill development. *Journal of Educational Computing Research*, *62*(4), 1010–1034.
8. Dautenhahn, K., Nehaniv, C. L., Walters, M. L., Robins, B., Kose-Bagci, H., Mirza, N. A., & Blow, M. (2009). KASPAR–a minimally expressive humanoid robot for human–robot interaction research. *Applied Bionics and Biomechanics*, *6*(3–4), 369–397.
9. Brooks, R. A. (1991). Intelligence without representation. *Artificial Intelligence*, *47*(1–3), 139–159.
10. Belpaeme, T., Kennedy, J., Ramachandran, A., Scassellati, B., & Tanaka, F. (2018). Social robots for education: a review. *Science Robotics*, *3*(21), eaat5954.
11. Robaczewski, A., Bouchard, J., Bouchard, K., & Gaboury, S. (2021). Socially assistive robots: the specific case of the NAO. *International Journal of Social Robotics*, *13*, 795–831.
12. Robinson, H., MacDonald, B., & Broadbent, E. (2014). The role of healthcare robots for older people at home: a review. *International Journal of Social Robotics*, *6*, 575–591.
13. Maranesi, E., Amabili, G., Cucchieri, G., Bolognini, S., Margaritini, A., & Bevilacqua, R. (2022). Understanding the acceptance of IoT and social assistive robotics for the healthcare sector: a review of the current user-centred applications for the older users. *Internet of Things for Human-Centered Design: Application to Elderly Healthcare*, *1*, 331–351.
14. Storjak, I., Krzic, A. S., & Jagust, T. (2022). Elementary school pupils' mental models regarding robots and programming. *IEEE Transactions on Education*, *65*(3), 297–308.
15. Alzubaidi, L., Zhang, J., Humaidi, A. J., Al-Dujaili, A., Duan, Y., Al-Shamma, O., Santamaría, J., Fadhel, M. A., Al-Amidie, M., & Farhan, L. (2021). Review of deep learning: concepts, CNN architectures, challenges, applications, future directions. *J Big Data*, *8*(1), 53. https://doi.org/10.1186/s40537-021-00444-8. Epub 2021 Mar 31. PMID: 33816053; PMCID: PMC8010506.
16. Ghahramani, Z. (2015). Probabilistic machine learning and artificial intelligence. *Nature*, *521*(7553), 452–459.
17. Villani, V., Pini, F., Leali, F., & Secchi, C. (2018). Survey on human–robot collaboration in industrial settings: safety, intuitive interfaces and applications. *Mechatronics*, *55*, 248–266.
18. Yan, T., Cempini, M., Oddo, C. M., & Vitiello, N. (2015). Review of assistive strategies in powered lower-limb orthoses and exoskeletons. *Robotics and Autonomous Systems*, *64*, 120–136.

19. Taye, M. M. (2023). Understanding of machine learning with deep learning: architectures, workflow, applications and future directions. *Computers*, *12*(5), 91.
20. Aflalo, T., Kellis, S., Klaes, C., Lee, B., Shi, Y., Pejsa, K., ... & Andersen, R. A. (2015). Decoding motor imagery from the posterior parietal cortex of a tetraplegic human. *Science*, *348*(6237), 906–910.
21. Abdi, J., Al-Hindawi, A., Ng, T., & Vizcaychipi, M. P. (2018). Scoping review on the use of socially assistive robot technology in elderly care. *BMJ Open*, *8*(2), e018815.
22. Turing, A. M. (1950). Mind. *Mind*, *59*(236), 433–460.
23. Newell, A., & Simon, H. A. (1972). *Human Problem Solving* (Vol. 104, No. 9). Prentice-Hall.
24. LeCun, Y., Bengio, Y., & Hinton, G. (2015). Deep learning. *Nature*, *521*(7553), 436–444.
25. Martin, J. H., & Jurafsky, D. (2020). Vector semantics and embeddings. *Speech Language Processing*, *1*, 1–31.
26. Szeliski, R. (2022). *Computer Vision: Algorithms and Applications*. Springer Nature.
27. Barto, A. G. (2021). Reinforcement learning: an introduction. by richard's sutton. *SIAM Review*, *6*(2), 423.
28. Wada, K., & Shibata, T. (2007). Living with seal robots—its sociopsychological and physiological influences on the elderly at a care house. *IEEE Transactions on Robotics*, *23*(5), 972–980.
29. Alimisis, D. (2019). Teacher training in educational robotics: the ROBOESL project paradigm. *Technology, Knowledge and Learning*, *24*(2), 279–290.
30. Nye, B. D., Graesser, A. C., & Hu, X. (2014). AutoTutor and family: a review of 17 years of natural language tutoring. *International Journal of Artificial Intelligence in Education*, *24*, 427–469.
31. Chou, C. H., Shrestha, S., Yang, C. D., Chang, N. W., Lin, Y. L., Liao, K. W., ... & Huang, H. D. (2018). miRTarBase update 2018: a resource for experimentally validated microRNA-target interactions. *Nucleic Acids Research*, *46*(D1), D296–D302.
32. Hidler, J., Nichols, D., Pelliccio, M., Brady, K., Campbell, D. D., Kahn, J. H., & Hornby, T. G. (2009). Multicenter randomized clinical trial evaluating the effectiveness of the Lokomat in subacute stroke. *Neurorehabilitation and Neural Repair*, *23*(1), 5–13.
33. Lucas, R. J., Peirson, S. N., Berson, D. M., Brown, T. M., Cooper, H. M., Czeisler, C. A., ... & Brainard, G. C. (2014). Measuring and using light in the melanopsin age. *Trends in Neurosciences*, *37*(1), 1–9.
34. Matarić, M. J. (2018). Integration of representation into goal-driven behavior-based robots. In *The Artificial Life Route to Artificial Intelligence* (pp. 165–186). Routledge.
35. Sharkey, S. W., Windenburg, D. C., Lesser, J. R., Maron, M. S., Hauser, R. G., Lesser, J. N., ... & Maron, B. J. (2010). Natural history and expansive clinical profile of stress (Tako-Tsubo) cardiomyopathy. *Journal of the American College of Cardiology*, *55*(4), 333–341.
36. Tiku, S., Pasricha, S., Notaros, B., & Han, Q. (2020). A hidden Markov model based smartphone heterogeneity resilient portable indoor localization framework. *Journal of Systems Architecture*, *108*, 101806.

37. Sparrow, P. R., & Makram, H. (2015). What is the value of talent management? Building value-driven processes within a talent management architecture. *Human Resource Management Review*, *25*(3), 249–263.
38. Li, J., & Pu, K. (2019). Development of organic semiconducting materials for deep-tissue optical imaging, phototherapy and photoactivation. *Chemical Society Reviews*, *48*(1), 38–71.
39. Wang, P., Chen, P., Yuan, Y., Liu, D., Huang, Z., Hou, X., & Cottrell, G. (2018, March). Understanding convolution for semantic segmentation. In *2018 IEEE Winter Conference on Applications of Computer Vision (WACV)* (pp. 1451–1460). IEEE.
40. Belpaeme, T., Kennedy, J., Ramachandran, A., Scassellati, B., & Tanaka, F. (2018). Social robots for education: A review. *Science robotics*, *3*(21), eaat5954.
41. Wan, S., Gu, Z., & Ni, Q. (2020). Cognitive computing and wireless communications on the edge for healthcare service robots. *Computer Communications*, *149*, 99–106.
42. Gabriel, O. T. (2023). *Data Privacy and Ethical Issues in Collecting Health Care Data Using Artificial Intelligence among Health Workers* (Master's Thesis, Center for Bioethics and Research).
43. Upreti, K., Vats, P., Nasir, M. S., Alam, M. S., Shahi, F. I., & Kamal, M. S. (2023, July). Managing the ethical and sociologically aspects of AI incorporation in medical healthcare in India: unveiling the conundrum. In *2023 IEEE World Conference on Applied Intelligence and Computing (AIC)* (pp. 293–299). IEEE.
44. Upreti, K., Kapoor, A., Hundekari, S., Upreti, S., Kaul, K., Kapoor, S., & Tiwari, A. (2024). Deep dive into diabetic retinopathy identification: a deep learning approach with blood vessel segmentation and lesion detection. *Journal of Mobile Multimedia*, *20*, 495–524.
45. Töberg, J. P., Ngonga Ngomo, A. C., Beetz, M., & Cimiano, P. (2024). Commonsense knowledge in cognitive robotics: a systematic literature review. *Frontiers in Robotics and AI*, *11*, 1328934.
46. Liu, S., Wang, L., & Gao, R. X. (2024). Cognitive neuroscience and robotics: Advancements and future research directions. *Robotics and Computer-Integrated Manufacturing*, *85*, 102610.

Chapter 2

A sustainable approach to cloud computing

A comprehensive analysis of load balancing techniques in cognitive environments

Uma Shankar, Rituraj Jain, GV Radhakrishnan, Akhilesh Tiwari, Shubham Mahajan, and Sumegh Tharewal

2.1 INTRODUCTION

One reason for the growth of cloud computing is that the internet is easily available and does not put strain on its operation. Cloud technology is used with a huge amount of data, and other high configuration technology or programs over the internet. Cloud technology provides an organization with the ability to save expenses by requiring fewer human resources, hardware, and costs. Cloud technology-based systems are easy to use because they are simple, flexible, and have well-designed architecture. Then there are the many services being migrated over the internet among many well-known and trusted companies such as Amazon, Yahoo, and Microsoft [1, 2].

Some of the issues that impact the success of cloud computing have been mentioned, however, there are also challenges. One of such problems is load balancing [3]. Load balancing is a type of distribution of the workload to all those nodes that are connected in the computer network in the same way. This provides greater resource utilization and better-performing execution while simultaneously helping to produce the maximum throughput by minimizing latency and response time. Load balancing helps the failed service to work by redirecting the traffic to a working service. Further, it also ensures fair distribution of all the computing service resources over the network.

2.2 OVERVIEW OF CLOUD COMPUTING

Internet realization representation in the form of a cloud is called a network diagram. Cloud computing is what is known as interconnected computers that access computing resources over the internet, and the "clouds" are the interconnected nodes of the hardware system located at different network points. This defines cloud computing as the use of technology-enabled

DOI: 10.1201/9781003569251-2

services through the internet, which connects people and organizations [4]. According to Jansen and Grance [5], the National Institute of Standards and Technology (NIST) defines cloud computing as a system model for delivery of a network of computing resources (server, memory, application, customized configuration, etc.) with easy access from cloud service providers.

Features that the NIST reports as a cloud system are on-demand self-service, broad network access, resource pooling, rich elasticity, and measurable service [6]. Infrastructure as a service (IaaS), platform as a service (PaaS), functions as a service (FaaS), and software as a service (SaaS) are terms to describe these cloud service types. These are the categories of the cloud system used according to services supplied to the cloud system. IaaS deals with offering hardware-related services. The infrastructure in this manner provides services to all infrastructure related to the compute, storage, load balancer, and firewall services [7, 8]. IaaS provides users with computing resources; PaaS provides a way to develop and manage versions of your software solution products. SaaS represents a mode of provisioning complete developed software solution applications over the internet on a subscription basis. SaaS providers will be necessary for infrastructure, operating systems, middleware, and data, as and when the customers want. FaaS is a serverless computing that runs the code on allocated resources only when the code is executed and not beforehand. Cloud service providers act as an infrastructure to the customers who are required to pay for what service or resource they use [8].

2.3 LOAD BALANCING IN CLOUD COMPUTING ENVIRONMENT

Earlier it was mentioned that the loading of traffic is the act of dividing or distributing the network system and server load among different nodes on the network system and servers. Additionally, it can also allow sending and receiving data without delay, and every resource should load the same amount of workload at any point in time. Load distribution will also facilitate the maximization of throughput, the reduction in response time, and the prevention of overload [7].

Load balancing is simply a process in cloud computing that makes even the distribution of workload throughout the cloud. Load balancing, utilized in their own cloud computing platform by the cloud service provider, provides an effective solution to end users. This means any number of requests for applications can be fulfilled by means of distributing them to any number of deployed applications at dispersed data centers [9]. A framework used under which all load balancing algorithms work for a cloud environment is shown in Figure 2.1 [7].

As the number of uses of the cloud system increases, the available resources reduces and delays will occur. In this scenario, there is the need for a load balancing process. Network traffic has not been intelligently

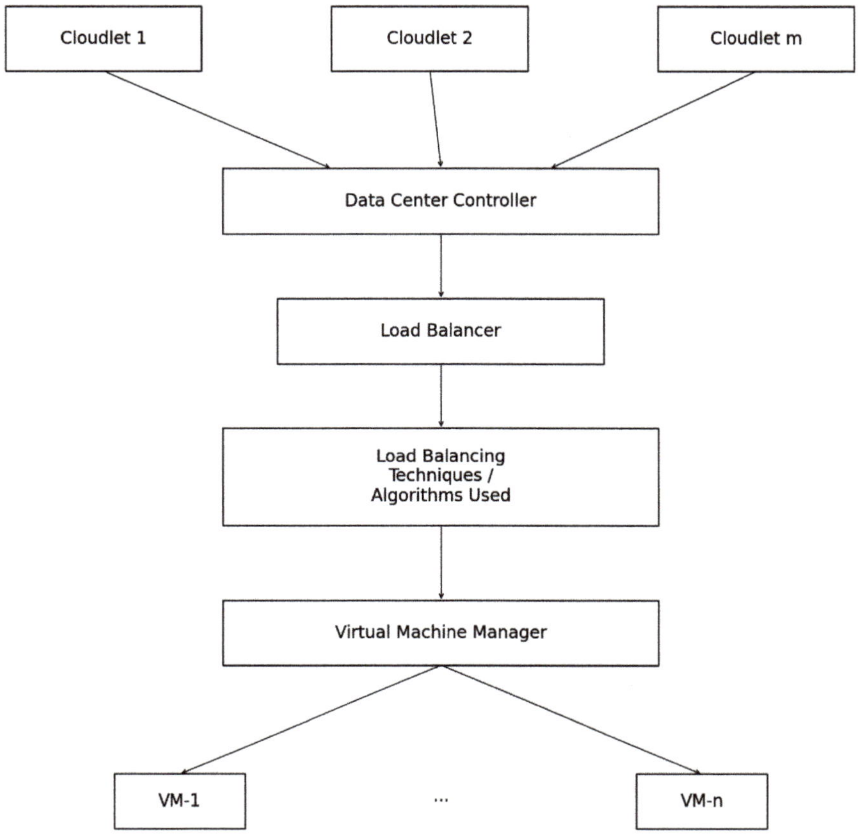

Figure 2.1 Load balancing algorithms execution [7]

managed, and some of the network nodes may become overloaded, while others are underloaded. In load balancing, it has to be avoided. When load balancing algorithms are applied to the cloud environment, however, there are so many other attributes to take into consideration, such as reliability, security, and throughput. The first main objective of the cloud computing load balancing algorithm is to improve the response time with proper load distribution over the system with no node being overloaded [10].

There are a number of aspects associated with the concept of load balancing in the cloud: throughput, migration time, resource utilization, scalability, fault tolerance, overhead, reaction time, and performance factor. "Total" is a term that is used to refer to the total number of requests or jobs in the system. This is how long it takes to move resources or jobs from one node to another. To maintain a record of resource utilization, the resource usage feature is used. One other valuable feature of the algorithm is that it can assign culpability with the display algorithm. Moreover, there is fault tolerance, that is, the capability of completing load balancing without any node of the cluster breaking. The amount of overhead that is associated with the overhead that is reported at the time of a load balancing method is realized if it is listed. The maintenance of the response time says that in any algorithm, some requests should be returned within a predetermined time. System performance is characterized and used from the point of view of efficiency.

Partitioning refers to dividing load balancing algorithms into those that do not contain an initiator and those that are caused by the condition of the system. Whether it is load balancing initiated by the sender or receiver can be divided into three categories: sender initiated, receiver initiated, and symmetric based. If the load balancing is done by the sender, then the load being initiated would be considered as done by the sender. However, the load is initiated by the receiver if the load balancing is performed at the receiver. The load balancing process starts symmetrically, during which the transmitter as well as the receiver manage to achieve a symmetrically balanced load distribution. In each algorithm, the information that is already noted about the system from the present state of service is used [11]. The reason for this is that the techniques for dynamic load balancing make use of the information that has been observed about the system.

2.4 FACTORS AFFECTING LOAD BALANCING IN CLOUD COMPUTING

In the event of problems, the cloud computing environment needs to implement load balancing. Each of the algorithms on the cloud system has a process to maximize resource consumption, minimize usage or response, and minimize the time of response. The problem of load balancing has to be handled in the context of the cloud computing environment, including the following aspects of handling: geographical dispersion of nodes, Single point of failure, heterogeneous nodes, storage management, algorithm complexity, and traffic analysis among other geographical regions [12].

2.5 STUDY OF AVAILABLE ALGORITHMS FOR LOAD BALANCING IN CLOUD ENVIRONMENT

The number of cloud computing load balancing algorithms defined is numerous. In this chapter, we look at various general well-defined algorithms (which perform certain tasks on a series of inputs) as well as some of the new ones (which are static or dynamic).

2.5.1 Static load balancing techniques

Static algorithms use information that becomes available in advance to the system. The process is not a preemptive one, so as soon as a task is assigned to a node then it cannot have it removed by being shifted to any other node. For this, the node performance and assigned work matrix are identified and workload distribution done upon the start of the balancing algorithm [13]. The four different static load balancing techniques applied are the round robin algorithm, opportunistic load balancing, threshold algorithm, and randomized algorithm. In the round robin algorithm, the load that is to be given to each node is equal. Each node may have a load index to keep the load index round robin. Every node is given an equal portion of the time. It means each node has only its time slice to complete the assigned task [14]. Thus, the objective of the opportunistic load balancing algorithm is to keep all the nodes busy. It is an algorithm that randomly distributes the tasks with no weight given to the current load of a node. It doesn't calculate the implementation time of the nodes, and hence performance is poor in the completion of tasks [15]. In the min–min algorithm, a job that requires the less time to completed and a node that requires less time to complete the assigned jobs are selected so that they can be mapped. The time required for the remaining tasks changes after taking one task on a node. Moreover, the tasks to be assigned are changed, at the task that was just assigned is removed from this list. All the tasks that are not assigned to the nodes [16, 17] will continue. For min–max, the uncompleted jobs are set in increasing order of their completion time. The first minimum completion time job on the uncompleted job list is determined and the maximum value is assigned to this machine. The mapped node and the mapped job will be finally selected. When the job is assigned to the selected node, the completion time of the assigned task is updated to the completion time of the chosen node, and the completion time can be up to the time spent by the first task to run all tasks that are assigned to that chosen node [18, 19].

Kumar and Raj [20] propose a new static load balancing algorithm. The active monitoring with the throttled load balancing technique is used as the basis of this approach. One of their techniques is to combine two lists of hash tables into one hash table. These tables contain the virtual machine keys along with their status (AVAILABLE or BUSY) and are used to hold the virtual machine keys together. Once the status check of the table is finished, we should assign the request to the virtual machine assigned. It will change to BUSY once the respective status has been allocated. If the answer cannot find a free node, it places the request line into the waiting queue. It is found that this continues until the method returns when the request is not met. Based on functionality, Table 2.1 presents the comparison of static load balancing techniques.

2.5.2 Dynamic load balancing techniques

In contrast to static load balancing, this study uses the given workload to resolve the distribution of the load across the nodes. We describe a method in which the load, that is, moving object, is carried from an overloaded node to an underloaded node on a basis of weighed estimation of the regions where all the loads are at. The strategies are selection strategy, transfer strategy, location strategy, and information strategy. Moreover, they determine

Table 2.1 Comparison of static load balancing algorithms based on performance parameters

Evaluation criteria	*Round robin*	*Min–min*	*Max–min*	*Threshold*	*Opportunistic*	*Randomized*
High throughput	✓	X	✓	X	X	✓
Low overhead	✓	X	✓	X	X	✓
Fault tolerance	X	X	X	✓	X	X
Supports migration	X	X	X	✓	X	X
Fast response time	✓	X	✓	✓	X	✓
Efficient resource utilization	✓	✓	✓	✓	✓	✓
Scalability	✓	X	X	✓	X	✓
Overall performance	✓	✓	✓	✓	✓	✓

which jobs will be transferred, the nodes responsible for those activities, and create a centralized information hub for load balancing. Based on your choice, dynamic load balancing may be performed in either a distributed or nondistributed fashion. A distributed dynamic load balancing approach, with the application of a scheduling job, can be used for the load balancing process. The load balancing process was distributed over all the nonshared nodes using this method. A distributed load balancing approach is therefore one that features no single node that is operated independently of the others and the nodes must come together to accomplish a single ultimate goal. Each of the nondistributed load balancing is made up of two distinct versions: a central and a semicentral. If the system partitioned in part into a number of clusters for the purpose of semidistributed dynamic load balancing, the load balancing will be centralized. Overall, it results in that, at the node level, it chooses a central node from the cluster to which users should be redirected and therefore the cluster gets load balanced [21]. This method is also limited to a single node in the whole system, so that only one node is able to fulfill the load by connecting with other nodes [22].

Finish time estimation for tasks in such load balancing processing is to be done, in which the current job size and amount of processing power on the node are considered [23]. We propose a novel approach to this estimation that takes two factors into account [24]. The first factor is the nodes that have completed their work, and another factor is the finish time of queued jobs. The upcoming jobs also need to be estimated. The enhanced processing time and response time also lead to an increase in power consumption and carbon emissions.

Another research suggests that the prediction on nodes could be made using machine learning-based prediction of the nodes' load achieved during the run of the program [25]. The authors define an approach of loading prediction to identify the machines underloaded and overloaded so that proactive load balancing can be initiated. There, the proposed model was simulated using a random forest approach in the CloudSim tool environment.

Swift also presented us with a new framework [26] for resource management and work balancing. From the cluster, the authors designed a workload monitoring and analysis system to determine overloaded and underloaded nodes. These implemented and used different algorithms such as split, merge, and pair to do the physical machine workload regulation, and the resource reallocate algorithm to do the virtual machine workload regulation. The designed framework can be easily deployed without the need for any modification at the user end. It was also said that the proposed framework scope is limited to a small number of nodes.

There is another study that provides a model for resource identification and communication in load balancing units [27]. Additionally, the more precise load balancing decision and resource identification decision are made by utilizing resource capacities and running processes' behavior. The load balancing units are different than the resource discovery units that exist for every node. The different load balancing units that have resources available use these message exchanges to avail themselves with sufficient information from the resources about the available resources so that more accurate decisions can be made. On top of getting that, scalability will be easier to come by and response time will increase, as will discovery time of resources.

An optimized round robin method [28], which improves the duration of tasks and the virtual machine's capacity of task execution, is presented. This method finds the node having minimum completion time considering the current load of a virtual machine and an executing job. Upon job completion, the load balancer allocates the load among the nodes. They take three steps in their approach. The first step is a static priority table for primary placement, and the second is a dynamic one that considers the load in designing virtual machines and a provisional completion time related to the current load and the estimated time duration of completion. Finally, the identical virtual machine node is used for task allocation at the last stage. In particular, the results of this method show that it is effective for heterogeneous and homogeneous tasks using heterogeneous resources (virtual machines, VMs).

Virtual machine scheduling problems [29] have been proposed as a combined approach based on the ant colony optimization and particle swarm optimization. Instead, it checks for maximum available memory each time the load of a server is accessed when a request comes in to be processed. With this historical information, it predicts the workload of the new input requests. Memory availability and size of the task are either accepted or rejected upon arrival for a new request. For this strategy, improvisation takes place in the resource computation time of tasks. According to the authors, this proposed technique's response time is faster than the traditional ant colony optimization technique only on homogeneous servers.

On the other hand, a task scheduling and load balancing method named DeMS is also provided [30], composed of the staged task migration (STM), querying and migrating task (QMT) and on demand scheduling algorithms. The first algorithm is employed in order to perform the time overhead communication between the master node and the slave node. The second part of the algorithm responsibility is workload balancing. The third one is for scheduling the tasks. In the heavy load situation, QMT improves the balance of workload by transferring the last arriving task load to the node that is underloaded. QMT and SMT both proved to work well for dependent and independent tasks but have the problem of high transmission and scheduling time, as stated by the authors.

Ghoneem and Kulkarni defined a modified active load balancing technique [31]. Like the existing technique, however, the author's method did allocate unevenly during peak time, but made a more efficient decision based on the reservation and allocation table as to where to bring the request to allocate it to the nodes. Although current methods increase response time, the author's method reduces user response time.

Chen et al. [32] proposed a load balancing technique inspired by honey bee behavior with the intention of maximizing the throughput by better balancing the traffic load between the nodes. The balance it offers between the tasks on the nodes is another plus: it balances the priorities of the tasks on the nodes in such a manner that minimizes waiting time. This applies to cases whereby each task is independent.

To achieve load balancing with more efficiency and higher effectiveness in a cloud environment, the three qualities of service, migration, and response time as load balancing parameters have been considered to formulate a modified bee colony algorithm [33]. This method is inspired by the honey bee's search for food. The goal is to make the deadline completion and the number of virtual machine migrations as minimal as possible. The improvements seen in the proposed method's results demonstrate improved QoS delivery to end users.

Wang et al. [34] introduced a technique called adaptive scheduling with quality-of-service satisfaction (AsQ) to improve resource usage and by delay extension, so the tasks will respond for a greater time in a hybrid cloud environment. The strategies have been recommended regarding the utilization rate, completion time, and execution time of private and public clouds, and then these parameters are applied to maximize the allocation of jobs on the private cloud and are evaluated. To improve resource utilization and reduce runtime in private clouds, their technology was used for multiple runtime executions and scheduling algorithms. Reduction of the rate of the transferred task is done with the employment of the minimum cost to reduce the rate of the transferred task for public cloud services. In the context of QoS, the outcomes show superior performance.

The overlapping computation request process with computer loading and server processing has raised the requirement of an integrated cloud load balance architecture [35] to solve that problem. It also provides a load balancing technique for virtual and physical web servers, with server load information, priority, and running capacity information simultaneously to improve response time.

The hybrid algorithm is composed of three different cloud brokering algorithms (round robin, throttled load balancing and performance optimized service broker, service proximity broker) for optimization of the system's performance [36]. The authors state that they are load aware and cost aware. It generated lower cost but also had a load aware algorithm that still provided lower processing time. The cost aware algorithms result in cost minimization. In the service broker algorithm, the server is selected based on user programmatically. When a data center is selected, it is by means of a service proximity broker algorithm. This is close to the client region selection. A throttled load balancer maintains a table of all virtual machine availability. All these algorithms were used in pairs and asked to process a request that reduces the processing time and response time to minimum values.

An agent based allocation algorithm (ABAA) for work scheduling in the cloud setting was presented that dealt with the emergent job scheduling issues [37]. Therefore, an equitable competitive rule and a dynamic adjustment rule for the customer balance and job variation, respectively, are employed. The method of the bidder is further improved by the buffer pool despite the fact that he does not possess the authority to assign the task for resource allocation. Approaches were used to assign the resources that did not disturb transmission and processing time.

To select the physical host for task deployment, the researcher introduced LB-BC, a Bayes and clustering method. Rather than choosing an immediate load balancing method, the authors defined the long-term process of load balancing as a whole. Specifically, the approach consists of constraints that are designed to choose the best host and compute posterior probabilities over physics nodes. After that, it produces a set of best clusters for the physical host by integrating the clustering process with balancing efforts, data variance, and requested jobs. The authors state that the proposed method aims at improving the performance of cloud, as well as data center designs, while enhancing throughput and reducing the incidence of task deployment failures.

Based on the genetic algorithm and ant colony optimization algorithm, the hybrid task scheduling algorithm has taken an action to speed up the user's task finish [39]. The genetic algorithm selects the most suitable optimal solution and the resource is searched by the ant colony algorithm, which will choose the optimal resource. The combination of both genetic and ant colony algorithms benefits the inherited properties from genetic such as load balancing and improved efficiency from ants.

A new cloud job scheduler is presented in which jobs are expected to announce their deadline by each tenant and a quantity of reserved bandwidth for each job is obtained by using the elastic feature to maximize total job rewards [40]. This method aims to reduce overall transmission time, completion time for jobs, and the cost of data. They reported that tasks deal with dynamic changes in the availability of resources, dependency of data fetching time on the size and location of the tasks of reduced tasks, and the impact from load on data access latency. Results from the experiment indicated that this job scheduler is both effective and efficient.

One study utilized virtual machine live migration in a collaborative agent-based load balancing method that balances the workloads over commodity and heterogeneous servers [41]. The method is agent based: the nodes to be migrated, destination nodes, migration strategies, acceptance polices for virtual machines, and load balancing program for choice of initial host are identified. Results of the authors' method show that the balance in load as well as the response time balance and resource utilization can be done in a distributed manner, with the overhead in migration increasing at the same time.

Load balancing of the multimedia streams is necessary with such a dynamic nature of the strategy. This context has suggested the Inter Cloud Manager (ICM) cluster-based job dispatching algorithm for the large-scale cloud environment [42]. The processing was done across two tasks: clustering and decision-making. The hello packet is used for broadcasting packets to the connected neighbor servers, thus clustering. Clustering is simply to manage complexity of platform, monitoring activities, and corresponding QoS of the host. The job request is forwarded, and the execution results are obtained under a network delay that occurs while forwarding the job request and receiving the execution results, and the decision of transferring the job request is taken based on the required execution time. The authors compared the experimental performance with the honey bee foraging algorithm (HFA), workload and client aware policy (WCAP), and ant colony algorithm. According to author, the experimental results of their method provide better response time.

Another hybrid method, named SRDQ, proposes a different combination of the features of shortest job first and round robin schedulers [43]. Dynamic task quantum is the idea to separate long and short tasks into two separate queues with each waiting for whose waiting time in the task would be balanced. Over the evaluation, the author also uses the CloudSim simulator. The shortest job first, round robin, and time slice priority-based round robin algorithms were executed on the different datasets. Simulation results showed that the waiting time, response time, and starvation of large tasks were reduced.

To maximize the resource utilization, load balancing based on multiagent-based architecture is used. Load shifting at the IaaS level is taken into consideration using the multiple agent based load balancing algorithm [44]. It first selects between two load balancing techniques, initially the sender initiated and later the receiver initiated. This ensures sticking to the service level agreement and never allowing the wait time of the tasks in the queue to increase. The design of the methods is specified with three agents: virtual machine monitor agent (VMM agent), data center monitor agent (DcM agent), and negotiator ant agent (NA agent). All the virtual machines and supporting information are loaded by the monitoring tool with VMM. The information policy is managed by the DcM agent and is the responsibility of the machines categorized for characteristics. In many cases, the DcM agent starts the NA agent, which is responsible for collecting the status of other virtual machines in other data centers. According to the authors, simulation result of their method is better than the improvised makespan and response time.

Improved load balancing and scalability is given through a class of scalable distributed loop self-scheduling schemes [45]. This model will be applicable to heterogeneous as well as homogeneous environments. Large-scale clustering has been used to apply the proposed scheme on the Mandelbrot set, adjoint convolution, matrix multiplication, and quick sort scientific computations, which also gives better load balancing with improvement in performance and scalability.

Furthermore, the resolution of the issues of the centralized controller in huge data centers is also addressed using another approach called functional small world, which leads to effective load balancing in a heterogeneous system environment [46]. In order to achieve higher efficiency, this approach breaks the network in particular regions with allocated controller, which, in addition, reduces the network diameter, communication overhead, time delays and number of nodes exchanging the information. In addition, the authors also defined another load balancing algorithm in which nodes calculated the average effective load and the defined the capacity. If the load is imbalanced, migration was performed in the form of controller migration to handle it in the dynamic load. According to the authors, their simulation of the method yielded improved throughputs, response times, communication overhead, and cost of movements over the original approach.

Establishing the scheduling architecture with multiple schedulers is shown to solve the issues of using large-scale parallel tasks and reduces their cost of execution. Thus, the cost-efficient scheduling method (CESM) method is based upon this aspect and determined by the use of the weighted random schedule policy for assigning devices to tasks [47] to avoid competition between tasks and reduce device load. Parameters such as cost and communication time, cost and execution time, and cost and busy state are weighed with tasks. A higher weight on this machine makes it a task executor machine, since it has lower cost. This scheduling method provides better results in execution time and cost, which is a benefit. The authors tested the effectiveness and consistency of the CESM with four experiments on the arrival time test, the device dependence test, the task structure test, and the device set test.

A self-organizing approach to load balancing in a network providing heterogeneous services has been proposed [48]. Self-organized techniques for clustering peers based on service types and creating and maintaining super peer topologies are proposed as they can provide a way to utilize a small fraction of the peer node base to satisfy all query masking goals. Different configurations and environmental conditions have been implemented by the authors using the PeerSim simulator and achieved reduced service response times.

The load balancing approach power aware load balancing algorithm (PALB) [49] is devised to work on IaaS cloud architecture. Using compute nodes status and utilization percentages, it determines the number of nodes operating. According to the authors, within the context of their proposed solution based on a job schedule and a size of the virtual machine request distributing approach, their scheme can be as much as 70–90% more energy efficient when compared to non-power-aware load balancing algorithms.

Examples of high computational client server applications with the need for real-time quality of service (QoS) are online games with multiplayer facility. Though maintaining the sessions of high number of resources is not efficient for maintaining QoS requirements, it is mandatory. Nae, Prodan, and Fahringer [50] proposed a new way for cost-efficient hosting of massively multiplayer online games (MMOG) sessions over cloud resources is proposed. It helps with the accurate use of resources, which is required for smooth execution, and also allocates resources for load and services. Once the resource allocation is determined for a play session, the game server keeps this load as determined by the load balance service. As per simulation results, the authors received reduced hosting costs.

VectorDot was proposed for balancing the load with a resource chain of multiple-layer data centers that are established with the resource constraints of hierarchy and multidimensionality [51]. Their method was evaluated on a number of synthetic and real data center testbeds. The method was presented to be end-to-end valid as per various conditions to solve the server switch overload problem efficiently.

The proposed central load balancing policy for virtual machines (CLBVM) evenly distributes the loads and enhances system performance [52]. Load balancing decisions belong to globally stated information with location rules and centralized information. The authors state that this method can't be used with fault tolerance (fault-tolerant mode) in the process.

Mehta, Kanungo, and Chandwani [53] proposed a new content aware load balancing policy, the workload and client aware policy (WCAP). Both the workload aware policy and the client aware policy are used to process the request distribution policy. This approach is also used in web servers and for internet distribution service purposes of load balancing, with a unique and special property (USP) parameter being introduced to define the special and unique property of the processing node and request. It also selects the suitable node for the processing of the request based on this parameter.

CARTON algorithms use the concepts of load balancing and distributed rate limiting in the cloud control [54]. The purpose is to reduce the load cost by distributing loads uniformly to each server, and also relatively distribute the resources. They find that their methods incur less overhead for computation and communication and all servers equally performed.

The ACCLB is an algorithm with the aim to find if we can find the smallest path between the requests and nodes in an open cloud computing network [55]. Upon a task starting, it moves along the pathway according to the ant phenomenon. It continues on the same path after finding an underloaded processor. However, if an overloaded processor is found, it moves back to the previous underloaded processor and so continues looking for another suitable path. The shared computation and storage over long distances also mean that this approach also gets optimal job scheduling. The authors analyzed their approach and built a prototype to quantify it.

Random walk sampling can reduce the effects of communication latency if the geographic awareness factor is used [56]. The load balancing method presented by the authors used biased sampling and a self-organized graph that depends only on locally available information to identify resources and to allocate the load. Node connectivity and biased sampling used for random sampling in load balancing can be identified using virtual graphs. Simulation of the proposed approach was also shown to be a reliable, scalable, and effective load balancer, although there is a trade-off between load balancing performance and reliable load balancing in the high load scenario.

The virtual machine scheduling problem has been defined by a genetic algorithm-based load balancing approach [57]. Using the current and historical state of the system and genetic algorithm, it computes the effects of the needed virtual machine resources on the deployed system. It then selects a solution for load balancing as well as reduced dynamic migration. Experimental results of the approach were said to have worked effectively for load balancing and resource utilization.

Two capable distributed load balancing algorithms, namely, deadline based distributed load balancing (DDLB) and capability-based distributed load balancing (CDLB), are proposed for distributed load balancing in a cloud environment [58]. The algorithm's objective is getting global balancing of the cloud system by redistributing more loads toward the nearest nodes of the overloaded server. Both algorithms equally distributed the load to storage servers in the cloud environment. Load balancing is performed for DDLB with the service time per client request as per the service rate and server queue, and for the CDBL with the server queue and service rate. Both algorithms reduced response times and request delays, and effectively utilized the server capabilities.

Inspired by nature, the hybridized BAT and ABC algorithm, termed the HBATAABC, algorithm has been proposed to find an improvement of the dynamic load balancing problem proposed [59]. Based on VM transfer strategy rules, the proposed algorithm performed better in the VM's data allocation system in the cloud data center. Th algorithm was evaluated based on performance throughput, average time, and network stability. The proposed algorithms yielded a decrease of 1.30% in network execution time, stability, and throughput.

A binary bird swarm optimization load balancing (BSO-LB) algorithm was proposed by Mishra and Majhi [60]. Removing required tasks from an overloaded VM can be done by using available resources, moving the tasks to an underloaded VM. Using the proposed method, each task has its own information and is tracked and updated. The proposed algorithms are shown to minimize makespan while maximizing resource utilization by adopting the correct function in the proposed algorithms. Additionally, their proposed algorithms improved response time, makespan, and resource use, and worked well on any cloud environment with independent, preemptive, nonpreemptive tasks.

A hybrid solution was proposed by using the advantages of the Harries hawks optimization (HHO) and pigeon-inspired optimization (PIO) approaches, and also improvised response time for each request sent by users [61]. To analyze underloaded and overloaded virtual machine balancing of new requests, the HHO approach concept is used. The PIO approach leads to an optimal response time. An improvement in performance and reduction in the response time were the result of the proposed approach. However, on their limitation of increasing the volume of cloud infrastructure and data centers, the authors state that the quality of the proposed system as well as the number of virtual machines will affect the number of virtual machines, which will result in higher system costs and reduced response time.

A new approach of DLB3M (dynamic load balancing max–min and max) algorithms has been defined [62]. It uses ant colony optimization and load balancing max–min and max algorithms to minimize execution time, efficiently use resources, and produce the best optimal solution for load balancing. The resource-utilizing capabilities of the server node is used to assign user tasks on the basis of the resource-utilizing capabilities. They are scalable and fault tolerant because they are compiled once for every job task entered in the cloud environment and dynamically explore the free node information. They have stated that their proposed system leads to less execution time, efficient resource utilization, response time, and good throughput [63–66]. Table 2.2 lists the general comparison of the studied dynamic load balancing approaches.

2.6 CONCLUSION

Cloud computing is one of the growing technologies involving the utilization of data and data sharing generated by large-scale computing and data centers exchanging information with each other. This means it keeps equilibrium among all of the nodes in the network as the network is traversed. An analysis of the problems and difficulties encountered during load balancing is done in this chapter through a comparative study method. An attempt has been made to conduct complete research on the methods of static and dynamic load balancing, their merits and demerits, and comparison of several parameters that determine the process of load balancing. A condition not discussed in this work is when no fault tolerance is implemented as a

Table 2.2 Comparative analysis of dynamic load balancing algorithms in cloud computing environments

Source	*Approach*	*Methodology*	*Advantages*	*Drawbacks*	*Simulation tools*
[24]	Forecasting task completion using node performance and current workload	Assigns tasks based on idle nodes and predicts finish times for pending and upcoming jobs	Enhanced processing speed and responsiveness	Higher energy usage and increased carbon footprint	CloudSim via Java
[25]	Predictive load balancing using machine learning	Identifies under- and overloaded nodes using ML algorithms	Optimized resource usage and fewer task migrations	Not validated on an actual cloud platform	CloudSim with Weka
[26]	Load management for rapid storage response	Applies operations like split, merge, and pair on physical machines; reallocates virtual resources dynamically	Faster task completion	Limited to small-scale environments	Not mentioned
[27]	Process- and resource-aware balancing	Distributed load balancing and discovery with communication between units on different nodes	Supports scaling and reduces job response time	Extended time needed for discovering resources	KARE Grid Sim
[28]	Balancing based on VM capacity and task duration	Selects the VM with the shortest predicted task finish time, based on real-time load	Performs well with varied tasks and system heterogeneity	Less efficient for uniform systems	CloudSim
[29]	Hybrid optimization using ACO and PSO	Uses historical workload data to predict demand and allocate tasks based on available memory	Improved task efficiency and resource use	High cost and limited to homogeneous systems	Implemented in C
[30]	DeMS: A multi-algorithm scheduling and balancing technique	Integrates on-demand scheduling with two-stage migration via QMT and STM methods	Effective for managing both dependent and independent tasks	Time-consuming scheduling and data transfer	Not provided

(*Continued*)

Table 2.2 (Continued)

Source	*Approach*	*Methodology*	*Advantages*	*Drawbacks*	*Simulation tools*
[31]	Enhanced active load distribution strategy	Utilizes a reservation and assignment table for optimized task processing decisions	Reduces user wait time and supports elasticity	Fixed task distribution across data centers	Not specified
[32]	Load distribution inspired by honey bee behavior	Mimics foraging patterns of bees for dynamic task allocation	Boosts overall system throughput	Effective only for nondependent tasks	Not specified
[33]	Adapted bee colony mechanism	Treats tasks like bee behavior; underutilized VMs act as food sources	Low latency and enhanced quality of service for users	High complexity and limited scalability	Not specified
[34]	ASQ: Adaptive scheduling based on QoS	Employs multiple scheduling techniques and runtime executions for efficient resource use	Enhances resource usage and reduces task delays	Ignores energy and operational cost considerations	CloudSim
[35]	CLB: Cloud load balancing framework	Manages overloads by factoring in server usage and task processing load	Offers strong scalability	May suffer from delayed responses under load	CloudSim
[36]	Integrated brokering and load balancing model	Combines round robin, throttled balancing, and advanced brokering methods for hybrid optimization	Decreases processing and response durations	Performance decline due to increased execution time	CloudAnalyst
[37]	ABAA: Agent-based allocation algorithm	Utilizes competitive fairness and dynamic buffer adjustment for task allocation	Effective in allocating resources optimally	Higher overhead in data transfer and task processing	CloudSim
[38]	LB-BC: Load balancing using Bayesian and clustering techniques	Selects optimal hosts using probabilistic modeling with limited constraints for global balancing	Boosts throughput, lowers failure rates, and enhances service delivery	Not validated under real-time conditions	CloudSim

(Continued)

Table 2.2 (Continued)

Source	*Approach*	*Methodology*	*Advantages*	*Drawbacks*	*Simulation tools*
[39]	Hybrid scheduling: Genetic and ant colony optimization	Merges genetic and ACO methods for resource selection and execution optimization	Accelerates task completion and improves load distribution	No implementation on a real cloud platform	CloudSim
[40]	Topology-aware transmission scheduling	Allocates jobs based on tenant-declared deadlines and reserved bandwidth using elasticity principles	Enhances cluster usage and reduces task completion time	Lacks bandwidth reservation model; untested under diverse networks	Not available
[41]	Collaborative agent-based load management	Migrates VMs dynamically to balance load across mixed server infrastructures using agent support	Improves response time and resource usage	Migration introduces additional system overhead	Java, JADE agent platform
[42]	ICM: Inter-cloud cluster-based job distribution	Employs hello packet broadcasting for job clustering and server interaction	Faster task responses through effective clustering	Network congestion may lead to packet loss	Not available
[43]	SRDO: SJF and round robin-based scheduling	Divides tasks by size into separate queues and adjusts execution with dynamic quantum	Decreases task wait and response duration	Risk of starvation for longer tasks	CloudSim
[44]	MA: Multi-agent load distribution strategy	Utilizes both sender- and receiver-initiated agent cooperation for load handling in IaaS	Improves overall task duration and system responsiveness	Lacks a self-termination mechanism for agents	CloudSim, Java Programming
[45]	Scalable loop self-scheduling model	Designed for large-scale clusters using scientific computing patterns like Mandelbrot and matrix operations	Enhances system scalability and ensures load distribution	Limited by lack of loop-level parallel implementation	FlexCloud, C++ Programming

(Continued)

Table 2.2 (Continued)

Source	*Approach*	*Methodology*	*Advantages*	*Drawbacks*	*Simulation tools*
[46]	Functional small world model	Divides the network into zones managed by dedicated controllers; node load is evaluated based on capacity averages	Increases system throughput and responsiveness, reduces communication and migration costs	High power consumption	Sim-Java
[47]	CESM: Cost-efficient scheduling method	Reduces contention between jobs by using a weighted random scheduling approach to map devices to tasks	Minimizes execution costs and device workload; improves runtime efficiency	Parameter optimization is lacking	Not available
[48]	Self-organizing load distribution	Uses self-organization for peer clustering based on service type, forming and maintaining super-peer structures	Decreases service latency	Method not validated on real systems	PeerSim
[49]	PALB: Power-aware load balancing	Makes scheduling decisions using compute node status and utilization data; adjusts active node count accordingly	Saves 70–90% of energy consumption	Assumes equal priority for all tasks, which is unrealistic	Not available
[50]	MMOG session scheduling	Allocates resources by accurately identifying session demands to support smooth multiplayer game hosting	Reduces infrastructure costs	Few QoS violations from minimal workload events	RuneScape, AWS EC2
[51]	Vectordot	Addresses multilevel and hierarchical resource limitations across data centers for balanced workload distribution	Valid under diverse conditions with minimal server or storage overloads	Simulation details not available	Not available

(*Continued*)

Table 2.2 (Continued)

Source	*Approach*	*Methodology*	*Advantages*	*Drawbacks*	*Simulation tools*
[52]	CLBVM: Centralized load balancing for VMs	Centralized decision-making using global and location-aware policies for load balancing among VMs	Scalable to support more nodes in live cloud setups	Lacks real-world traffic testing and fault tolerance analysis	Not available
[53]	WCAP: Workload and client-aware hybrid policy	Merges workload-driven request routing with client-based service policies in a decentralized manner	Simple, efficient, and improves system-wide performance	Slightly slower than round robin or latency-based methods	Not available
[54]	CARTON algorithm	Distributes load evenly across servers and allocates resources equitably	Reduces overhead in both communication and computation	Lacks redundancy for control packet loss	Not available
[55]	ACCLB: Ant colony and complex network theory	Determines shortest routing paths between tasks and compute nodes using optimization techniques	Supports optimal task scheduling	Does not clearly specify effect of dynamic parameter variations	Not available
[56]	Geo-aware random walk sampling	Implements load balancing via self-organized virtual graphs enhanced with location-aware random sampling	Ensures reliable and scalable performance	May sacrifice performance for reliability in high-load situations	Not available
[57]	GA-based VM scheduling	Applies genetic algorithm using current and historical data to predict VM resource needs and balance loads	Increases utilization and balances loads effectively	Potential for imbalance under some conditions	ISF® OpenNebula

(*Continued*)

Table 2.2 (Continued)

Source	*Approach*	*Methodology*	*Advantages*	*Drawbacks*	*Simulation tools*
[58]	CDLB and DDLB: Capability-based and deadline-based distributed load balancing	CDLB leverages server queue lengths and service rates for efficient load distribution; DDLB incorporates client-specific service times alongside server metrics for task scheduling	Decreases response delays, boosts server use, and increases successful task completions	Not specified	Java, NetBeans 7.0
[59]	HBATAABC: Hybrid BAT and artificial bee colony algorithm	Implements VM task transfer strategies to optimize data placement, improving performance and stability within cloud data centers	Achieved notable reduction in execution time and better throughput stability	Not specified	CloudSim, Origin Software
[60]	BSO-LB: Bird swarm optimization load balancing	Relocates tasks from overburdened VMs to underutilized ones, updates task status iteratively based on resource availability	Enhances response time, makespan, and resource efficiency; supports varied task types	Not specified	CloudSim
[61]	HHO-PIO: Hybrid Harries hawks and pigeon-inspired optimization	Uses HHO to assess VM load for new request balancing; PIO fine-tunes decisions for minimal task delay	Increases system efficiency and lowers response time	Larger data centers may degrade performance quality	CloudSim, Java, NetBeans IDE
[62]	DLB3M: Dynamic max–min-based load balancing	Assigns tasks to servers based on real-time resource capabilities; dynamically monitors and allocates to idle nodes	Faster task execution, efficient resource use, and improved throughput	Lacks full adaptability to all execution scenarios in distributed systems	Visual Studio 2015, Newtonsoft. Json, DevExpress 16.1.4, WebClient, C#

complete solution for virtual machine migration. However, to reduce the power consumption and for green computing of the cloud environment, we have to adopt more advanced approaches that provide a lower power state and the ability to do more.

REFERENCES

1. V. Reddy, K. Surya, M. Praveen, B. Lokesh, A. Vishal, and K. Akhil (2016). Performance analysis of load balancing algorithms in cloud computing environment. *Indian Journal of Science and Technology, 9*. https://doi.org/10.17485/ijst/2016/v9i18/90697.
2. A. Sidhu, A., and S. Kinger (2005). Analysis of load balancing techniques in cloud computing. *International Journal Of Computers & Technology*, 4(2), 737–741. https://doi.org/10.24297/ijct.v4i2C2.4194.
3. Zenon Chaczko, Venkatesh Mahadevan, Shahrzad Aslanzadeh, and Christopher Mcdermid (2011). Availability and load balancing in cloud computing. In *2011 International Conference on Computer and Software Modeling*, IPCSIT vol. 14. Singapore: ACSIT Press.
4. Farhan Bashir Shaikh, and S. Haider (2011). Security threats in cloud computing. In *2011 International Conference for Internet Technology and Secured Transactions*, Abu Dhabi, pp. 214–219.
5. Wayne Jansen, and Timothy Grance (2011). Guidelines on security and privacy in public cloud computing. National Institute of Standards and Technology Gaithersburg. https://doi.org/10.6028/NIST.SP.800-144.
6. Peter Mell, and Timothy Grance (2011). The NIST definition of cloud computing. *NIST Special Publication 800-145*.
7. N. S. Raghava, and Deepti Singh (2014). Comparative study on load balancing techniques in cloud computing. *Open Journal of Mobile Computing and Cloud Computing,* 1(1), 18–25.
8. Blair Felter (2020). Th*e Different Types of Cloud Computing and How They Differ.* https://www.vxchnge.com/blog/different-types-of-cloud-computing (accessed on 12 Oct, 2020).
9. B. Mondal, K. Dasgupta, and P. Dutta (2012). Load balancing in cloud computing using stochastic hill climbing-a soft computing approach. *Procedia Technology*, 4, 783–789. https://doi.org/10.1016/j.protcy.2012.05.128.
10. S. Moharana. (2013). Analysis of load balancers in cloud computing. *International Journal of Computer Science and Engineering*, 2, 101–108.
11. David Escalante, and Andrew J. Korty (2011). Cloud services: policy and assessment. *EDUCASE Review*, 46(4), 60–61.
12. Pawan Kumar, and Rakesh Kumar (2019). Issues and challenges of load balancing techniques in cloud computing: a survey. *ACM Comput. Surv*, 51(6), 35. https://doi.org/10.1145/3281010.
13. Daniel Grousa, and Anthony T. (2005). Non-cooperative load balancing in distributed systems. *Journal of Parallel and Distributing Computing*, 2005, 1022–1034. https://doi.org/10.1016/j.jpdc.2005.05.001.
14. H. Abubakar, and R. Usman (2004). Evaluation of load balancing strategies. *National Conference on Emerging Technologies*, 6(6), 32–37.

15. M. S. Jyoti, and R. Mishra (2019). Cloud computing and load balancing in cloud computing –survey In *2019 9th International Conference on Cloud Computing, Data Science & Engineering (Confluence)*, Noida, India, pp. 51–55. https://doi.org/10.1109/CONFLUENCE.2019.8776948.
16. Maimit A. Patel, and Rutvik Mehta (2015). A comparative study of heuristic load balancing in cloud environment. *International Journal of Advance Engineering and Research Development,* 2(1), 1–24.
17. G. Liu, J. Li J., and J. Xu (2013). An improved min-min algorithm in cloud computing. In: Z. Du (eds) *Proceedings of the 2012 International Conference of Modern Computer Science and Applications. Advances in Intelligent Systems and Computing*, vol 191. Springer, Berlin, Heidelberg. https://doi.org/10.1007/978-3-642-33030-8_8.
18. T. Kokilavani, and D. I. George Amalarethinam (2011).Load balanced Min-Min algorithm for static meta task scheduling in grid computing. *International Journal of Computer Applications,* 20(2). https://doi.org/10.5120/2403-3197.
19. G. Ritchie, and J. Levine (2003). A fast, effective local search for scheduling independent jobs in heterogeneous computing environments. *Proceedings of the 22nd Workshop of the UK Planning and Scheduling Special Interest Group*, Glasgow, United Kingdom.
20. A. Kumar, and A. Raj (2015). A new static load balancing algorithm in cloud computing. *International Journal of Computer Applications*, 132, 13–18. https://doi.org/10.5120/ijca2015907285.
21. S. Begum, and C. S. R. Prashanth (2013). Review of load balancing in cloud computing. *IJCSI International Journal of Computer Science*, 10(1), 1694–1784.
22. Shiny (2013). Load balancing in cloud computing: a review. *IOSR Journal of Computer Engineering (IOSR-JCE)*, 15(2), 22–29.
23. Ali Mohammad Alakeel (2010). A guide to dynamic load balancing in distributed computer systems. *International Journal of Computer Science and Network Security*, 10(6), 153–160.
24. N. K. Chien, N. H. Son, and H. Dac Loc (2016). Load balancing algorithm based on estimating finish time of services in cloud computing. In *2016 18th International Conference on Advanced Communication Technology (ICACT),* Pyeongchang, pp. 228–233. https://doi.org/10.1109/ICACT.2016.7423340.
25. Anju Bala, and Inderveer Chana (2016). Prediction-based proactive load balancing approach through VM migration. *Engineering with Computers*, 32(4), 581–592. https://doi.org/10.1007/s00366-016-0434-5.
26. Zhenhua Wang, Haopeng Chen, Ying Fu, Delin Liu, and Yunmeng Ban (2015). Workload balancing and adaptive resource management for the swift storage system on cloud. *Future Generation Computer Systems,* 51, 120–131. https://doi.org/10.1016/j.future.2014.11.006.
27. M. N. Arab, and M. Sharifi (2014). A model for communication between resource discovery and load balancing units in computing environments. *Journal of Supercomputing*, 68, 1538–1555. https://doi.org/10.1007/s11227-014-1124-y.
28. D. Devi, and V. Uthariaraj (2016). Load balancing in cloud computing environment using improved weighted round robin algorithm for nonpreemptive

dependent tasks. *The Scientific World Journal*, 2016, 1–14. https://doi.org/10.1155/2016/3896065.
29. K. M. Cho, P. W. Tsai, C.W. Tsai, et al. (2015). A hybrid meta-heuristic algorithm for VM scheduling with load balancing in cloud computing. *Neural Computing and Applications*, 26, 1297–1309. https://doi.org/10.1007/s00521-014-1804-9.
30. Yu Liu, Changjie Zhang, Bo Li, and Jianwei Niu (2015). DeMS: A hybrid scheme of task scheduling and load balancing in computing clusters. *Journal of Network and Computer Applications,* 83, 213–220. https://doi.org/10.1016/j.jnca.2015.04.017.
31. Mohammad Ghoneem, and Lalit Kulkarni (2017). An adaptive MapReduce scheduler for scalable heterogeneous systems. In: S. Satapathy, V. Bhateja, and A. Joshi (eds) *Proceedings of the International Conference on Data Engineering and Communication Technology. Advances in Intelligent Systems and Computing*, vol. 469. Springer, Singapore. https://doi.org/10.1007/978-981-10-1678-3_57.
32. S. -L. Chen, et al. (2016). CLB: a novel load balancing architecture and algorithm for cloud services. *Computers and Electrical Engineering*, 58, 154–160. http://dx.doi.org/10.1016/j.compeleceng.2016.01.029.
33. K. R. Remesh Babu, and P. Samuel (2016). Enhanced bee colony algorithm for efficient load balancing and scheduling in cloud. In: V. Snášel, A. Abraham, P. Krömer, M. Pant, and A. Muda (eds) *Innovations in Bio-Inspired Computing and Applications. Advances in Intelligent Systems and Computing*, vol. 424. Springer, Cham. https://doi.org/10.1007/978-3-319-28031-8_6.
34. W. J. Wang, Y. S. Chang, W. T. Lo, et al. (2013). Adaptive scheduling for parallel tasks with QoS satisfaction for hybrid cloud environments. *Journal of Supercomputing*, 66, 783–811. https://doi.org/10.1007/s11227-013-0890-2.
35. Dhinesh Babu, and Venkata Krishna (2013). Honey bee behavior inspired load balancing of tasks in cloud computing environments. *Applied Soft Computing Journal* 13(5), 2292–2303.
36. Ranesh Kumar Naha and Mohamed Othman (2016). Cost-aware service brokering and performance sentient load balancing algorithms in the cloud. *Journal of Network and Computer Applications* 75(C), 47–57. https://doi.org/10.1016/j.jnca.2016.08.018.
37. C. Chen, X. Zhu, W. Bao, L. Chen, and K. M. Sim (2013). An agent-based emergent task allocation algorithm in clouds. In *2013 IEEE 10th International Conference on High Performance Computing and Communications & 2013 IEEE International Conference on Embedded and Ubiquitous Computing*, Zhangjiajie, pp. 1490–1497. https://doi.org/10.1109/HPCC.and.EUC.2013.210.
38. J. Zhao, K. Yang, X. Wei, Y. Ding, L. Hu, and G. Xu (2016) A Heuristic clustering-based task deployment approach for load balancing using Bayes theorem in cloud environment. In *IEEE Transactions on Parallel and Distributed Systems*, vol. 27(2), 305–316. https://doi.org/10.1109/TPDS.2015.2402655.
39. Z. Wu, S. Xing, S. Cai, Z. Xiao, and Z. Ming (2017). A genetic-ant-colony hybrid algorithm for task scheduling in cloud system. *In Lecture Notes in Computer Science,* 183–193. https://doi.org/10.1007/978-3-319-52015-5_19.

40. H. Shen, L. Yu, L. Chen, and Z. Li (2016). Goodbye to fixed bandwidth reservation: job scheduling with elastic bandwidth reservation in clouds. In *2016 IEEE International Conference on Cloud Computing Technology and Science (CloudCom)*, Luxembourg City, pp. 1–8. https://doi.org/10.1109/CloudCom.2016.0017.
41. J. O. Gutierrez-Garcia, and A. Ramirez-Nafarrate (2015). Agent-based load balancing in Cloud data centers. *Cluster Computing,* 18, 1041–1062. https://doi.org/10.1007/s10586-015-0460-x.
42. B. Kang, H. and Choo (2016). A cluster-based decentralized job dispatching for the large-scale cloud. *EURASIP Journal on Wireless Communications and Networking,* 2016, 25. https://doi.org/10.1186/s13638-016-0523-6.
43. S. Elmougy, S. Sarhan, and M. Joundy (2017). A novel hybrid of shortest job first and round Robin with dynamic variable quantum time task scheduling technique. *Journal of Cloud Computing,* 6, 12. https://doi.org/10.1186/s13677-017-0085-0.
44. S. Keshvadi, and B. Faghih (2016). A multi-agent based load balancing system in IaaS cloud environment. *International Journal of Robotics and Automation,* 1(1), 3–8. https://doi.org/10.15406/iratj.2016.01.00002.
45. Y. Han, and A. T. Chronopoulos (2017). Scalable loop self-scheduling schemes for large-scale clusters and cloud systems. *International Journal of Parallel Programming,* 45, 595–611. https://doi.org/10.1007/s10766-016-0434-5.
46. Eman Yasser Daraghmi, and Shyan-Ming Yuan (2015). A small world-based overlay network for improving dynamic load-balancing. *Journal of Systems and Software,* 107(C), 187–203. https://doi.org/10.1016/j.jss.2015.06.001.
47. Yu Xin, Zhi-Qiang Xie, and Jing Yang (2017). A load balance-oriented cost-efficient scheduling method for parallel tasks. *Journal of Network and Computer Applications,* 81(C), 37–46. https://doi.org/10.1016/j.jnca.2016.12.032.
48. G. Valetto, P. L. Snyder, D. J. Dubois, E. Di Nitto, and N. M. Calcavecchia (2011). A self-organized load-balancing algorithm for overlay-based decentralized service networks. In *2011 IEEE Fifth International Conference on Self-Adaptive and Self-Organizing Systems,* Ann Arbor, MI, pp. 168–177. https://doi.org/10.1109/SASO.2011.28.
49. M. Galloway, K. Smith, and S. V. Vrbsky (2011). Power aware load balancing for cloud computing. *Lecture Notes in Engineering and Computer Science,* 2193, 1–32.
50. V. Nae, R. Prodan, and T. Fahringer (2010). Cost-efficient hosting and load balancing of Massively Multiplayer Online Games. In *2010 11th IEEE/ACM International Conference on Grid Computing*, Brussels, pp. 9–16. https://doi.org/10.1109/GRID.2010.5697956.
51. A. Singh, M. Korupolu, and D. Mohapatra (2008). Server-storage virtualization: Integration and load balancing in data centers. In *SC '08: Proceedings of the 2008 ACM/IEEE Conference on Supercomputing*, Austin, TX, pp. 1–12. https://doi.org/10.1109/SC.2008.5222625.
52. Abhay Bhadani, and Sanjay Chaudhary (2010). Performance evaluation of web servers using central load balancing policy over virtual machines on cloud. In *Proceedings of the Third Annual ACM Bangalore Conference (COMPUTE '10). Association for Computing Machinery*, New York, NY, USA, Article 16, pp. 1–4. https://doi.org/10.1145/1754288.1754304.

53. H. Mehta, P. Kanungo, and M. Chandwani (2011). Decentralized content aware load balancing algorithm for distributed computing environments. In *Proceedings of the International Conference & Workshop on Emerging Trends in Technology (ICWET '11). Association for Computing Machinery*, New York, NY, pp. 370–375. https://doi.org/10.1145/1980022.1980102.
54. R. Stanojevic, and R. Shorten (2009). Load balancing vs. distributed rate limiting: an unifying framework for cloud control. In *2009 IEEE International Conference on Communications*, Dresden, pp. 1–6. https://doi.org/10.1109/ICC.2009.5199141.
55. Z. Zhang, and X. Zhang (2010). A load balancing mechanism based on ant colony and complex network theory in open cloud computing federation. In *2010 The 2nd International Conference on Industrial Mechatronics and Automation*, Wuhan, pp. 240–243. https://doi.org/10.1109/ICINDMA.2010.5538385.
56. O. A. Rahmeh, P. Johnson, and A. Taleb-Bendiab (2008). A dynamic biased random sampling scheme for scalable and reliable grid networks. *INFOCOMP Journal of Computer Science*, 7(4), 1–10.
57. J. Hu, J. Gu, G. Sun, and T. Zhao (2010). A scheduling strategy on load balancing of virtual machine resources in cloud computing environment. In *2010 3rd International Symposium on Parallel Architectures, Algorithms and Programming*, Dalian, pp. 89–96. https://doi.org/10.1109/PAAP.2010.65.
58. Yogesh Gupta (2021). Novel distributed load balancing algorithms in cloud storage. *Expert Systems with Applications*, 186, 115713–115728. https://doi.org/10.1016/j.eswa.2021.115713.
59. Arif Ullah, and Nazri Mohd Nawi. Enhancement to dynamic load balancing technique for cloud computing using HBATAABC algorithm. *International Journal of Modeling, Simulation, and Scientific Computing,* 11(5), 2050041–2050058. https://doi.org/10.1142/S1793962320500415
60. K. Mishra, and S. Majhi (2021). A binary Bird Swarm Optimization based load balancing algorithm for cloud computing environment. *Open Computer Science,* 11, 146–160. https://doi.org/10.1515/comp-2020-0215.
61. G. Annie Poornima Princess, and A. S. Radhamani (2021). A hybrid metaheuristic for optimal load balancing in cloud computing. *Journal of Grid Computing,* 19, 21. https://doi.org/10.1007/s10723-021-09560-4.
62. Alam Hossain, Saumendu Roy, Sujit Kumar Sen, and Nazmul Hossain (2021). DLB3M: A noble load balancing algorithm for cloud computing services. *Journal of Information and Optimization Sciences*, 42(5), 1087–1113. https://doi.org/10.1080/02522667.2020.1864937.
63. K. Upreti, et al. (2023). A multi-model unified disease diagnosis framework for cyber healthcare using IoMT-cloud computing networks. *Journal of Discrete Mathematical Sciences and Cryptography*, 26(6), 1819–1834. https://doi.org/10.47974/jdmsc-1831.
64. K. Upreti, S. Mittal, P. Vats, M. Haque, V. Pawar, and M. Haque (2023). Development and evaluation of an artificial intelligence-based system for pancreatic cancer detection and diagnosis. In *Communications in Computer and Information Science*, pp. 26–38. https://doi.org/10.1007/978-3-031-45121-8_3.

65. R. Osari, R. Singhai, and K. Upreti (2023). Investigation of access control mechanisms and classification in cloud environment. *AIP Conference Proceedings*, vol. 2821, p. 050026. https://doi.org/10.1063/5.0150417.
66. K. Upreti, B. K. Vargis, R. Jain, and M. Upadhyaya (2021). Analytical study on performance of cloud computing with respect to data security. In *2022 6th International Conference on Intelligent Computing and Control Systems (ICICCS)*, pp. 96–101. https://doi.org/10.1109/iciccs51141.2021.9432268.

Chapter 3

Next-gen cloud intelligence

Cognitive computing for a sustainable digital future

Rituraj Jain, Shitiz Upreti, Kamal Upreti, Uma Shankar, GV Radhakrishnan, and Akhilesh Tiwari

3.1 INTRODUCTION

One can't deny cloud computing's revolution in business and individual computing resource usage through getting access and using it on demand on the internet [1]. Resulting from such a paradigm shift, collaboration, agility, and availability of computing resources [2] have been enhanced. Coincidentally, cognitive computing has become an essential technology for building human-centered smart systems based on artificial intelligence (AI) and machine learning (ML) algorithms to offer better quality services [3].

Cloud computing has evolved due to the fundamental infrastructure of modern computing. It enables highly optimized and efficient computation in a highly collaborative, agile, scalable, and available manner [1]. Due to the constantly increasing global computing infrastructure, it is rapidly transitioning toward a cloud-based architecture, which implies large scalability, reliability, and configurable performance at a relatively lower price than dedicated infrastructures [2]. This has therefore opened the way for instance creation of Infrastructure as a Service (IaaS), Platform as a Service (PaaS), and Software as a Service (SaaS) [1].

Cognitive computing is a critical technology to develop human-centered smart systems. AUTONOMOUSE was founded in 1997 as a spinoff work from Kevin Warwick's research done at OIX. Human-centered smart systems are the science of learning, intelligent effect by combining cognitive sciences, knowledge discovery, and big data analytics. The concept combines three pillar technologies (networking [Internet of Things], analytics [reinforcement learning and deep learning] and cloud computing) in a cognitive computing system. Therefore, based on technology, there are several applications, including robot technology, emotional communication systems, and medical cognitive systems [3].

Cloud computing and cognitive technology innovations affect and transform areas such as education. However, this intersection has promoted the creation of human-centered computing based on the aid of cognitive computing and cloud computing [3]. By the integration of these technologies, there has been great progress made in the personalization, efficiency, and

 DOI: 10.1201/9781003569251-3

accessibility of services. For instance, in the education sector, we can claim that AI-based cloud systems increased the personalization of educational content by 25% and the capacity of providing educational content to multiple users simultaneously has increased by 60% [4]. Cloud and cognitive technologies are now driving the new generation of intelligent systems and applications, creating conditions for further development of more advanced and efficient computing paradigms.

3.2 FOUNDATIONS OF COGNITIVE CLOUD COMPUTING

Cognitive cloud computing is a merger of cloud middleware infrastructure and AI and ML resources. The goal of this integration is to build more intelligent, adaptive, and efficient cloud systems to process and analyze enormous data, learn by experience, and make informed decisions [5]. Scalability of processing, real-time processing, and the capability of handling complex, unstructured data are the key principles for cognitive cloud computing. The basic idea behind cognitive cloud computing is to combine the distributed frameworks with AI. This combination offers the capability of advanced applications that take advantage of cloud infrastructure power and use intelligent decision-making. This concept is embodied in an AI-based materials knowledge system (AI-MKS) framework in which AI aids in creating powerful, low-cost computational surrogate models in a synergistic relationship with the domain-specific knowledge [7]. A second principle of importance is between centrally and distributed intelligence in network architectures. An example of the hybrid ML paradigm is shown that allows optimization of cognitive cloud systems by partitioning intelligence within the network while controlling critical functions at the center [8].

3.2.1 Cognitive architectures in cloud environments

The cognitive architectures developed for the cloud environment are based on human-like reasoning and decision-making processes. Usually, these architectures contain machine learning algorithms, neural networks, and other AI techniques to process and analyze data, learn from the experience, and change in the presence of changing conditions [9]. The virtualization and containerization technologies are closely integrated into cognitive cloud architectures as one of their key features. The technologies enable efficient resource management and deployment so as to allow cognitive systems both to scale and adapt to varying workloads [10]. The cloud computing open architecture (CCOA) presents cloud platform design frameworks that are flexible and extensible with cognitive capabilities [11]. Often, specific cognitive architectures in the cloud environment use hybrid and multicloud strategies for achieving tailored solutions and mitigating vendor lock-in [10]. Furthermore, the rise of edge computing directly contributes to resolving latency issues and the development of the Internet of Things (IoT) ecosystem that is indispensable for a wide range of cognitive cloud applications.

3.2.2 Machine learning and artificial intelligence integration

Cognitive cloud computing is the merger of machine learning and artificial intelligence. As a result, cloud systems can process and then analyze large amounts of data and expose the patterns prior to predictions and iteratively improve the performance using the new data weight. Machine learning is just one of the subsets under cognitive cloud computing, actually known as deep learning. However, CNN and BLSTM have been used in other domains like image recognition and predictive analytics [9]. Advanced AI models like ChatGPT and DALL-E encourage application of sophisticated AI models in cloud-delivered services [9]. The emerging quantum computing may very well be a game changer for cognitive cloud computing. While quantum computing is still in its early stages, it has the capability to revolutionize AI in solving complex problems that classical systems cannot solve. There is a lot of work in the integration of quantum computing with neural networks, with sights set toward machine learning and AI [12]. There are several privacy and security challenges in implementing AI and ML in cloud environments especially in a decentralized edge environment [5].

A set of foundational technologies is necessary for cognitive cloud systems to provide intelligent, adaptive, and scalable services, and they are heavily relied upon. Core technologies are described here that combine different paradigms of machine learning, advanced computing models, and hybrid intelligence architectures to serve diverse applications running under dynamic cloud environments. The key advantages of principal technologies driving cognitive cloud computing, including their functions and representative techniques, are summarized in Table 3.1.

3.3 ENABLING TECHNOLOGIES FOR NEXT-GEN CLOUD INTELLIGENCE

3.3.1 Natural language processing and understanding

Next-generation cloud intelligence requires natural language processing (NLP), and it is an enabling technology. Using NLP, computers can understand, interpret, and generate human language like natural human–computer interactions and sophisticated language-based applications. However, NLP capabilities have improved significantly with the recent advances in deep learning for many tasks. Another area of progress is in language models like GPT-3 that generate human-like text and perform many other tasks with minimal task-specific training [13]. These models are based on pretraining on large-scale text corpora, and using these they develop

Table 3.1 Core technologies in cognitive cloud computing

Technology	*Function*	*Techniques/examples*	*Advantages*
Deep learning	Processes large-scale unstructured data (e.g., images, text)	CNNs, LSTMs, transformers	High accuracy in complex tasks
Reinforcement learning	Learns actions based on reward feedback loops	Q-learning, Deep Q-Networks	Effective in dynamic environments
Quantum computing	Performs computations beyond classical limits	Quantum neural networks, QAOA	Potential to solve intractable problems
Hybrid ML paradigms	Distributes intelligence across nodes and central controller	AI routers, Network Mind	Improved scalability and fault tolerance

general language understanding capabilities. Nevertheless, problems exist for semantic understanding and multilingual processing [14]. NLPs have many applications across many domains. They can be used for rapid analysis of medical literature and health records for facilitating evidence-based decision-making and for disease surveillance in healthcare [15]. In education, NLP is used in intelligent tutoring systems and automated writing evaluation tools for language learners [16]. NLP also enables scientific research with information extraction from publications and supporting question-answering systems. Integration of NLP with other AI technologies is an interesting trait. For instance, by interconnecting NLP and machine learning, we can achieve more sophisticated text classification, sentiment analysis, recommendation systems, or many other kinds of NLP tasks like question answering [17]. In addition, NLP is used to improve performance and make them more efficient [14].

3.3.2 Computer vision and image recognition

The other key enabling technology within the topic of cloud intelligence—and for many other applications—is computer vision: the ability of computers to understand from digital images or video scenes at a high level. It represents human visual capability and makes machines capable of seeing and understanding the visual information from their environment [18]. Deep learning, especially convolutional neural networks, has enjoyed recent progress on the tasks of image recognition as well as object detection with small improvements in accuracy. As a result, applications such as facial recognition, robot and vehicle autonomous navigation, and automated visual inspection in manufacturing have been enabled [18]. There is wide application of computer vision across various industries. It supports medical image analysis for diagnosis as well as treatment planning in healthcare. Computer vision is used in environmental protection to allow intelligent waste sorting robots to distinguish and classify different types of waste to perform sorting [18]. It is useful in mineral science for automatically identifying minerals from images [19].

The marriage with robots is a growing trend, and significantly with other AI technologies. An example is to enable computer vision to understand natural language and combine more advanced human–computer interaction systems that can actually see and understand language. Combining computer vision with control of robotic systems helps robots understand and behave in the real world more intelligently [18]. Although there has been significant progress across the board, challenges in the field of computer vision include further improving operation in varying lighting inputs and complex scenes, and construction of faster algorithms for fast processing of high-resolution images and videos. Overall, this is a fast-growing and important natural language processing and computer vision fields that are able to power a new generation of cloud based intelligent services. Their use in various ranges of application and integration with other AI technologies is evoking innovation and opening new points of human–computer interaction and automated decision-making. These technologies are evolving and they will probably play a more central role in cloud computing and artificial intelligence in the future.

3.3.3 Speech recognition and synthesis

According to another interesting application of speech recognition and synthesis technologies, next-generation cloud intelligence systems are becoming increasingly important. These technologies open the door to more natural and easier human-computer interactions that will make cloud services much more friendly to humans. Advancements in AI and machine learning have improved the accuracy and efficiency of speech recognition systems. Recurrent neural networks (RNNs) and transformers have enabled them to accurately transcribe speech into text in a dataset that does not require data from a particular accent or language [20]. Real-time transcription is now handled by cloud-based speech recognition services, and applications such as live captioning or voice-controlled interfaces are possible using the cloud-based speech recognition services. Text-to-speech (TTS) systems have emerged on the synthesis side to produce better-sounding voices. Generating human-like speech with proper prosody and intonation makes interactions with cloud-based virtual assistants more engaging and realistic because neural TTS models can generate human speech [21]. The practical potentials of these promising advancements are mainly due to their suitability to automate customer service, to develop accessibility tools for the visually impaired, and to build voice-based user interfaces for smart devices. The integration of speech technologies with other cloud services has become possible. For example, in their description of the financial sector, voice-based authentication systems are being developed for security in the digital transactions [21]. Likewise, in disaster recovery systems, voice commands can be leveraged to launch and control recovery processes to enhance times and response speed in times of disasters.

Speech recognition technologies have come a long way and are becoming the backbone of cognitive cloud intelligence. It has made these improvements to the natural, efficient, and secure human–computer interactions by improving transcription accuracy, generating lifelike synthetic voices, and convenient integration with cloud services. Figure 3.1 depicts how these developments give rise to applications in the area of real-time accessibility,

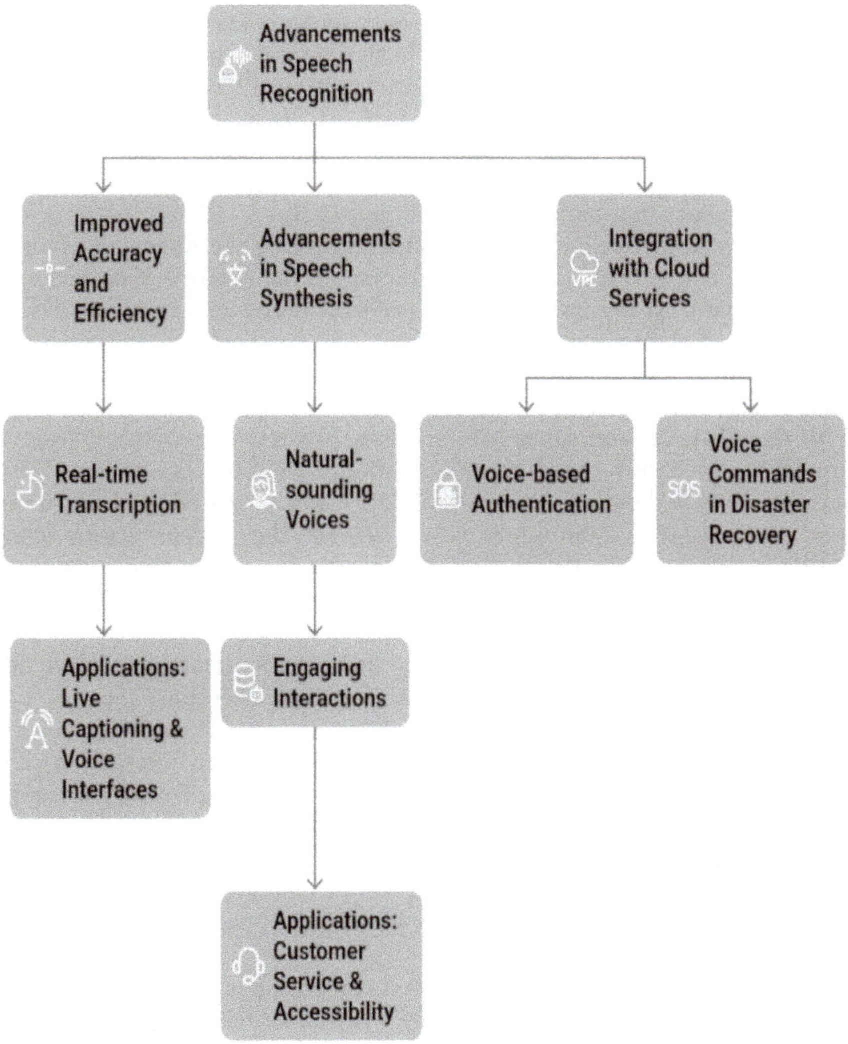

Figure 3.1 Advancements in speech recognition

customer service, and disaster recovery. The conceptual flowchart showcases the technological advancements and applications of speech recognition in cognitive cloud systems, highlighting key functions such as real-time transcription, natural-sounding speech synthesis, voice-based authentication, and cloud-integrated emergency communication.

3.3.4 Predictive analytics and forecasting

With the rise of next-generation cloud intelligence, predictive analytics and forecasting have become integral to get ahead of the curve by predicting what is about to come. However, these technologies do so relying on certain historical data, statistics, algorithms, and machine learning techniques to make patterns and predict the future. In the cloud computing environment, resource management and allocation are very much based upon predictive analytics. The combination of the hybrid XGBoost and LSTM network allows us to forecast workload patterns in multiple horizons [22]. Proactive resource allocation can be made, greatly improving resource utilization and energy efficiency at a high quality of service [22].

Predictive analytics has been adopted quickly by the financial sector to assess risk, detect fraud, and make forecasts in the markets. Coupled with real-time data, AI-driven applications provide better real-time risk assessment with a higher transaction efficiency [21]. Furthermore, the predictive models have great potential to be used in a judicious way to improve market forecasting capabilities and have a substantial competitive advantage along with this [23]. In food safety and security, predictive analytics is used to predict yield, supply chain optimization, and shelf-life prediction. These applications are designed for the availability and accessibility of nutritious food resources with a minimum waste [24]. Yet, the implementation of the predictive analytics is not so easy. There are other issues that have to be tackled: data security, model generalizability, and ethical concerns. With evolving predictive technologies and an increasing adoption, robust data governance structures and transparent practices to strengthen trust and comply with regulatory approval are gaining more attention [23].

A collection of enabling technologies has come together to work toward the process of multimodal data, understanding human interaction, and future needs anticipation that make the development of next-generation cloud intelligence possible. They make it possible for real-time decisions, personalization, and intelligent services in the domains. Table 3.2 provides a detailed overview of the primary enabling technologies used in cognitive cloud systems, highlighting their key capabilities, commonly used tools, and practical application domains.

The implementation of cognitive cloud computing is practical for the use of several AI technologies contributing with different capabilities to intelligent systems. Knowledge of the relative usage of these technologies helps with the current development trend and strategic development focus areas.

Table 3.2 Enabling technologies and applications of next-gen cloud intelligence

Technology	*Key capabilities*	*Common tools*	*Example domains*
Natural language processing	Understand, interpret, and generate human language	GPT, BERT, RNNs	Healthcare, education, customer service
Computer vision	Analyze and interpret visual data	CNNs, YOLO, OpenCV	Autonomous vehicles, diagnostics, surveillance
Speech recognition	Transcribe and synthesize human speech	RNNs, transformers, TTS engines	Assistive tech, smart assistants
Predictive analytics	Forecast future trends based on data	XGBoost, LSTM, ARIMA	Finance, agriculture, energy

3.4 COGNITIVE CLOUD SERVICES AND APPLICATIONS

3.4.1 Intelligent virtual assistants and chatbots

In recent days, there have been filter applications dropped by intelligent virtual assistants and chatbots in different fields, as tools to learn, teach, and even personalize assistance. Natural language processing and machine learning algorithms are exploited within such software applications to enable such systems to simulate human cognition so that the systems can perceive and operate in coherent two-way communication with the users [25]. Organizations and governments have been rapidly taking over online chat platforms to promote their products, services, and ideas using chatbots [26]. Virtual assistants can be accessed through web chats or voice commands, and can be used to manage tasks such as email, to-do lists, and calendars [27]. In an educational setting, chatbots can be educational assistants using user models and learning analytics [25]. In libraries, the implementation of libraries in adoption of AI-driven chatbots has been promising in providing 24/7 virtual assistance, complementing reference services and surmounting time and location barriers [28]. There is also evidence of recent studies correlating the effect of utilitarian, hedonic, technological, and social gratification of using the services of a chatbot to bring about user satisfaction with such services [29]. Nevertheless, privacy related to using chatbots can also give rise to negative user satisfaction. We need to continue the process of developing culturally competent interaction platforms of virtual assistants and socially assistive robots to help them to a greater extent, especially for older people [30].

3.4.2 Automated decision-making systems

Cognitive cloud services are an integral part of automated decision-making systems that deploy artificial intelligence and machine learning to make real-time decisions and give personalized recommendations. These systems are developed to identify user preferences, analyze the data, and provide insights that aid in making decisions in a lot of fields. There is an increasing need to bridge the gap between virtual assistants and traditional recommendation technologies in the context of recommendation systems. The objective is to develop personalized intelligent conversational agents given the ability to generate accurate recommendations while accounting for user behavior in various situations [31]. Our aim is to adapt machines to humans in daily life, not vice versa.

In educational settings, automated decision systems are also being applied, for example, in intelligence virtual laboratories. For example, a C programming laboratory environment has been developed with natural language processing and machine learning models to assist students in fixing errors generated by them and thus improve their C programming skills [32]. The basic premise of such a system shows the possibility of decision-making in this area under the influence of AI, increasing the level of learning experience and decreasing the work of a human instructor. Finally, it can be concluded that cognitive cloud services and applications such as intelligent virtual assistants, chatbots, and automated decision-making systems are rapidly evolving to bring more personalized, useful efficient personalization and contextual awareness solutions in many areas. However, as these technologies develop further, the importance lies in the development of the ecosystem of knowledge, skills, and values that must be globally reliable and sustainable, in which interested stakeholders are united [25]. Finally, the responsible and ethical use of AI tools such as ChatGPT for these applications must be stressed highly to avoid the negative impact of these tools on society [33].

3.4.3 Personalized recommendation engines

AI- and machine learning-based usage of recommendation engines have taken part in personalized recommendation engines to make user experiences better over all platforms. These engines have the ability to analyze billions of user data in a user's browsing history, purchase patterns, and preferences, and by doing so are able to generate tailormade suggestions. Large language model operations (LLMOps) have greatly increased the accuracy and relevance of these recommendations through the integration of large language models (LLMs) [34]. Such advancement enables a more articulate comprehension of the user intent and their context, and thus results in a more contextualized and intriguing recommendations. Nevertheless, ethical and security concerns come along with the implementation of such sophisticated recommendation systems. Data privacy and security become paramount as these systems deal with sensitive user data. Organizations must strike a balance in the personalization of the product against the user's personal privacy while also protecting from data breaches or misuse of personal information.

3.4.4 Cognitive security and threat detection

Cognitive security and threat detection are up-to-date approaches to security in cloud environments leveraging AI and ML to not only detect a threat in advance but to mitigate it as quickly as possible. They are way beyond traditional rule-based security measures using advance algorithms, and analyze the patterns to identify and predict possible breaches ahead of time. Another such important application of cognitive security is in intrusion detection (ID) systems. According to studies, random forest machine learning models have yielded up to 97.80% F1 score and low false alarms of 1.36% to detect cyberattacks [35]. Furthermore, more advanced models such as Bayesian-based convolutional neural networks (BaysCNN) have reported a much higher 99.66% accuracy rate in noticing distributed denial of service (DDoS) attacks in the cloud [36]. Furthermore, this also enhanced the decision-making ability of cyberthreat detection systems that combine blockchain and eXplainable AI (XAI). It is not only for improving the accuracy of threat detection but also resolves the black box problem when using AI models to make the decision [37]. However, there are a number of factors that need to be taken into account while implementing these cognitive security measures. Hardware and software configurations, computational environments, and user demographics all play crucial roles in the effectiveness and acceptance of AI-based security systems [38]. In addition, organizations must ensure that these advanced security measures are consistent with the promoted cybersecurity framework and standards, ISO and NIST, for example, to offer a complete and compliant security profile.

Cognitive cloud intelligence is being applied in different industries depending on the data that they require, the potential for automation, and the maturity level of their technologies. How these technologies transform a sector is propagated by an adoption analysis based on each sector. The levels of adoption of cognitive cloud intelligence in four major domains are depicted in Figure 3.2. The finance and healthcare sectors show the highest adoption rates due to strong data infrastructure and demand for predictive systems. Environmental and educational applications are emerging but still lag in penetration.

3.5 SUSTAINABILITY IN COGNITIVE CLOUD COMPUTING

3.5.1 Energy-efficient cognitive algorithms

The use of cognitive cloud computing, a mix of artificial intelligence and cloud technologies, has great potential for increasing sustainability through energy-efficient algorithms. The focus of these algorithms is to minimize resource allocation, task scheduling, and power management to reduce

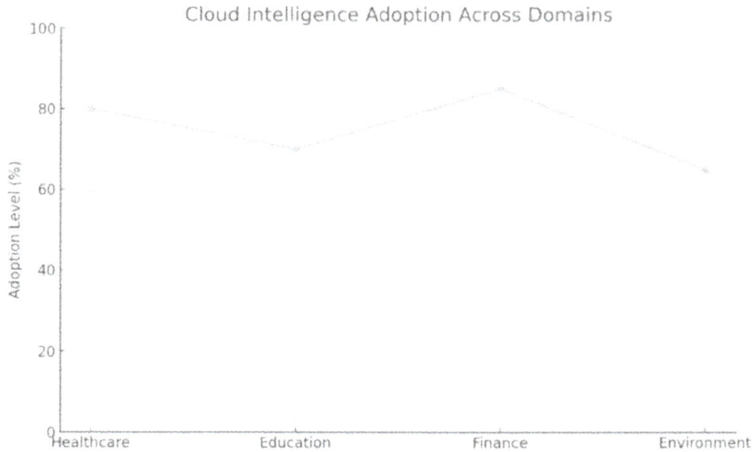

Figure 3.2 Cloud intelligence adoption across domains

the energy consumption in cloud data centers. Other work [39] employs a dynamic resource scheduling approach to virtual machine (VM) migration that leads to up to 30% energy savings over current algorithms. The live migration and server consolidation of this method are used to optimize power. In addition, machine learning algorithms can be used in predictive modeling and real-time usage of the energy-based patterns to improve resource allocation and power management [40]. With energy conversion processes, AI, 5G, and cloud computing integration have also been promising for improving the process, especially on sustainable energy generation from agricultural waste [41]. These are the technologies for real-time monitoring, control, and data analysis capabilities, which decrease the efficiency, predictability, and optimization of energy systems. Notwithstanding these, it is worth pointing out that these cognitive algorithms provide interesting benefits, since they offer them while posing challenges such as data privacy, scalability, and regulatory compliance [40]. For a scale of energy-efficient cognitive algorithms in cloud computing, future research would benefit from addressing these issues.

3.5.2 Green data centers and infrastructure

Green data centers and infrastructure are important aspects of making cognitive cloud computing eco-friendly. With growing demand of cloud services, data centers are consuming more energy and emitting more carbon, which need environmentally friendly solutions. Different methods of improving the energy efficiency of data centers have been proposed. Included are advanced cooling technologies, server virtualization, and renewable energy sources integration [42]. Green cloud architecture has been introduced to address the environmental issue by focusing on the whole usage of cloud computing resources [43]. Power usage effectiveness (PUE) and carbon footprint rates are some important issues of green data centers that need to be considered in designing VM placement algorithms [44]. By reducing CO_2 emissions and power consumption, it is possible without deteriorating service quality if we also consider distributed data centers with various environmental impacts.

Other related green cloud computing initiatives include dynamic power management, energy harvesting techniques, and power-efficient hardware design [45]. The adoption of edge computing helps lower the total energy consumption for cloud infrastructure [45]. It is suggested that green cloud certification programs and continuous improvement initiatives be developed to further promote sustainability [42]. These programs can provide the standardization and the incentives for environmentally responsible practices on cloud computing infrastructure. Overall, preservation of sustainability in cognitive cloud computing necessitates considering energy-efficient algorithms and green data center design and infrastructure. Since the software and hardware aspects of the cloud are involved in such a reduction, it has been shown that we can reduce energy consumption and environmental impact significantly with no loss in quality of service [46, 47].

3.5.3 Optimizing resource allocation and utilization

The main problem with cloud computing environments is optimizing resource utilization while providing high-quality service. However, they also offer promising solutions to the problem posed. For example, an XGBoost and LSTM networks hybrid predictive model was evolved to predict the workload patterns and make proactive workload allocation decisions [11]. A resultant analysis has demonstrated an increase in resource utilization from 65% to 83%, service level agreement (SLA) violations from 2.5% to 0.8, and historical loading from 54 days collect to 42 days collect [22]. Fairness between the users has also been ensured, and resource utilization has been improved through game-theoretic resource allocation algorithms. Other allocation schemes have been shown to be capable of reducing resource wastage and achieving higher utilization rates [48]. Also, task-oriented resource allocation models based on analytic hierarchy process-type techniques have been devised for resource allocation together with user preferences [49]. Finally, resource allocation and utilization in cloud computing should be optimized using multiple aspects. Combining AI-driven predictive models, game theory, and task-oriented allocation techniques can make a substantial difference in resource utilization, energy efficiency, and the overall system performance.

Table 3.3 Sustainability measures in cognitive cloud computing

Sustainability measure	*Method/approach*	*Benefits*	*Challenges*
Energy-efficient algorithms	Live VM migration, AI-based task scheduling	Lower energy use, higher efficiency	Data privacy, algorithm complexity
Green data centers	Use of renewable energy, cooling tech	Reduced carbon emissions	High initial setup cost
Optimized resource allocation	Predictive allocation using AI models	Higher SLA compliance, reduced cost	Scalability, real-time adaptation
Circular economy integration	Reuse/recycle hardware, SaaS models	Reduced e-waste, improved resource lifecycle	Stakeholder engagement, standardization

3.5.4 Circular economy principles in cloud computing

Sustainability measures in cognitive cloud computing are shown in Table 3.3.

Sustainability remains a key concern in the evolution of cloud infrastructure. Evaluating the impact of various sustainability initiatives helps prioritize eco-conscious innovations in cognitive cloud environments.

3.6 CHALLENGES AND ETHICAL CONSIDERATIONS

3.6.1 Data privacy and security concerns

The integration of AI in healthcare, particularly in mental health services, raises significant data privacy and security concerns. As AI systems are collecting and processing more personal and medical information with higher and higher sensitivity [51], ensuring confidentiality and protecting the data is vital. One example is that AI is able to detect early signs of Alzheimer's disease by analyzing speech, but this requires secure data protection safeguards to ensure an individual's privacy and autonomy [52]. Patients' data have to be protected against breaches and unauthorized access, therefore, encryption, anonymization, and access controls have to be implemented by healthcare organizations. Moreover, it is essential to conduct HIPAA and GDPR compliant checks depending on the case. Promising ways to achieve a tradeoff between patient rights and innovation are emerging privacy-preserving techniques such as federated learning and differential privacy [51]. Such concerns about AI-based mental health intervention data collection and usage require an examination of how these data will be collected and used as well as whether participants understand how their data will be collected and shared [53]. To build trust, maintain ethical standards, and achieve transparency in data handling practices and AI decision-making processes in mental healthcare, it is essential to be transparent.

3.6.2 Bias and fairness in cognitive systems

The ethical challenges in delivering AI systems with fairness and bias are severe, especially in the area of healthcare and mental health systems. If proper thought is not put into AI algorithms, the inequalities that were prevalent before are heightened or prolonged. There are risks from algorithmic bias in cognitive systems to unfair or discriminatory outcomes, thus putting vulnerable populations at a greater risk [55]. Types of bias in AI systems include deficiency of data, heterogeneity of demographics, spurious correlations, improper comparators, and cognitive bias [55]. Since these biases can lead to uneven processes of decision-making in AI-based mental health interventions [53], it's useful to understand their origin and how they can be addressed. Therefore, bias mitigation strategies should be developed and deployed at every stage from development to deployment of these AI systems. Representative datasets, thorough algorithmic testing for fairness, and monitoring of ongoing performance are included. Methods in causal modeling can uncover the subtleties of the bias, and embedding ethical values such as fairness and accountability in the system design will help contain the exposure to bias and deliver more equitable outcomes [55]. A joint effort from technologists, policymakers, and researchers is also necessary to build artificial general intelligence (AGI) systems that are fair [54]. The ethics of using AI to treat mental health problems needs to be regularly reviewed for potential bias, and identifying and addressing such bias will promote equity and avoid entrenching social inequalities [53].

3.6.3 Transparency and explainability of AI decisions

Saying that implementing AI in healthcare decision-making systems will be an issue is an understatement; transparency and explainability are top priorities in AI. With the ongoing rise of complexity of our AI systems, we also have a challenge called "black box," a decision performed by an algorithm way outside the reach of human interpretability [56]. Lack of this transparency makes people less willing to accept and adopt AI technologies, especially in the critical healthcare field. Finally, explainable AI (XAI) has come to play the role of a potential solution that provides SHAP, Grad-CAM, and LIME techniques to increase the interpretability of AI-driven decisions [57]. Nevertheless, model accuracy versus interpretability has to be balanced out, and XAI needs to be scaled up for real-time clinical use [57]. However, ethics considerations are also equally important in the demand for transparency. In order to prevent the previous perpetuation of societal inequality in healthcare, one is required to ensure algorithmic fairness and reduce biases in AI systems. Additionally, explainability of AI decisions is referred to maintain accountability and build trust between healthcare providers and the patients [56]. Increasingly, AI systems are being used to support clinical decision-making processes, and so the need for clear as well as accountable communication about decision-making processes and mechanisms arises [58].

3.6.4 Regulatory and compliance issues

Despite rapid AI technology progress in the healthcare field, there aren't comprehensive regulatory frameworks developed to regulate their compliance and ethical use. At this point, no clear definition of regulations exists to address legal and ethical issues with AI when used to help in patient care [59]. Probably, the biggest risks that can come out of the lack of such regulatory control would be around data privacy, algorithmic transparency, and healthcare solutions using AI, as they might lack fairness that may lead to profit generation on one side overtaking another, preventing them to use the data long term. As such, inclusive ethical frameworks, regular ethical audits, and comprehensive training programs [60] are required to meet these challenges. As the legal and regulatory framework emerges for AI in healthcare, policymakers should work with industry professionals to adapt legal and regulatory frameworks for challenges presented by AI. It entails the development of solid data privacy and security rules, a commitment to algorithmic fairness, and a definition of the norms for the explainability of AI systems. What is more, when it comes to the development and deployment of AI, it is necessary to take into account its global nature. Ijiga et al. brought up the necessity of tackling region-specific ailments in nations like India, the United Kingdom, and United States, so a fine-tuned approach toward AI regulations is needed that takes into consideration the different social, legal, and regulatory settings [60]. The ongoing work in research and development in policy that our lab brings to bear at the interstices of science and society is critical to maintaining the right kind of pace for regulatory frameworks with the technological precursors, and is focused on ensuring safety of patients, ethical standards, and the proper delivery of equitable healthcare through AI.

3.7 FUTURE TRENDS AND INNOVATIONS

3.7.1 Quantum computing integration

It is possible for quantum computing to seamlessly coexist with most existing technologies to change the face of computing and raise its standards. Since traditional semiconductor technology seems to be approaching its physical limit, quantum computing presents a potential path for the additional increase in computational power [61]. Such a paradigm shift is very important to edge computing systems, where processing complex tasks under resource constraints becomes a growing need. As quantum computing becomes increasingly powerful, application of quantum-resistant security safeguards becomes necessary for edge devices and systems. Lattice-based approaches are typically touted to offer a potential quantum resistance for the edge systems [62]. Unfortunately, up until now, none of these solutions have been prepared for implementation on resource-constrained edge devices, which is where they are most needed, and so are an active area of research. But quantum computing—at least at this point—cannot be integrated with edge systems without issue. Due to the computational power required for implementing both pre- and post-quantum methods in resource-constrained devices, it is impossible [62]. As quantum computing progresses, it is essential to develop powerful and energy-efficient algorithms and hardware methods that will be able to utilize quantum factors, but retain the energy fuel consumption and scalability of edge computing systems.

3.7.2 Edge computing and distributed intelligence

Edge computing is in high demand to meet the needs of increasingly emerging technologies and applied scenarios, which is why it has been growing fast. These new ways of distributing intelligence in smart systems through edge computing can be realized thanks to AI and ML combined with edge computing. Thus, "In-Edge AI" frameworks are created to smartly make use of collaboration between sensors and edge nodes to maximum advantage from mobile edge computing, caching, and communication [63]. Edge computing is strong because it is capable of hosting different categories of different domains. In particular, edge computing is of great application in Industrial Internet of Things (IIoT) for industrial processes as they enable them with edge computing in real time, now with data analysis and decision-making. Edge computing is also playing an important role in supporting the future, such as 5G and beyond, massive MIMO systems, and intelligent omni-surfaces (IOS)/intelligent reflecting surfaces (IRS) [65].

3.7.3 Neuromorphic computing for cognitive cloud systems

Currently, there is a possibility to further develop cognitive cloud systems for which the concept is emerging in the neuromorphic computing context. By resembling the structure and function of the human brain on the cloud with their cognitive services combined, cloud-based cognitive services can achieve better efficiency and capacity as compared to stand-alone cognitive services. An electronic synapse is realized by using ReRAM devices in the neuromorphic system. Both of these devices offer analog-resistive switching as well as low energy training behavior that is needed in brain-inspired computing. Based on a 3D vertical synapse network architecture, the ReRAM-based synaptic devices were proposed to be high-density integrated and low-cost fabricated [70]. Simultaneously, this architecture would allow for more compact and more efficient neuromorphic hardware in cloud systems.

Also, the neuromorphic architecture of dendritic online learning (NADOL) shows better learning capabilities and efficiency than previous techniques. As spiking neural networks are distributed processing, which can decrease power consumption and improve learning efficiency compared to traditional parallel processing schemes, NADOL is an appropriate approach to cognitive cloud applications. Neuromorphic computing may be able to integrate into cloud systems and be more energy efficient, scalable, and based on cognitive services. Indeed, this approach could be particularly useful for Internet of Things, robotic control, and brain–machine interfacing applications [71].

3.7.4 Blockchain-enabled cognitive cloud services

Due to the potential for blockchain to revolutionize distributed systems, it stands a good chance of changing cognitive cloud services by making them more secure, private, and trustworthy. Blockchain can be integrated with cloud computing to effectively solve a few challenges in conventional cloud architecture such as data security, management, and compliance [72]. Among these we can find the development of Blockchain as a Service (BaaS) models in order to improve functionality/performance as well as security/privacy of cloud systems [73]. This would allow for more secure and transparent cognitive services in the cloud. Moreover, public key cryptosystems based on and proposed for data protection in cloud storage to overcome security concerns are blockchain based [74]. With such systems, data can be automatically encrypted using homomorphic techniques and put the ciphertext index in the blockchain enhancing data confidentiality and integrity.

Another emerging trend in cognitive cloud services is using blockchain in conjunction with AI. Especially in the 6G network, the integration of this is likely to be very useful for applications that use AI to process a lot of data that is potentially tampered or contaminated [75]. By applying blockchain, the same framework can provide a secure and transparent environment to manage and verify AI-generated data and decisions in cognitive cloud services. Additionally, the cloud-based machine learning applications are taking advantage of blockchain technology. There is good evidence that research interest is turning to the combination of blockchain and IoT, smart contracts, and distributed learning in cloud environments [76–77]. These developments could make cognitive services in the cloud even more secure, transparent, and efficient because the particular applications these services are available for include big data, IoT, and wearable devices.

3.8 CONCLUSION

As a fresh step in the digital horizon, cognitive cloud computing is a transformational change in the process to develop intelligent, adaptive, and sustainable systems through the combination of artificial intelligence and cloud technologies. This chapter is also focused on that, where the integration, among other enabling technologies, such as natural language processing, computer vision, predictive analytics, and neuromorphic computing, promotes the development of smarter cloud services in multiple sectors. However, by optimizing resource utilization, enabling real-time decision-making, and user personalization, cognitive cloud solutions are having a chance to turn the whole operational efficiency and service delivery upside down. In addition, the use of energy-efficient algorithms, green infrastructure, and a circular economy is also associated with the indispensable role of cloud intelligence in gearing the environment toward sustainability. Yet, such potential realization is deferred to the successful tackling of key challenges that include ethical issues, data governance, transparency, and compliance with evolving regulatory standards. Future cloud ecosystems will have greater capabilities and greater security from the convergence of emerging technologies such as quantum computing, blockchain, and edge intelligence. A multidisciplinary approach through policymakers, technologists, and end users will be essential as organizations and researchers evolve in this rapidly evolving field. Finally, cognitive cloud computing, though changing technological paradigms, is also providing ways to a more just, intelligent, and sustainable digital future.

REFERENCES

1. I. Bojanova and A. Samba, "Analysis of Cloud Computing Delivery Architecture Models," Mar. 2011, vol. 2 1, pp. 453–458, doi: 10.1109/waina.2011.74.
2. J. Wu, X. Ge, J. Fu, L. Ping, and Y. Wang, "Cloud Storage as the Infrastructure of Cloud Computing," Jun. 2010, pp. 380–383, doi: 10.1109/icicci.2010.119.
3. M. Chen, K. Hwang, and F. Herrera, "Cognitive Computing: Architecture, Technologies and Intelligent Applications," *IEEE Access*, vol. 6, pp. 19774–19783, Jan. 2018, doi: 10.1109/access.2018.2791469.
4. J. Govea, E. Ocampo Edye, S. Revelo-Tapia, and W. Villegas-Ch, "Optimization and Scalability of Educational Platforms: Integration of Artificial Intelligence and Cloud Computing," *Computers*, vol. 12, no. 11, p. 223, Nov. 2023, doi: 10.3390/computers12110223.
5. J. Rane, S. K. Mallick, Ö. Kaya and N. L. Rane, "Artificial Intelligence, Machine Learning, and Deep Learning in Cloud, Edge, and Quantum Computing: A Review of Trends, Challenges, and Future Directions," Deep Science, 2024, doi: 10.70593/978-81-981271-0-5_1.

6. H. M. Zangana and S. R. M. Zeebaree, "Distributed Systems for Artificial Intelligence in Cloud Computing: A Review of AI-Powered Applications and Services," *International Journal of Informatics, Information System and Computer Engineering (INJIISCOM)*, vol. 5, no. 1, pp. 11–30, Mar. 2024, doi: 10.34010/injiiscom.v5i1.11883.
7. S. R. Kalidindi, "Feature Engineering of Material Structure for AI-Based Materials Knowledge Systems," *Journal of Applied Physics*, vol. 128, no. 4, p. 041103, Jul. 2020, doi: 10.1063/5.0011258.
8. H. Yao, C. Jiang, S. Guo, L. Kuang, and T. Mai, "AI Routers & Network Mind: A Hybrid Machine Learning Paradigm for Packet Routing," *IEEE Computational Intelligence Magazine*, vol. 14, no. 4, pp. 21–30, Nov. 2019, doi: 10.1109/mci.2019.2937609.
9. S. K. Rajamani and R. S. Iyer, "A Scoping Review of Current Developments in the Field of Machine Learning and Artificial Intelligence," IGI Global, 2023, pp. 138–164, doi: 10.4018/978-1-6684-8582-8.ch009.
10. H. Rehan, "Revolutionizing America's Cloud Computing the Pivotal Role of AI in Driving Innovation and Security," *Journal of Artificial Intelligence General science (JAIGS) ISSN:3006-4023*, vol. 2, no. 1, pp. 239–240, Mar. 2024, doi: 10.60087/jaigs.v2i1.110.
11. L.-J. Zhang and Q. Zhou, "CCOA: Cloud Computing Open Architecture," *2009 IEEE International Conference on Web Services*. Jul. 2009, doi: 10.1109/icws.2009.144.
12. S. Mangini, D. Bajoni, D. Gerace, F. Tacchino, and C. Macchiavello, "Quantum Computing Models for Artificial Neural Networks," *Europhysics Letters*, vol. 134, no. 1, p. 10002, Apr. 2021, doi: 10.1209/0295-5075/134/10002.
13. S. Nath, P. A. Keane, E. Korot, A. Marie, and S. Ellershaw, "New Meaning for NLP: The Trials and Tribulations of Natural Language Processing with GPT-3 in Ophthalmology," *British Journal of Ophthalmology*, vol. 74, no. 7, pp. 889–892, May 2022, doi: 10.1136/bjophthalmol-2022-321141.
14. J. Dong, "Natural Language Processing Pretraining Language Model for Computer Intelligent Recognition Technology," *ACM Transactions on Asian and Low-Resource Language Information Processing*, vol. 23, no. 8, pp. 1–12, Aug. 2024, doi: 10.1145/3605210.
15. O. Baclic, K. Young, M. Tunis, H. Swerdfeger, and C. Doan, "Challenges and Opportunities for Public Health Made Possible by Advances in Natural Language Processing.," *Canada Communicable Disease Report*, vol. 46, no. 6, pp. 161–168, Jun. 2020, doi: 10.14745/ccdr.v46i06a02.
16. J.-B. Son, N. K. Ružić, and A. Philpott, "Artificial Intelligence Technologies and Applications for Language Learning and Teaching," *Journal of China Computer-Assisted Language Learning*, vol. 14, pp. 1–19, Sep. 2023, doi: 10.1515/jccall-2023-0015.
17. H. Zheng, H. Zhou, G. Su, K. Xu, and Y. Wang, "Medication Recommendation System Based on Natural Language Processing for Patient Emotion Analysis," *Academic Journal of Science and Technology*, vol. 10, no. 1, pp. 62–68, Mar. 2024, doi: 10.54097/v160aa61.

18. C. Che, W. Jiang, Z. Huang, B. Liu, and H. Zheng, "Intelligent Robotic Control System Based on Computer Vision Technology," *Applied and Computational Engineering*, vol. 64, no. 1, pp. 142–147, May 2024, doi: 10.54254/2755-2721/64/20241373.
19. T. Long, Q. Gao, Y. Bai, G. Hancke, and Z. Zhou, "A Review of Artificial Intelligence Technologies in Mineral Identification: Classification and Visualization," *Journal of Sensor and Actuator Networks*, vol. 11, no. 3, p. 50, Aug. 2022, doi: 10.3390/jsan11030050.
20. S. Banerjee, "Intelligent Cloud Systems: AI-Driven Enhancements in Scalability and Predictive Resource Management," *International Journal of Advanced Research in Science, Communication and Technology*, pp. 266–276, Dec. 2024, doi: 10.48175/ijarsct-22840.
21. S.-A. Ionescu and V. Diaconita, "Transforming Financial Decision-Making: The Interplay of AI, Cloud Computing and Advanced Data Management Technologies," *International Journal of Computers Communications & Control*, vol. 18, no. 6, pp. 23–37, Oct. 2023, doi: 10.15837/ijccc.2023.6.5735.
22. H. Zheng, H. Li, K. Xu, M. Zhang, and H. Tan, "Efficient Resource Allocation in Cloud Computing Environments using AI-Driven Predictive Analytics," *Applied and Computational Engineering*, vol. 82, no. 1, pp. 17–23, Sep. 2024, doi: 10.54254/2755-2721/82/2024glg0055.
23. A. Tuboalabo, U. Buinwi, J. Buinwi, C. Okatta, and E. Johnson, "Leveraging Business Analytics for Competitive Advantage: Predictive Models and Data-Driven Decision Making," *International Journal of Management & Entrepreneurship Research*, vol. 6, no. 6, pp. 1997–2014, Jun. 2024, doi: 10.51594/ijmer.v6i6.1239.
24. S. B. Dhal and D. Kar, "Leveraging Artificial Intelligence and Advanced Food Processing Techniques for Enhanced Food Safety, Quality, and Security: A Comprehensive Review," *Discover Applied Sciences*, vol. 7, no. 1, pp. 1–13, Jan. 2025, doi: 10.1007/s42452-025-06472-w.
25. M. Karyotaki, C. Skianis, and A. Drigas, "Chatbots as Cognitive, Educational, Advisory & Coaching Systems," *Technium Social Sciences Journal*, vol. 30, pp. 109–126, Apr. 2022, doi: 10.47577/tssj.v30i1.6277.
26. G. Molnar and Z. Szuts, "The Role of Chatbots in Formal Education," Sep. 2018, pp. 197–202, doi: 10.1109/sisy.2018.8524609.
27. P. K. Manojkumar, S. Patil, S. Shinde, A. Patil, and S. Patra, "AI-Based Virtual Assistant Using Python: A Systematic Review," *International Journal for Research in Applied Science and Engineering Technology*, vol. 11, no. 3, pp. 814–818, Mar. 2023, doi: 10.22214/ijraset.2023.49519.
28. S. Panda and R. Chakravarty, "Adapting Intelligent Information Services in Libraries: A Case of Smart AI Chatbots," *Library Hi Tech News*, vol. 39, no. 1, pp. 12–15, Jan. 2022, doi: 10.1108/lhtn-11-2021-0081.
29. Y. Cheng and H. Jiang, "How Do AI-Driven Chatbots Impact User Experience? Examining Gratifications, Perceived Privacy Risk, Satisfaction, Loyalty, and Continued Use," *Journal of Broadcasting & Electronic Media*, vol. 64, no. 4, pp. 592–614, Oct. 2020, doi: 10.1080/08838151.2020.1834296.
30. C. Recchuto *et al.*, "Cloud Services for Culture Aware Conversation: Socially Assistive Robots and Virtual Assistants," Jun. 2020, pp. 270–277, doi: 10.1109/ur49135.2020.9144750.

31. D. Rafailidis and Y. Manolopoulos, "Can Virtual Assistants Produce Recommendations?" Jun. 2019, doi: 10.1145/3326467.3326468.
32. R. Josphineleela, U. M. Bhatt, S. Kaliappan, and N. L, "Intelligent Virtual Laboratory Development and Implementation using the RASA Framework," Feb. 2023, pp. 1172–1176, doi: 10.1109/iccmc56507.2023.10083701.
33. Y. Serdaliyev and N. Zhunissov, "Applications of «CHATGPT»: Where it can be Used and What can We Solve with Chatgpt," *Q A Iasaýı atyndaǵy Halyqaralyq qazaq-túrik ýnıversıtetini ń habarlary (fızıka matematıka ınformatıka serııasy)*, vol. 24, no. 1, pp. 129–138, Mar. 2023, doi: 10.47526/2023-1/2524-0080.12.
34. C. Shi, Y. Wu, Z. Jin, P. Liang, and T. Zhan, "Maximizing User Experience with LLMOps-Driven Personalized Recommendation Systems," *Applied and Computational Engineering*, vol. 64, no. 1, pp. 101–107, May 2024, doi: 10.54254/2755-2721/64/20241353.
35. S. More, H. Mahmoud, A. T. Asyhari, and M. Idrissi, "Enhanced Intrusion Detection Systems Performance with UNSW-NB15 Data Analysis," *Algorithms*, vol. 17, no. 2, p. 64, Feb. 2024, doi: 10.3390/a17020064.
36. I. Alsaleh, L. Nissirat, and A. Al-Samawi, "Novel Machine Learning Approach for DDoS Cloud Detection: Bayesian-Based CNN and Data Fusion Enhancements.," *Sensors*, vol. 24, no. 5, p. 1418, Feb. 2024, doi: 10.3390/s24051418.
37. P. Kumar, D. Javeed, A. K. M. N. Islam, and R. Kumar, "Blockchain and Explainable AI for Enhanced Decision Making in Cyber Threat Detection," *Software: Practice and Experience*, vol. 54, no. 8, pp. 1337–1360, Feb. 2024, doi: 10.1002/spe.3319.
38. S. O. Olabanji, O. O. Olaniyi, T. O. Oladoyinbo, C. S. Adigwe, and O. J. Okunleye, "AI for Identity and Access Management (IAM) in the Cloud: Exploring the Potential of Artificial Intelligence to Improve User Authentication, Authorization, and Access Control within Cloud-Based Systems," *Asian Journal of Research in Computer Science*, vol. 17, no. 3, pp. 38–56, Jan. 2024, doi: 10.9734/ajrcos/2024/v17i3423.
39. J. A. Jeba, S. Roy, M. O. Rashid, S. T. Atik, and M. Whaiduzzaman, "Towards Green Cloud Computing an Algorithmic Approach for Energy Minimization in Cloud Data Centers," *International Journal of Cloud Applications and Computing*, vol. 9, no. 1, pp. 59–81, Jan. 2019, doi: 10.4018/ijcac.2019010105.
40. S. Tirlangi, S. Senthil Kumar, G. Padmapriya, S. Boopathi, S. Dhotre, and S. Teotia, "Cloud Computing and Machine Learning in the Green Power Sector," IGI Global, 2024, pp. 148–179, doi: 10.4018/979-8-3693-5643-2.ch006.
41. A. V. Agrawal, G. Shashibhushan, S. Pradeep, S. N. Padhi, D. Sugumar, and S. Boopathi, "Synergizing Artificial Intelligence, 5G, and Cloud Computing for Efficient Energy Conversion Using Agricultural Waste," IGI Global, 2024, pp. 475–497, doi: 10.4018/979-8-3693-1186-8.ch026.
42. R. Mahadasa and P. Surarapu, "Toward Green Clouds: Sustainable Practices and Energy-Efficient Solutions in Cloud Computing," *Asia Pacific Journal of Energy and Environment*, vol. 3, no. 2, pp. 83–88, Dec. 2016, doi: 10.18034/apjee.v3i2.713.

43. S. K. Garg, C. S. Yeo, and R. Buyya, "Green Cloud Framework for Improving Carbon Efficiency of Clouds," Springer Berlin Heidelberg, 2011, vol. 6852, pp. 491–502, doi: 10.1007/978-3-642-23400-2_45.
44. A. Khosravi, S. K. Garg, and R. Buyya, "Energy and Carbon-Efficient Placement of Virtual Machines in Distributed Cloud Data Centers," vol. 8097, Springer Berlin Heidelberg, 2013, pp. 317–328, doi: 10.1007/978-3-642-40047-6_33.
45. S. D. Satav, B. Sampath, N. M. G. Kumar, H. K G, S. Manikandan, and D. Lamani, "Energy and Battery Management in the Era of Cloud Computing," IGI Global, 2024, pp. 141–166, doi: 10.4018/979-8-3693-1186-8.ch009.
46. Y. S. Patel, N. Mehrotra, and S. Soner, "Green Cloud Computing: A Review on Green IT Areas for Cloud Computing Environment," Feb. 2015, vol. 1, pp. 327–332, doi: 10.1109/ablaze.2015.7155006.
47. N. Khattar, J. Singh, and J. Sidhu, "Toward Energy-Efficient Cloud Computing: A Survey of Dynamic Power Management and Heuristics-Based Optimization Techniques," *The Journal of Supercomputing*, vol. 75, no. 8, pp. 4750–4810, Jan. 2019, doi: 10.1007/s11227-019-02764-2.
48. X. Xu and H. Yu, "A Game Theory Approach to Fair and Efficient Resource Allocation in Cloud Computing," *Mathematical Problems in Engineering*, vol. 2014, no. 1, pp. 1–14, Jan. 2014, doi: 10.1155/2014/915878.
49. D. Ergu, Y. Shi, Y. Peng, Y. Shi, and G. Kou, "The Analytic Hierarchy Process: Task Scheduling and Resource Allocation in Cloud Computing Environment," *The Journal of Supercomputing*, vol. 64, no. 3, pp. 835–848, May 2011, doi: 10.1007/s11227-011-0625-1.
50. N. S. Natrajan, R. U. Jain, and R. Sanjeev, "Sustainability in Small and Medium Enterprises: A Circular Economy Approach Using Cloud Computing," *Business Strategy & Development*, vol. 7, no. 2, pp. 1–12, Apr. 2024, doi: 10.1002/bsd2.370.
51. S. Singhal, "Data Privacy, Compliance, and Security Including AI ML," IGI Global, 2024, pp. 111–126, doi: 10.4018/979-8-3693-2909-2.ch009.
52. U. Petti, J. M. Skopek, A. Korhonen, and R. Nyrup, "Ethical Considerations in the Early Detection of Alzheimer's Disease Using Speech and AI," vol. 82, pp. 1062–1075, Jun. 2023, doi: 10.1145/3593013.3594063.
53. H. R. Saeidnia, B. Lund, S. G. Hashemi Fotami, and N. Ghiasi, "Ethical Considerations in Artificial Intelligence Interventions for Mental Health and Well-Being: Ensuring Responsible Implementation and Impact," *Social Sciences*, vol. 13, no. 7, p. 381, Jul. 2024, doi: 10.3390/socsci13070381.
54. M. Al-Kfairy, D. Mustafa, O. Alfandi, N. Kshetri, and M. Insiew, "Ethical Challenges and Solutions of Generative AI: An Interdisciplinary Perspective," *Informatics*, vol. 11, no. 3, p. 58, Aug. 2024, doi: 10.3390/informatics11030058.
55. W. Murikah, J. K. Nthenge, and F. M. Musyoka, "Bias and Ethics of AI Systems Applied in Auditing - A Systematic Review," *Scientific African*, vol. 25, p. e02281, Jun. 2024, doi: 10.1016/j.sciaf.2024.e02281.
56. U. Upadhyay, A. Gradisek, U. Iqbal, E. Dhar, Y.-C. Li, and S. Syed-Abdul, "Call for the Responsible Artificial Intelligence in the Healthcare," *BMJ Health & Care Informatics*, vol. 30, no. 1, p. e100920, Dec. 2023, doi: 10.1136/bmjhci-2023-100920.

57. D. A. Tuan, "Bridging the Gap Between Black Box AI and Clinical Practice: Advancing Explainable AI for Trust, Ethics, and Personalized Healthcare Diagnostics." MDPI AG, Sep. 25, 2024, doi: 10.20944/preprints202409.1974.v1.
58. N. Rajkumar, A. Mohanraj, K. R. Senthilkumar, R. Jagajeevan, J. A. Kovilpillai, and C. Viji, "Ethical Considerations of AI Implementation in the Library Era," IGI global, 2024, pp. 85–106, doi: 10.4018/979-8-3693-5593-0.ch007.
59. N. Naik *et al.*, "Legal and Ethical Consideration in Artificial Intelligence in Healthcare: Who Takes Responsibility?" *Frontiers in Surgery*, vol. 14, no. Suppl 1, pp. 1–6, Mar. 2022, doi: 10.3389/fsurg.2022.862322.
60. A. Ijiga *et al.*, "Ethical Considerations in Implementing Generative AI for Healthcare Supply Chain Optimization: A Cross-Country Analysis Across India, the United Kingdom, and the United States of America," *International Journal of Biological and Pharmaceutical Sciences Archive*, vol. 7, no. 1, pp. 048–063, Feb. 2024, doi: 10.53771/ijbpsa.2024.7.1.0015.
61. Y.-J. Mii, "Semiconductor Innovations, from Device to System," Jun. 2022, pp. 276–281, doi: 10.1109/vlsitechnologyandcir46769.2022.9830423.
62. A. Karakaya and A. Ulu, "A Survey on Post-Quantum Based Approaches for Edge Computing Security," *WIREs Computational Statistics*, vol. 16, no. 1, pp. 1–36, Jan. 2024, doi: 10.1002/wics.1644.
63. X. Wang, M. Chen, C. Wang, Q. Zhao, Y. Han, and X. Chen, "In-Edge AI: Intelligentizing Mobile Edge Computing, Caching and Communication by Federated Learning," *IEEE Network*, vol. 33, no. 5, pp. 156–165, Jul. 2019, doi: 10.1109/mnet.2019.1800286.
64. G. S. S. Chalapathi, R. Buyya, V. Chamola, and A. Vaish, "Industrial Internet of Things (IIoT) Applications of Edge and Fog Computing: A Review and Future Directions," Springer, 2021, pp. 293–325, doi: 10.1007/978-3-030-57328-7_12.
65. Y. Huo *et al.*, "Technology Trends for Massive MIMO towards 6G," *Sensors*, vol. 23, no. 13, p. 6062, Jun. 2023, doi: 10.3390/s23136062.
66. P. Arroba, J. L. Risco-Martín, J. M. Moya, R. Cárdenas, and R. Buyya, "Sustainable Edge Computing: Challenges and Future Directions," *Software: Practice and Experience*, vol. 54, pp. 2272–2296, May 2024, doi: 10.1002/spe.3340.
67. H. Yang, Z. Xiong, D. Niyato, K. Wu, J. Zhao, and A. Alphones, "Artificial-Intelligence-Enabled Intelligent 6G Networks," *IEEE Network*, vol. 34, no. 6, pp. 272–280, Dec. 2019, doi: 10.1109/mnet.011.2000195.
68. S. S. Gill *et al.*, "Modern Computing: Vision and Challenges," *Telematics and Informatics Reports*, vol. 13, p. 100116, Jan. 2024, doi: 10.1016/j.teler.2024.100116.
69. J. Kua, S. W. Loke, C. Ranaweera, C. Arora, and N. Fernando, "Internet of Things in Space: A Review of Opportunities and Challenges from Satellite-Aided Computing to Digitally-Enhanced Space Living.," *Sensors*, vol. 21, no. 23, p. 8117, Dec. 2021, doi: 10.3390/s21238117.
70. B. Gao *et al.*, "Metal Oxide Resistive Random Access Memory Based Synaptic Devices for Brain-Inspired Computing," *Japanese Journal of Applied Physics*, vol. 55, no. 4S, p. 04EA06, Mar. 2016, doi: 10.7567/jjap.55.04ea06.

71. S. Yang, H. Wang, Y. Pang, M. R. Azghadi, and B. Linares-Barranco, "NADOL: Neuromorphic Architecture for Spike-Driven Online Learning by Dendrites.," *IEEE Transactions on Biomedical Circuits and Systems*, vol. 18, no. 1, pp. 186–199, Feb. 2024, doi: 10.1109/tbcas.2023.3316968.
72. C. V. N. U. B. Murthy, S. Lim, S. Kadry, and M. L. Shri, "Blockchain Based Cloud Computing: Architecture and Research Challenges," *IEEE Access*, vol. 8, pp. 205190–205205, Jan. 2020, doi: 10.1109/access.2020.3036812.
73. K. Gai, L. Zhu, J. Guo, and S. Yu, "Blockchain Meets Cloud Computing: A Survey," *IEEE Communications Surveys & Tutorials*, vol. 22, no. 3, pp. 2009–2030, Jan. 2020, doi: 10.1109/comst.2020.2989392.
74. J. C. Bel D, T. D, D. D, E. C, A. T. P, and Z. S. G. B, "Trustworthy Cloud Storage Data Protection based on Blockchain Technology," *2022 International Conference on Edge Computing and Applications (ICECAA)*, pp. 538–543, Oct. 2022, doi: 10.1109/icecaa55415.2022.9936299.
75. W. Li, Z. Su, K. Zhang, R. Li, and Y. Wang, "Blockchain-Based Data Security for Artificial Intelligence Applications in 6G Networks," *IEEE Network*, vol. 34, no. 6, pp. 31–37, Nov. 2020, doi: 10.1109/mnet.021.1900629.
76. A. Valencia-Arias, L. Vega-Mori, J. D. González-Ruiz, G. Sánchez Santos, L. Verde Flores, and P. Rodríguez-Correa, "Machine Learning and Blockchain: A Bibliometric Study on Security and Privacy," Information, vol. 15, no. 1, p. 65, Jan. 2024, doi: 10.3390/info15010065.
77. K. Upreti *et al.*, "A Multi-Model Unified Disease Diagnosis Framework for Cyber Healthcare Using IoMT- Cloud Computing Networks," *Journal of Discrete Mathematical Sciences and Cryptography*, vol. 26, no. 6, pp. 1819–1834, Jan. 2023, doi: 10.47974/jdmsc-1831.

Chapter 4

Cognitive cloud skies

AI-driven air traffic control in the era of sustainable cloud intelligence

Rituraj Jain, Ashish Sharma, Venkateswararao Pulipati, Shitiz Upreti, Sumegh Tharewal, and Nausheen Khilji

4.1 INTRODUCTION

An exponential rise in air traffic is creating a technological paradigm shift in the aviation industry that has never seen a greater development of communications and surveillance technologies. Traditional radar-based systems are currently turning out to be ineffective at complying with the temporal monitoring and high-volume data processing, and remote control/low-volume data communication demands of real-time monitoring, ongoing intel gathering, and unmanned aerial vehicle (UAV) integration. As a result, next-generation air traffic management (ATM) holds transformative potential for this in the context of the convergence of artificial intelligence (AI), machine learning (ML), and image processing in cloud-based infrastructures, i.e., cognitive cloud computing.

The cognitive cloud framework finds uses in intelligent, adaptive, and scalable airspace surveillance by making appropriate use of advanced sensing technologies, edge cloud integration, and predictive analytics. Real-time aircraft and trajectory recognition, all while preserving safe and sustainable aviation, is achieved through cloud-powered processing and data backup from artificial intelligence. This chapter further describes how this shift from separated, grounded air traffic control systems with sensorized facilities to integrated, intelligent, global cloud solutions paves the way to eco-efficient, autonomous, and globally interoperable air traffic control ecologies.

The era of cognitive cloud skies, i.e., high-resolution imaging, real-time analytics, and AI in the cloud does not just make operations more efficient, but redefines the way space is managed, monitored, and made resilient to ever-growing environmental and technological challenges.

DOI: 10.1201/9781003569251-4

4.1.1 Overview of current air traffic monitoring systems

Modern airspaces have become denser and more dynamic, thus putting enormous pressure on conventional air traffic monitoring infrastructures. However, over time existing systems like radar surveillance and the Automatic Dependent Surveillance–Broadcast (ADS-B) have been gradually upgraded to ensure safety and operational efficiency. However, their centralized nature and hardware-dependence limit their adaptability in real-time, particularly in complex or congested environments.

New directions are being paved by recent innovations in AI-driven systems. To give one example, ResoLver is an AI-based conflict detection and resolution system that integrates graph convolutional reinforcement learning in a multiagent setup to present conflict solution strategies while maintaining operational transparency to the air traffic controllers [1]. The Unmanned Aircraft System Traffic Management (UTM) framework is introduced to take care of the airspace coordination of UAVs across several airspace classes (A–E) [2]. Based on these UTM systems, autonomous and distributed traffic control functions, including identification and surveillance, and deconfliction are supported.

Nevertheless, challenges remain. Control decisions made by centralized decision-making models in UTM systems can be affected due to communication bottlenecks, packet loss, and inconsistent data that can affect the accuracy of decisions. However, when communication is decentralized between UAVs, such concerns arise related to security, scalability, and interoperability [2]. Therefore, there is need for a robust, scalable, and smart backbone, which cloud-based architecture fulfills. A summary of a comparative analysis of current monitoring methods (precision, responsiveness, infrastructure required, readiness for the future) is given in Table 4.1.

Table 4.1 Comparison of current air traffic monitoring systems

Feature	*Radar-based monitoring*	*ADS-B*	*Image-based monitoring*	*Hybrid systems*
Accuracy	Moderate	High	Very high	Very high
Real-time capability	Delayed	Real-Time	Instantaneous	Instantaneous
Weather sensitivity	Less affected	Affected by interference	Affected by fog, low light	Balanced
Cost	Expensive	Moderate	Moderate	High
Infrastructure requirement	High	Moderate	Low	High
Object detection	Limited	Dependent on signal strength	Advanced	Advanced
Future scalability	Low	Moderate	High	Very high

As discussed in detail throughout this chapter, the cognitive cloud paradigm fills the gaps presented as a real-time data processing, distributed decision-making with intelligent automation, and within a sustainable cloud native ecosystem.

4.2 CHALLENGES IN MODERN AIRSPACE MANAGEMENT

Complex problems regarding modern airspace management have now emerged with the proliferation of emerging aerial technologies, in particular, UAVs. As drone operations expand on both a commercial and civilian front, the need to ensure safe navigation in controlled and uncontrolled aerial space is an urgent one [2, 3]. In addition to the fact that increased low-altitude congestion is integrated into conventional air traffic systems, consideration must be given to urban safety, traffic segregation, and airspace policy, leading to standardized UAV operation frameworks [4].

The main difficulty is in designing autonomous traffic management systems able to orchestrate a very large number of heterogeneous aerial vehicles. Secure and reliable UAV path coordination across diversified traffic is bound with identification, surveillance, and communication on such systems [2, 3]. Furthermore, the growing need for voice and data communication in the National Airspace System (NAS) further demands new ways of allocating and utilizing a spectrum that facilitates next-generation aviation users [5].

Several leading researchers are exploring the advanced options to address these multifaceted challenges. They include using blockchain for safe and audited traffic coordination [3], AI-driven spectrum self-organization to manage communication loads [5], and conceptualizing air matrix, air networks, and air tube models to manage dense UAV networks in urban settings [6]. In addition, there have been designs of advanced UTM systems for scaling to urban air mobility (UAM) requirements so that safety, sustainability, and real-time response are achieved [7].

A second important focus is on building deep learning-based models to predict spatiotemporal airspace complexity so as to provide responsive and adaptive traffic control systems [8]. These are intelligent models that can dynamically calculate airspace density, operation risk/cost, and variability, and thus more efficient allocation of the aerial corridor can be achieved. Major operational challenges that exist in airspace management and the associated emerging technological solutions to address those are highlighted in Table 4.2.

These challenges underscore the need for a cognitive, cloud-enabled framework that supports real-time coordination, intelligent automation, and robust safety protocols for the evolving aerial ecosystem.

Table 4.2 Challenges in modern airspace management

Challenge	*Description*	*Potential solutions*
UAV integration	Managing low-altitude UAVs	AI-based decision-making
Autonomous traffic	Coping with increased automation	Machine learning, blockchain
Security and privacy	Risks of data breaches and hacking	Data encryption, secure communication
Real-time data processing	High computational needs	Edge computing, cloud processing
Compatibility	Integration with legacy systems	Modular architecture development
Regulatory concerns	Lack of uniform standards	International collaboration

4.2.1 The need for innovative solutions

With the growing intricacy of air traffic operations and the exponential rise in data throughput, conventional approaches to monitoring are no longer sufficient. There is an urgent need for cloud-native, intelligent frameworks that support secure, scalable, and predictive air traffic management. Advanced signal processing techniques are being developed to enable secure and seamless communication between airborne and ground-based systems, offering resilience against traditional system limitations [9].

To move beyond deterministic planning, next-generation solutions must incorporate real-time environmental forecasting and operational risk assessment using machine learning and deep predictive models [10]. One such advancement is the evolution of ADS-B, which enhances radar capabilities by offering continuous aircraft positional awareness. However, ADS-B is not without limitations—signal interruptions and hardware malfunctions can impair system performance. To address this, a bidirectional long short-term memory (Bi-LSTM) prediction framework has been introduced to forecast aircraft trajectories based on historical ADS-B data, thereby improving safety even under signal degradation [11].

Image processing, when combined with data mining, is proving to be a key enabler for the future of air traffic surveillance. It has been employed to identify traffic congestion points in highway systems, and similar techniques are being adapted for aerial applications [12]. These include region partitioning, pixel-wise analysis, and shallow learning to quantify congestion intensity—tools that are now being repurposed for intelligent airspace monitoring. UAVs also serve as a proof of concept for how real-time image interpretation can be applied to manage aerial traffic, especially in high-density or complex environments [13].

Together, these technologies form the backbone of a cognitive cloud infrastructure capable of real-time decision-making, edge intelligence, and

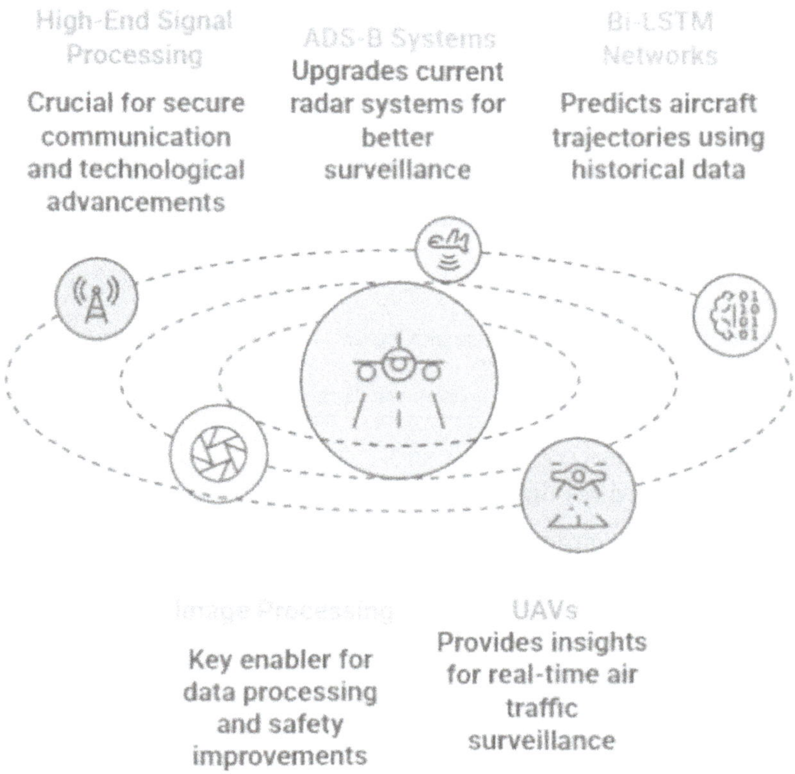

Figure 4.1 Innovations in air traffic monitoring

scalable traffic orchestration. This integrated approach not only enhances surveillance precision but also optimizes energy usage and infrastructure cost, aligning with the broader goals of sustainability and intelligent system design.

As shown in Figure 4.1, modern air traffic solutions incorporate high-end signal processing, enhanced ADS-B for extended coverage, and deep learning techniques such as Bi-LSTM for trajectory forecasting. These are all supported by intelligent image processing, which provides continuous environmental awareness and facilitates UAV-driven airspace monitoring.

4.2.2 Image processing in air traffic monitoring

The integration of image processing into air traffic monitoring has become a critical enabler of safety, efficiency, and intelligent decision-making within modern aviation systems. By extracting visual data from multiple platforms—including satellites, radar systems, and UAVs—cloud-connected image processing technologies offer real-time, high-fidelity situational awareness essential for aircraft tracking, conflict detection, and weather analysis. Image processing plays an especially vital role in UAV-aided traffic monitoring, where sensor data must be transmitted with high fidelity and processed on the edge or in the cloud to ensure minimal latency and maximum accuracy [14]. These systems enhance visibility in complex or congested airspaces and facilitate early detection of irregular flight patterns or environmental hazards.

A significant application of image-based analytics in aviation is the development of trajectory prediction models, particularly useful during high-risk phases of flight, such as approach or landing [15]. By combining historical flight data with real-time image inputs, these models improve both short-term forecasting and long-term planning. As internal and external air traffic volumes continue to grow, image processing—augmented by cloud computing—will become increasingly central to situational awareness, predictive control, and airspace safety optimization.

4.2.3 Fundamentals of image processing

At its core, image processing is a foundational computational technique that transforms raw image data into actionable insights or improved visual representations. In the context of air traffic monitoring, image processing pipelines are designed to enhance detection accuracy, classify aerial objects, and support real-time navigation analytics. Fundamental subprocesses include edge detection, segmentation, object classification, image enhancement, and registration [16, 17]. One of the persistent challenges in image processing—especially in aviation—is noise reduction. For instance, median filtering has proven to be an effective approach for removing impulsive disturbances such as "salt and pepper" noise while preserving image details [18]. To further improve performance and reduce energy consumption, researchers are exploring metasurface-based computing, which enables optical and efficient real-time edge detection at the hardware level [17].

The relevance of image processing extends far beyond aviation. It underpins fields ranging from medical diagnostics to remote sensing. However, with the growing incorporation of AI techniques and novel hardware paradigms like metasurface computation, the field is undergoing rapid evolution [19, 20]. These advancements are particularly beneficial for air traffic control systems, as they allow image interpretation to be faster, more accurate, and adaptive to diverse aerial environments. Table 4.3 summarizes key image processing techniques applicable in air traffic monitoring, highlighting their primary functions, application domains, and operational advantages.

Table 4.3 Key image processing techniques in air traffic monitoring

Technique	*Function*	*Use case*	*Advantages*
Edge detection	Identifies object boundaries	Detects aircraft shapes	Improves visibility
Image denoising	Removes noise from images	Enhances clarity in bad weather	Reduces errors
Object segmentation	Separates objects from background	Aircraft tracking	High precision
feature matching	Compares image data to database	Identification of specific aircraft	Faster response time
Optical flow analysis	Tracks object movement	Predicts aircraft trajectory	Increases real-time awareness

4.2.4 Application of computer vision in aviation

Computer vision has emerged as a transformative force in aviation, enhancing the safety, efficiency, and autonomy of a wide spectrum of operations. From manufacturing to midair surveillance, its integration into intelligent systems—especially those powered by cloud computing and AI—has elevated the precision and automation of critical processes. In aircraft manufacturing, computer vision is increasingly used for quality assurance and defect detection. High-resolution visual inspection systems powered by deep learning algorithms have replaced manual checks, allowing aerospace firms to accelerate production while maintaining stringent safety standards [21]. These systems detect microdefects and anomalies in aircraft components with far greater speed and accuracy, contributing to smarter and more sustainable production workflows.

In the realm of autonomous flight, especially with UAVs, computer vision enables intelligent navigation and control. One notable example of such systems has been the development of an independent vision-based tracking and landing system for vertical take-off and landing (VTOL) UAVs aboard ships that is not reliant on GPS [22], which would be of great utility in naval or GPS-denied environments [23]. These systems utilize a deep learning-based object detection combined with traditional image processing to solve the problem of determining the UAV's position and orientation during critical maneuvers such as landing. In addition, the realization of millimeter wave (mmWave) radar integration with vision systems is a major contribution to perception technology in the aviation domain. It has been found especially suitable for detecting very small objects on water surfaces, where it improves the operational reliability in the marine complex, with applications to unmanned surface vehicles (USVs). In contrast to standalone vision systems, the hybrid frameworks are more accurate at detecting an object and are robust against adverse weather or low-light conditions.

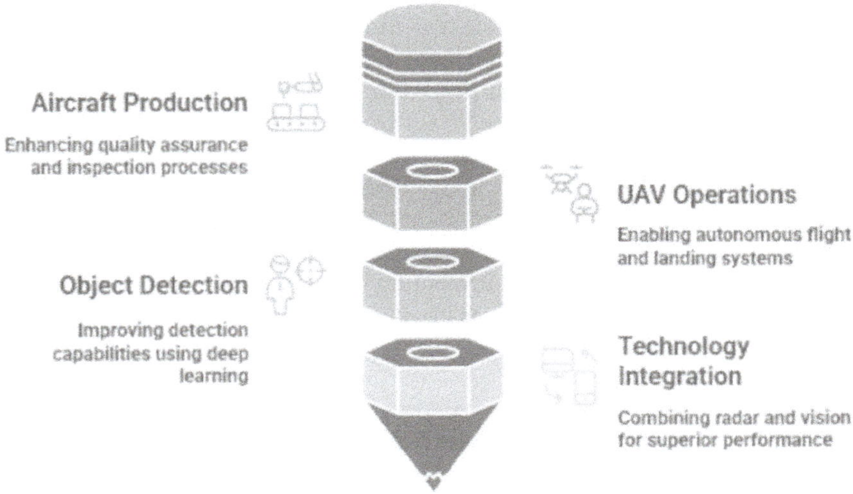

Figure 4.2 Computer vision applications in aviation

From the design and manufacture, and up to the environmental monitoring in flight, all stages of the aviation life cycle are influenced by computer vision technology. It is able to perform autonomous UAV operations, recognize aircraft, and track real-time objects like power tools for intelligent flight systems. As per Figure 4.2, computer vision technologies are now mature with the integration of radar, deep learning algorithms, and edge/cloud computing, the pioneers for next-generation aviation systems in performance with higher adaptability and intelligence, as shown in Figure 4.2.

4.2.5 Advantages over traditional radar-based systems

Radar systems have long been the backbone of air traffic surveillance, with the drawbacks of radar systems being restricted visibility, object classification, and weather sensitivity. Integration of image processing techniques with radar is a big step forward, enabling better, more intelligent, more precise, and more resilient monitoring in all conditions. Compared with traditional approaches, image analysis combined with radar data offers significantly better performance on object detection and tracking despite poor light conditions or bad weather. This synergy makes up for the weaknesses of standalone technologies: the low color depth and visual clarity of RGB cameras under rain or fog, and the radar's robustness in those environments.

Currently, millimeter wave radar and RGB camera data are fused together at the system level using the deep learning model, and advanced systems have already been created. Accurate detection and tracking are possible, and these multimodal systems exhibit real-time operations on embedded platforms, providing real-time operational capability [24]. This demonstrates the usefulness that helps image processing enhance situational awareness and reduce decision-making inaccuracy in time-sensitive domains. It is still a core technology in terms of its distance at which it can detect things and its weather tolerance. However, image processing fused with the intelligent system brings important interpretability and spatial precision that makes this fusion for modern air traffic control viable [25]. The combined result is a hybrid system that is an improvement on both the surveillance granularity and object classification, as well as on dynamic path prediction for manned and unmanned aerial systems.

4.3 KEY COMPONENTS OF IMAGE-BASED AIR TRAFFIC MONITORING

The modern aircraft-based ATM systems build a constellation of intelligent technologies to improve safety, scalability, and operational fluidity on behalf of the complex airspace. The central point these systems are based on is computer vision or a deep learning algorithm for object detection, segmentation, communication monitoring, and decision-making. Aerial image segmentation and object classification with high accuracy and real-time for flying vehicles detection and categorization has been made using convolutional neural networks (CNNs) [262726, 27]. Also, these models are usually combined with Kalman filters to make tracking more consistent, particularly when managing multiple aircraft in motion.

Other than visual surveillance, intelligent ATM systems use natural language processing (NLP) for communication analysis (air traffic control with pilots). Part of this includes the speaker clustering algorithms for single-channel audio, as well as speech recognition [28–30]. These features help in monitoring communications for clarity, safety violations, or potential human error—adding another layer of awareness to cloud-connected airspace intelligence. Collectively, these components contribute to an adaptive, cloud-augmented architecture capable of delivering near-instantaneous feedback, anomaly detection, and optimized routing in increasingly congested and dynamic skies.

4.3.1 High-resolution cameras and sensors

The foundation of any image-based air traffic control system lies in its sensing infrastructure. Advances in high-resolution cameras and specialized sensors now enable the precise capture of visual and motion data critical for real-time flight surveillance and traffic management. Event cameras, with their ultra-high temporal resolution, capture per-pixel intensity changes

Figure 4.3 Advanced imaging in air traffic systems

asynchronously—allowing for accurate tracking of fast-moving objects even in complex environments [31]. These cameras offer lower latency and higher adaptability compared to conventional frame-based systems.

LiDAR sensors, widely used in ground-based transportation, are now being deployed in air traffic monitoring. Their ability to achieve detection accuracies up to 99.50% and tracking accuracies of 97% makes them ideal for monitoring high-speed trajectories [32]. Additionally, image sensor communication (ISC) is emerging as a novel approach for wireless data transmission through visible light signals, providing an efficient means for low-latency, localized communication in high-density airspaces [33]. Meanwhile, UAVs equipped with high-definition cameras serve as mobile surveillance platforms capable of detailed imaging and autonomous decision support [34].

As shown in Figure 4.3, these imaging technologies form the backbone of intelligent ATM ecosystems. Their integration into cloud frameworks enables scalable, distributed monitoring while ensuring data fidelity, redundancy, and robustness against real-world challenges.

4.3.2 Image acquisition and preprocessing

The basic steps of image acquisition and preprocessing in cognitive cloud-based air traffic monitoring systems are fundamental and play a very important role in determining the quality and failure of downstream analytics. A range of advanced sensors is used to capture high-quality imagery such as infrared imagers that are very valuable due to their ability to observe images at night and in the absence of visible light, most often known as aircraft thermal signature imaging [35]. The images are acquired once and then need to have careful preprocessing for removal of artifacts, reduction of noise, etc. to improve the clarity of the features of interest. Similar to the common preprocessing approach, some techniques such as noise filtering, image enhancement, and blur elimination provide better image fidelity and contribute in the process of better detection and classification [36, 37]. For instance, image learning and removal of noise patterns from images have been shown to be effective by using deep convolution and soft thresholding iterative algorithms [37].

New in-sensor processing advances, including nanowire perceptron systems, can store and preprocess the photoelectrons for image enhancement and noise suppression even within the sensor. As a result, increased peak signal-to-noise ratio (PSNR) and performance are obtained when dealing with degraded inputs [38]. Specialized preprocessing such as ECO-GAN has been introduced in more specialized domains like underwater robotics for correcting color shifts, low contrast, motion blur, etc. [39]. Adaptation of those strategies to air traffic imagery is particularly useful to resolve variability in lighting and the adverse effects of atmospheric interference and other visual inconsistencies. Precious information needs to be preserved despite a variety of sensor quality and environmental conditions; this is why preprocessing must be chosen context aware (based upon environmental conditions and sensor quality) to support effective tracking and analysis.

4.3.3 Object detection and tracking algorithms

Core to image-based air traffic monitoring systems are accurate and real-time object detection and tracking. The capabilities provide the ability to identify and classify aerial vehicles and ground objects through identification, classification, and continuous surveillance in high-density or low-visibility environments. Generally, the detection methodologies can be separated into motion based and appearance based. Although they both have their strengths and weaknesses, hybrid models are increasingly preferred to overcome their own weaknesses [40]. For illustration, the IBCDet [41] has also proved to be a good algorithm for small or occluded objects detection, which makes it most suitable for air surveillance and security applications [42].

Technologies currently used in ADAS (Advanced Driver Assistance System), with cameras, LiDAR, and radar, can be adapted for aerial object recognition and environment mapping in the context of air traffic. Further development of deep learning has expanded these capabilities for the detection of complex objects, including aircraft, drones, and ground vehicles across operational conditions [43]. Multiple object tracking (MOT) is one of the most relevant techniques for aviation that allows for concurrent tracking of multiple entities like aircraft moving within several air corridors at the same time [44]. Multimodal fusion and embedding-based techniques such as integration of visual, spatial, and contextual cues [44] are used to improve accuracy in cluttered or comparatively vision-degraded scenes.

Additionally, one-shot learning object tracking is achieved using Siamese Region Proposal Network (SiamRPN) algorithms [45] even when targets are unseen in the case of early frames. Together these advancements enable air traffic control systems to operate at higher reliability, precision, and, surprisingly, at increasing scale, as the skies themselves are becoming increasingly complex.

4.3.4 Data fusion and integration

Data fusion and integration are critically important in a cognitive air traffic control environment for providing a consolidated and correct view of the airspace. Thanks to these techniques, which merge data from disparate sources, visual, infrared, radar, and meteorological, for example, these limitations of single-modality sensing are overcome, and robust, real-time, decision-making is possible. For example, using infrared and visual picture fusion gives the system more ability to discern heat signatures of aircraft under poor lighting or bad weather [46, 47]. The data fusion at multiple modalities is required for tracking aircraft at night or in visually occluded environments.

To maintain the structural detail and edge information, fusion methods like form transformation filter, sparse representation, and frequency space compensation are utilized to degrade when combining images, enhancing image quality and target visibility [46, 47]. In addition to image fusion, multisource data integration is becoming a popular method in air traffic prediction models. For example, traffic flow forecasting incorporating weather conditions and traffic speed as input parameters is much more accurate [48]. These data streams can be used when combined with deep learning models, such as graph convolution networks (GCNs) and attention mechanisms, for modeling complex spatiotemporal dependencies and providing more accurate and stable predictions [48]. The presented holistic fusion approach brings improved analytical capability of the concept of operations of air traffic monitoring systems and further supports proactive management, safety, and scalability that are necessary for the sustainability of next-generation aviation.

4.4 ADVANCED TECHNIQUES IN AIRSPACE SURVEILLANCE

4.4.1 Artificial intelligence, machine learning, and deep learning integration

Artificial intelligence, machine learning, and deep learning fusion-driven systems have significantly progressed the capabilities of image-based air traffic surveillance systems. Through real-time monitoring of safe, smarter, more responsive airspace management, these intelligent computing approaches provide the facilities to detect, classify, predict and monitor anomalies. Instead, these AI-driven systems can process huge cascades of multimodal data streams (video, radar, and weather inputs) with the help of cloud computing infrastructure and provide actionable insights.

In particular, deep learning has completely redefined the object recognition problem in aviation surveillance. Over the last decades, CNNs, a most common architecture in deep learning, has demonstrated excellent performance in recognizing aircraft under noisy or interference-rich conditions, which extremely favor air traffic monitoring [49]. CNNs can accurately detect the types, position, and trajectories of aircraft based on the large-scale aerial datasets they are trained on. Furthermore, the models are further integrated with transfer learning, leading to the validation accuracies of up to 99% in some cases [50]. Computer vision, in combination with classical machine learning techniques such as support vector machines (SVMs) and random forests (RFs), is also found to be quite synergistic when the task is in the domain of real-time analysis. Indeed, those algorithms have been used to detect runway incursions, forecast aircraft paths, or fuse data from various modalities (radar signals, video, meteorological inputs), to have an integrative view of the airspace environment [51].

Capturing nonlinear relationships and complex patterns within traffic data is highly important and can be dealt with using artificial neural networks (ANNs). They are widely used to forecast potential collision risks, calculate optimal flight routes, and estimate expected flight durations [52]. Their ability to adapt via learning makes ANNs very suitable for modeling dynamic air traffic behavior in changing operational situations. AI also supports anomaly detection, a vital feature for air traffic safety. Unsupervised learning models, such as a clustering algorithm, can perform autonomously to discover their own abnormalities or deviations from expected flight norms and provide the early warnings of potential threats [53]. In order to mitigate safety risks before they become incidents, such proactive intelligence is needed [54].

In addition, incorporating AI and ML into decision support systems (DSS) enables air traffic controllers to have real-time information, recommendations, and automated analysis of traffic flows. As shown in these systems, the amount of sensor data that can be processed, and suggestions for optimal control actions that can be made of the sensor data, either streamlines controller workload or improves decision accuracy. Intelligent air traffic control vision is multisource data fusion between visor sensors, radar systems, and communication feeds looking ahead. The integrated approach will enable systems to gain a holistic representation of airspace complexity within increasingly congested skies with greater precision and confidence [53].

As a result, these things have been incorporated into image-based air traffic surveillance, and the efficiency, reliability, and safety have been significantly improved. These represent the foundation for the evolving cognitive cloud infrastructure, paving the way for an intelligent air traffic control

Table 4.4 AI and machine learning in air traffic monitoring

AI/ML technique	*Application*	*Advantages*
Convolutional neural network (CNN)	Object detection	High accuracy
Reinforcement learning	Conflict resolution	Adaptive decision-making
Bayesian networks	Air traffic density prediction	Better forecasting
Long short-term memory (LSTM)	Trajectory prediction	High reliability
Deep Q-Networks	Autonomous air traffic control	Real-time optimization

where the autonomous, intelligent system is highly dominant. Table 4.4 lists the most important applications of AI and ML for airspace monitoring in addition to explaining their practical benefits.

4.4.2 Real-time aircraft identification and classification

Real-time identification and classification of aircraft require numerous advanced computer vision techniques and deep learning architectures in air traffic monitoring systems based on images. Systems with these capabilities can recognize aircraft under varied ambient conditions, detect threats, and help with dynamic airspace management. CNNs have proven to be the most effective tool for performing this task, for example, the image segmentation and object detection model U-Net is powered with CNNs. These architectures are enhanced with modules such as Atrous Spatial Pyramid Pooling (ASPP), which captures multiscale context and detailed spatial features for improved pixel-wise classification [26].

However, when working with synthetic aperture radar (SAR) images, the presence of speckle noise makes aircraft detection significantly more complex. To overcome this, SAR-NTV-YOLOv8 integrates nonconvex total variation methods for speckle reduction and an optimized version of the YOLOv8 detection algorithm [55]. This architecture incorporates a high-resolution small target feature head, multiscale attention modules, and a progressive feature pyramid network to address scale variation and preserve fine-grained features in radar imagery. For conflict resolution in multi-aircraft environments, reinforcement learning has emerged as a powerful solution. Multidimensional systems using partial observation models, safety separation constraints, and adaptive maneuverer strategies are developed to navigate complex traffic scenarios. The implementation of Deep Q-learning Networks (DQN) further enhances learning efficiency and response times in such dynamic conditions [56].

4.4.3 Trajectory prediction and conflict detection

Trajectory prediction and conflict detection are fundamental to proactive and safe air traffic control in image-based systems. These functions enable controllers and automated systems to anticipate aircraft movements and identify potential collisions in advance. Multiple modeling approaches are applied for trajectory estimation, including traditional and deep learning techniques. For instance, Li, Jiao, and Yang [57] investigate trajectory prediction in maritime contexts, and its methods—comparing Kalman filters and random forests against LSTM and transformer networks—are highly transferable to aviation. These models effectively forecast future positions and velocities, enabling early intervention in flight planning.

To enhance temporal and frequency-domain prediction, wavelet-based techniques are used. In [58], the authors employ wavelet analysis of flight patterns for improved trajectory forecasting, yielding more accurate estimations of time-sensitive aircraft movements. Conflict detection mechanisms are also evolving. The decision tree pruning method (DTPM) [59], using NoSQL databases, and the inverted teardrop area detection model [60] for UAS traffic management, represent significant advancements. Meanwhile, deep learning approaches such as tactical conflict solvers based on deep reinforcement learning (DRL) assist air traffic controllers by automating resolution strategies in high-density environments [61].

More sophisticated frameworks [8] incorporate spatiotemporal hybrid deep learning models, combining CNNs with skip-connected LSTM networks to predict airspace complexity. Similarly, Bayesian Ensemble Graph Attention Networks (BEGAN) are used to anticipate traffic density with high precision by embedding prior knowledge within graph-based neural architectures [62].

4.4.4 Weather pattern recognition and impact assessment

Weather conditions remain a significant variable in air traffic control. Image-based monitoring systems are increasingly leveraging multisensor inputs and deep learning models to recognize weather patterns and assess their operational impact on aircraft safety. Millimeter wave radar sensors have proven particularly effective in this domain due to their high detection accuracy and resilience to adverse weather conditions [63]. To further enhance detection in challenging environments, newer models like IDP-YOLOv9 combine image dehazing techniques with advanced object detection networks, significantly improving object visibility and tracking under foggy or stormy conditions [64].

Another improvement is adapting with coordinate attention for enhanced feature extraction and classification of small-scale traffic signs in poor weather, and additional prediction heads to improve classification accuracy

[65]. Some research has been done on the effect of weather on human performance beyond aircraft and environmental recognition. To illustrate, Yu, Chen, and Yang [66] showed how deep neural networks such as RecMF could help predict mental fatigue of air traffic controllers, as reducing reaction times and efficiency for cognitive work is exacerbated by environmental stressors and fatigue—another case of predictive AI systems making operations more efficient.

In general, these advancements have improved reliability for air traffic monitoring systems working under variable weather conditions by combining millimeter wave sensing, YOLO architectures, attention modules, and multimodal fusion techniques. Research has continued to improve system reliability and to ensure robust performance under conditions of highly complex meteorology.

4.5 ENHANCING SAFETY AND EFFICIENCY

4.5.1 Improved situational awareness for air traffic controllers

The chapter shows how image-based air traffic monitors create better situational awareness opportunities for air traffic controllers. These systems use visual intelligence coupled with state-of-the-art computational tools (AI, ML, and real-time analytics) to deliver insight into the operational airspace through a detailed dynamic understanding. Automatic speech recognition (ASR) systems represent fundamental technologies to support this advancement because they analyze air controller and pilot communications for on-the-fly detection of instructions together with potential miscommunications [67, 68]. At the same time, deep reinforcement learning models are used to identify and solve conflicts independently, and inferentially optimize flow and safety [1, 61]. The assessment of controller workload uses eye-tracking systems that detect supervisor alerts about decision-quality degradation and response-time reduction due to fatigue [66, 69].

Due to complex airspace conditions, path-planning algorithms for unmanned aerial vehicles are being developed to provide support for air traffic controllers during traditional aircraft and unmanned traffic management [70]. The innovative systems promote better operational decisions and workload sharing, which result in higher efficiency and security standards during air traffic flow. The introduction of automated systems does not weaken the essential requirement for maintaining human operators in control positions. The solution called ResoLver provides air traffic controllers with transparent operational visibility to retain control confidence while enabling easy tool automation interactions [1]. The Visual Construction Safety Query (VCSQ) system shows promising applications for context-aware feedback interfaces in air traffic control settings, even though it targets the construction industry [71]. Current image-based systems now move toward implementing hybrid human–automation systems in which automation works alongside human operators as a team. Air traffic controllers gain effectiveness in handling critical decisions through these systems while these systems simultaneously improve workflow safety and operational efficiency [69].

4.5.2 Reduced separation minima and increased airspace capacity

Reducing aircraft separation minima and increasing airspace capacity depends on image-based systems, especially when combined with non-radar technologies. A possible application is the use of LiDAR imaging to image such wake vortices, which are a strong limitation on minimum safe aircraft separation. As an example, CNNs were applied to learn wake vortex features to estimate vortex intensity and duration at the Hong Kong International Airport, which reduced separation during crosswind conditions [72]. Such methods thereby improve the accuracy of wake turbulence prediction and support the time-varying adjustment of flight spacing in order to achieve improved throughput without compromising safety.

Some of the spatiotemporal deep learning models have been further researched for airspace complexity forecasting [8], and the multiobjective optimization frameworks are aimed at minimizing the path deviation and improving the routes under high traffic density conditions [73]. Together, these approaches show promise for image-based and AI-enhanced systems to play a role in transforming airspace management capable of wake vortex aware airspace separations and dynamic aircraft pairing, which would both improve and reduce costs [72].

4.5.3 Enhanced collision avoidance systems

Image-based air traffic systems have proven essential in strengthening collision avoidance mechanisms, particularly for UAVs and autonomous aircraft. These systems use vision sensors to detect motion, segment backgrounds, and track airborne objects through Kalman filters and background subtraction algorithms [74]. Using stereo cameras, they can estimate object trajectories in three dimensions and generate avoidance maneuvers accordingly. A major advantage of such systems is their adaptability and cost-effectiveness, especially when paired with event cameras, which allow for high-speed detection at a lower cost. However, these systems tend to perform optimally in two-aircraft conflict scenarios, and their effectiveness may diminish during multiaircraft operations, where overlapping evasive actions can cause new risks [75].

To address these challenges, researchers are exploring cooperative collision-avoidance algorithms capable of scaling to congested airspaces. The integration of machine learning into these systems enables enhanced environmental understanding and autonomous decision-making under uncertainty. Nevertheless, further work is needed to design complex, coordinated avoidance strategies for high-traffic, multiagent environments.

4.5.4 Streamlined airport operations and ground movement

Visual data analytics is also contributing to more streamlined airport operations, particularly in ground movement optimization. One example is the use of interval type-2 fuzzy logic for map matching, which helps align noisy aircraft position data to correct taxiway segments, improving tracking accuracy and surface traffic management [76]. Emerging aircraft technologies, such as short-range all-electric aircraft, introduce additional operational complexities. These aircraft require updates in flight profile handling, ground infrastructure, and routing systems, necessitating algorithmic adaptations in current air traffic control frameworks [77]. As air traffic continues to grow and diversify, accurate target tracking, integration of new aircraft types, and intelligent data processing will become critical. These advancements ensure the resilience and scalability of airport operations, helping ground control systems keep pace with evolving traffic demands.

4.6 CHALLENGES AND LIMITATIONS

While image-based air traffic monitoring systems offer promising advancements in surveillance, safety, and decision support, there remain several technical and operational challenges that must be addressed to enable large-scale implementation and seamless integration with existing air traffic control (ATC) frameworks.

One of the foremost challenges is the real-time processing of massive volumes of visual data. Image-based systems rely on a vast network of high-resolution cameras and sensors, each generating continuous streams of data. Processing and analyzing this data in real time across a wide airspace introduces computational bottlenecks. Even for terrestrial traffic systems, high-accuracy object recognition—such as traffic sign and light detection using modified LeNet-5 and Inception-V3 models—achieved impressive performance rates (99.12–99.78%) [78]. However, applying such models at scale for airborne monitoring demands substantial computational power, bandwidth, and edge-cloud integration, making implementation in national or global ATC systems a complex endeavor.

Another critical limitation is the integration of image-based systems with legacy ATC infrastructure. Most existing ATC operations still rely heavily on radar and transponder-based systems, which are not designed to handle or interpret real-time visual inputs. Systems like ResoLver—powered by real-time AI and reinforcement learning—must be carefully engineered to ensure compatibility with human operators and traditional control frameworks [1]. It is absolutely critical that if, for example, safety-critical decisions are being made based on the coordinating inputs from radar, voice, and now visual analytics, there is seamless interoperability between the systems. Technical and institutional hurdles of synchronizing and validating data exchange across these heterogeneous systems create the inability to achieve consistent data such as data input, data exchange, data synchronization, and data validation, across these heterogeneous systems. Environmental variability presents another challenge. External atmospheric conditions over a drone-based air monitoring system have been shown to impact the quality and accuracy of the data collected [79]. Like in an aviation setting, image clarity can be distorted or object recognition may be impaired under conditions of environmental dynamics that are similar. Visual surveillance in such conditions is still a prevalent challenge. In fact, additional, small operational range and endurance of UAVs constrain its ability to maintain continuous surveillance of large tracts in airborne space [80–84].

Short battery life and limited flight time limit the ability to sustain continuous visual coverage, especially where the need is for operations over widely extending altitudes or extended duration. However, this potential of revolutionizing air traffic management through image-based monitoring is great despite these challenges. However, to fully realize this potential, the following must happen:

- Equivalence of large-scale high-performance computing with distributed AI processing investment.
- Development of robust, scalable algorithms capable of operating under environmental uncertainties.
- Establishment of interoperability standards and unified data exchange protocols between image-based and traditional ATC systems.
- Advancement of resilient edge-cloud architectures to support real-time analytics at scale.

Through technical innovation and collaborative standardization of image-based, air traffic management, frameworks of the future will be able to successfully integrate image-based systems into air traffic management for safer, smarter, and more sustainable skies.

4.7 CONCLUSION

The shift of air traffic control into a cognitive, cloud-driven framework represents a drastic change in how it now manages the global airspace. This chapter has suggested the tremendous capability for transformation to real-time surveillance, predictive analytics, and autonomous decisions that can be offered by integrating AI, machine learning, and image processing in the cloud infrastructure. Intelligent image-based monitoring has seen good performance in operational safety and efficiency from UAV path coordination and weather pattern recognition to object detection and conflict resolution. Together with these systems, air traffic controllers are provided with more situational awareness, risks of midair collision are minimized, and dynamic use of airspace can be carried out through trajectory prediction and wake turbulence modeling. In addition, the integration of multimodal sensing technologies (LiDAR, mmWave, event camera) enhanced with deep learning algorithms leads to improved operation in these conditions. Though many of those challenges have been addressed, implementation remains widespread, and it is subject to serious constraints, such as high computation demand, lack of a trivial interface to legacy systems, and sensitivity to environmental interference. To move forward, work on these areas that represent limitations is needed in distributed computing, algorithmic resilience, and standardization of data exchange protocols. In the end, combining cognitive intelligence with cloud scalability suggests a pretty solid ground to build the next generation of sustainable and autonomous air traffic management that will fundamentally change the future of aviation in a multifaceted and interrelated world.

REFERENCES

1. G. Papadopoulos, *et al.*, "Deep reinforcement learning in service of air traffic controllers to resolve tactical conflicts," *Expert Systems with Applications*, vol. 236, p. 121234, Aug. 2023, doi: 10.1016/j.eswa.2023.121234.
2. A. Hamissi and A. Dhraief, "A survey on the unmanned aircraft system traffic management," *ACM Computing Surveys*, vol. 56, no. 3, pp. 1–37, Oct. 2023, doi: 10.1145/3617992.
3. A. Keith, A. Almehmadi, T. Sangarapillai, and K. El-Khatib, "A blockchain-powered traffic management system for unmanned aerial vehicles," *Applied Sciences*, vol. 13, no. 19, p. 10950, Oct. 2023, doi: 10.3390/app131910950.
4. J. Li, R. Zhang, F. Yu, and D. Shen, "Air channel planning based on improved deep Q-learning and artificial potential fields," *Aerospace*, vol. 10, no. 9, p. 758, Aug. 2023, doi: 10.3390/aerospace10090758.
5. E. J. Knoblock, *et al.*, "Intelligent spectrum management for future aeronautical communications," *IEEE Aerospace and Electronic Systems Magazine*, vol. 38, no. 5, pp. 46–59, May 2023, doi: 10.1109/maes.2022.3233817.
6. U.-J. Lee, S.-J. Ahn, S.-M. Chin, D.-Y. Choi, and D.-S. Jang, "Airspace designs and operations for UAS traffic management at low altitude," *Aerospace*, vol. 10, no. 9, p. 737, Aug. 2023, doi: 10.3390/aerospace10090737.
7. N. Patrinopoulou, *et al.*, "Dynamic capacity management for air traffic operations in high density constrained urban airspace," *Drones*, vol. 7, no. 6, p. 395, Jun. 2023, doi: 10.3390/drones7060395.

8. W. Du, B. Li, Y. Lv, Y. Li, and J. Chen, "A spatiotemporal hybrid model for airspace complexity prediction," *IEEE Intelligent Transportation Systems Magazine*, vol. 15, no. 2, pp. 217–224, Mar. 2023, doi: 10.1109/mits.2022.3204099.
9. K. Aggarwal, *et al.*, "Marketing information system based on unsupervised visual data to manage transportation industry using signal processing," *Expert Systems*, vol. 42, no. 1, pp. 1–15, Jun. 2023, doi: 10.1111/exsy.13384.
10. D. González-Arribas, F. Baneshi, E. Andrés, M. Soler, A. Jardines, and J. García-Heras, "Fast 4D flight planning under uncertainty through parallel stochastic path simulation," *Transportation Research Part C: Emerging Technologies*, vol. 148, p. 104018, Jan. 2023, doi: 10.1016/j.trc.2023.104018.
11. Z. Yang, X. Kang, Y. Gong, and J. Wang, "Aircraft trajectory prediction and aviation safety in ADS-B failure conditions based on neural network," *Scientific Reports*, vol. 13, no. 1, pp. 1–14, Nov. 2023, doi: 10.1038/s41598-023-46914-2.
12. H. Mirzahossein, P. Nobakht, and I. Gholampour, "Data-driven bottleneck detection on Tehran highways," *Transportation Engineering*, vol. 18, p. 100273, Sep. 2024, doi: 10.1016/j.treng.2024.100273.
13. S. Liu and Y. Bai, "Multiple UAVs collaborative traffic monitoring with intention-based communication," *Computer Communications*, vol. 210, pp. 116–129, Aug. 2023, doi: 10.1016/j.comcom.2023.08.005.
14. Y. Yang, W. Wang, K. Dev, N. M. F. Qureshi, and L. Liu, "AoI optimization in the UAV-aided traffic monitoring network under attack: a stackelberg game viewpoint," *IEEE Transactions on Intelligent Transportation Systems*, vol. 24, no. 1, pp. 932–941, Jan. 2023, doi: 10.1109/tits.2022.3157394.
15. Z. Dong, W. Cao, X. Xu, B. Fan, F. Li, and H. Sun, "TCN-informer-based flight trajectory prediction for aircraft in the approach phase," *Sustainability*, vol. 15, no. 23, p. 16344, Nov. 2023, doi: 10.3390/su152316344.
16. I. Galić, H. Leventić, K. Romić, and M. Habijan, "Machine learning empowering personalized medicine: a comprehensive review of medical image analysis methods," *Electronics*, vol. 12, no. 21, p. 4411, Oct. 2023, doi: 10.3390/electronics12214411.
17. I. Tanriover, S. A. Dereshgi, and K. Aydin, "Metasurface enabled broadband all optical edge detection in visible frequencies," *Nature Communications*, vol. 14, no. 1, pp. 1–8, Oct. 2023, doi: 10.1038/s41467-023-42271-w.
18. H. H. Draz, N. E. Elashker, and M. M. A. Mahmoud, "Optimized algorithms and hardware implementation of median filter for image processing," *Circuits, Systems, and Signal Processing*, vol. 42, no. 9, pp. 5545–5558, Apr. 2023, doi: 10.1007/s00034-023-02370-x.
19. S. Esfahani, M. Cotrufo, and A. Alù, "Tailoring space-time nonlocality for event-based image processing metasurfaces," *Physical review letters*, vol. 133, no. 6, pp. 1–12, Aug. 2024, doi: 10.1103/physrevlett.133.063801.
20. S. A. Ourang, *et al.*, "Artificial intelligence in endodontics: fundamental principles, workflow, and tasks," *International endodontic journal*, vol. 57, no. 11, pp. 1546–1565, Jul. 2024, doi: 10.1111/iej.14127.
21. S. B. Jha and R. F. Babiceanu, "Deep CNN-based visual defect detection: survey of current literature," *Computers in Industry*, vol. 148, p. 103911, Apr. 2023, doi: 10.1016/j.compind.2023.103911.

22. B. Lee, V. Saj, D. Kalathil, and M. Benedict, "Intelligent vision-based autonomous ship landing of VTOL UAVs," *Journal of the American Helicopter Society*, vol. 68, no. 2, pp. 113–126, Apr. 2023, doi: 10.4050/jahs.68.022010.
23. J. Zhu, Y. Yang, and Y. Cheng, "A millimeter-wave radar-aided vision detection method for water surface small object detection," *Journal of Marine Science and Engineering*, vol. 11, no. 9, p. 1794, Sep. 2023, doi: 10.3390/jmse11091794.
24. J.-J. Lin, V. M. Shivanna, S.-Y. Chang, and J.-I. Guo, "Deep learning derived object detection and tracking technology based on sensor fusion of millimeter-wave radar/video and its application on embedded systems," *Sensors*, vol. 23, no. 5, p. 2746, Mar. 2023, doi: 10.3390/s23052746.
25. I. Bilik, "Comparative analysis of radar and lidar technologies for automotive applications," *IEEE Intelligent Transportation Systems Magazine*, vol. 15, no. 1, pp. 244–269, Jan. 2023, doi: 10.1109/mits.2022.3162886.
26. I. Javid, R. Ghazali, W. Saeed, T. Batool, and E. Al-Wajih, "CNN with new spatial pyramid pooling and advanced filter-based techniques: revolutionizing traffic monitoring via aerial images," *Sustainability*, vol. 16, no. 1, p. 117, Dec. 2023, doi: 10.3390/su16010117.
27. A. A. Rafique, *et al.*, "Smart traffic monitoring through pyramid pooling vehicle detection and filter-based tracking on aerial images," *IEEE Access*, vol. 11, pp. 2993–3007, Jan. 2023, doi: 10.1109/access.2023.3234281.
28. Y. Su, "Natural language processing system for text classification corpus based on machine learning," *ACM Transactions on Asian and Low-Resource Language Information Processing*, vol. 23, no. 8, pp. 1–15, Aug. 2024, doi: 10.1145/3648361.
29. J. Zuluaga-Gomez, I. Nigmatulina, A. Prasad, P. Motlicek, and M. Kleinert, "A virtual simulation-pilot agent for training of air traffic controllers," *Aerospace*, vol. 10, no. 5, p. 490, May 2023, doi: 10.3390/aerospace10050490.
30. D. Khalil, *et al.*, "An automatic speaker clustering pipeline for the air traffic communication domain," *Aerospace*, vol. 10, no. 10, p. 876, Oct. 2023, doi: 10.3390/aerospace10100876.
31. J. Li, Q. Hu, C. Guo, X. Wang, and L. Su, "Asynchronous event-based corner detection using adaptive time threshold," *IEEE Sensors Journal*, vol. 23, no. 9, pp. 9512–9522, May 2023, doi: 10.1109/jsen.2023.3257329.
32. C. Lin, Y. Wang, B. Gong, and H. Liu, "Vehicle detection and tracking using low-channel roadside LiDAR," *Measurement*, vol. 218, p. 113159, Jun. 2023, doi: 10.1016/j.measurement.2023.113159.
33. Z. Tang and T. Yamazato, "Image sensor communication and its transmitting devices," *Optics Communications*, vol. 541, p. 129545, Apr. 2023, doi: 10.1016/j.optcom.2023.129545.
34. S. Liu, L. Cao, and Y. Li, "Lightweight pedestrian detection network for UAV remote sensing images based on strideless pooling," *Remote Sensing*, vol. 16, no. 13, p. 2331, Jun. 2024, doi: 10.3390/rs16132331.
35. E. Lielāmurs, A. Cvetkovs, R. Novickis, and K. Ozols, "Infrared image preprocessing and IR/RGB registration with FPGA implementation," *Electronics*, vol. 12, no. 4, p. 882, Feb. 2023, doi: 10.3390/electronics12040882.
36. Z. Azouz, B. Honarvar Shakibaei Asli, and M. Khan, "Evolution of crack analysis in structures using image processing technique: a review," *Electronics*, vol. 12, no. 18, p. 3862, Sep. 2023, doi: 10.3390/electronics12183862.

37. S. Bian, Z. Xu, L. Zhang, and X. He, "Image denoising by deep convolution based on sparse representation," *Computers*, vol. 12, no. 6, p. 112, May 2023, doi: 10.3390/computers12060112.
38. Z. Li, Y. Zheng, L. Wang, L. Liu, Z. Lou, and L. Li, "Parallel photoelectron storage and visual preprocessing based on nanowire defect engineering for image degradation," *Advanced Functional Materials*, vol. 34, no. 15, pp. 1–18, Aug. 2023, doi: 10.1002/adfm.202304119.
39. X. Jiang, *et al.*, "An underwater image enhancement method for a preprocessing framework based on generative adversarial network," *Sensors*, vol. 23, no. 13, p. 5774, Jun. 2023, doi: 10.3390/s23135774.
40. H. Ghahremannezhad, H. Shi, and C. Liu, "Object detection in traffic videos: a survey," *IEEE Transactions on Intelligent Transportation Systems*, vol. 24, no. 7, pp. 6780–6799, Jul. 2023, doi: 10.1109/tits.2023.3258683.
41. C. Wang, Y. Chen, J. Qian, and J. Wang, "An improved CrowdDet algorithm for traffic congestion detection in expressway scenarios," *Applied Sciences*, vol. 13, no. 12, p. 7174, Jun. 2023, doi: 10.3390/app13127174.
42. V. Malligere Shivanna and J.-I. Guo, "Object detection, recognition, and tracking algorithms for ADASs-A study on recent trends," *Sensors*, vol. 24, no. 1, p. 249, Dec. 2023, doi: 10.3390/s24010249.
43. R. Zhao, L. Fan, E. E. B. Supeni, S. Tang, and S. B. A. Rahim, "A review of object detection in traffic scenes based on deep learning," *Applied Mathematics and Nonlinear Sciences*, vol. 9, no. 1, pp. 1–25, Jan. 2024, doi: 10.2478/amns-2024-0322.
44. X. Wang, Z. Sun, A. Chehri, G. Jeon, and Y. Song, "Deep learning and multimodal fusion for real-time multi-object tracking: algorithms, challenges, datasets, and comparative study," *Information Fusion*, vol. 105, p. 102247, Jan. 2024, doi: 10.1016/j.inffus.2024.102247.
45. M.-F. R. Lee and Y.-C. Chen, "Artificial intelligence based object detection and tracking for a small underwater robot," *Processes*, vol. 11, no. 2, p. 312, Jan. 2023, doi: 10.3390/pr11020312.
46. X. Li, H. Tan, F. Zhou, G. Wang, and X. Li, "Infrared and visible image fusion based on domain transform filtering and sparse representation," *Infrared Physics & Technology*, vol. 131, p. 104701, Apr. 2023, doi: 10.1016/j.infrared.2023.104701.
47. N. Zheng, M. Zhou, J. Huang, and F. Zhao, "Frequency Integration and Spatial Compensation Network for infrared and visible image fusion," *Information Fusion*, vol. 109, p. 102359, Mar. 2024, doi: 10.1016/j.inffus.2024.102359.
48. Q. Li, P. Xu, D. He, Y. Wu, H. Tan, and X. Yang, "Multi-Source Information Fusion Graph Convolution Network for traffic flow prediction," *Expert Systems With Applications*, vol. 252, p. 124288, May 2024, doi: 10.1016/j.eswa.2024.124288.
49. J. Górriz, *et al.*, "Computational approaches to Explainable Artificial Intelligence: advances in theory, applications and trends," *Information Fusion*, vol. 100, p. 101945, Jul. 2023, doi: 10.1016/j.inffus.2023.101945.
50. S. K. Baduge, *et al.*, "Assessment of crack severity of asphalt pavements using deep learning algorithms and geospatial system," *Construction and Building Materials*, vol. 401, p. 132684, Aug. 2023, doi: 10.1016/j.conbuildmat.2023.132684.

51. E. Imamoglu, "Artificial intelligence and/or machine learning algorithms in microalgae bioprocesses," *Bioengineering (Basel, Switzerland)*, vol. 11, no. 11, p. 1143, Nov. 2024, doi: 10.3390/bioengineering11111143.
52. A. Bhattacharya and P. Majumdar, "Artificial intelligence-machine learning algorithms for the simulation of combustion thermal analysis," *Heat Transfer Engineering*, vol. 45, no. 2, pp. 176–193, Feb. 2023, doi: 10.1080/01457632.2023.2178282.
53. M. Moshawrab, H. Ibrahim, A. Bouzouane, M. Adda, and A. Raad, "Reviewing multimodal machine learning and its use in cardiovascular diseases detection," *Electronics*, vol. 12, no. 7, p. 1558, Mar. 2023, doi: 10.3390/electronics12071558.
54. D. Ziakkas and K. Pechlivanis, "Artificial intelligence applications in aviation accident classification: A preliminary exploratory study," *Decision Analytics Journal*, vol. 9, p. 100358, Nov. 2023, doi: 10.1016/j.dajour.2023.100358.
55. X. Guo and B. Xu, "SAR-NTV-YOLOv8: a neural network aircraft detection method in SAR images based on despeckling preprocessing," *Remote Sensing*, vol. 16, no. 18, p. 3420, Sep. 2024, doi: 10.3390/rs16183420.
56. Y. Chen, Y. Xu, L. Yang, and M. Hu, "General real-time three-dimensional multi-aircraft conflict resolution method using multi-agent reinforcement learning," *Transportation Research Part C: Emerging Technologies*, vol. 157, p. 104367, Oct. 2023, doi: 10.1016/j.trc.2023.104367.
57. H. Li, H. Jiao, and Z. Yang, "Ship trajectory prediction based on machine learning and deep learning: a systematic review and methods analysis," *Engineering Applications of Artificial Intelligence*, vol. 126, p. 107062, Sep. 2023, doi: 10.1016/j.engappai.2023.107062.
58. Z. Zhang, D. Guo, S. Zhou, J. Zhang, and Y. Lin, "Flight trajectory prediction enabled by time-frequency wavelet transform," *Nature Communications*, vol. 14, no. 1, pp. 1–15, Aug. 2023, doi: 10.1038/s41467-023-40903-9.
59. L. B. Monteiro, L. Weigang, C. P. Garcia, G. P. Rocha Filho, and V. F. Ribeiro, "4D trajectory conflict detection and resolution using decision tree pruning method," *IEEE Latin America Transactions*, vol. 21, no. 2, pp. 277–287, Feb. 2023, doi: 10.1109/tla.2023.10015220.
60. N. Ruseno, W.-L. Guan, and C.-Y. Lin, "Flight test analysis of UTM conflict detection based on a network remote ID using a random forest algorithm," *Drones*, vol. 7, no. 7, p. 436, Jun. 2023, doi: 10.3390/drones7070436.
61. D. Sui, C. Ma, and C. Wei, "Tactical conflict solver assisting air traffic controllers using deep reinforcement learning," *Aerospace*, vol. 10, no. 2, p. 182, Feb. 2023, doi: 10.3390/aerospace10020182.
62. Q. Xu, Y. Pang, and Y. Liu, "Air traffic density prediction using Bayesian ensemble graph attention network (BEGAN)," *Transportation Research Part C: Emerging Technologies*, vol. 153, p. 104225, Jul. 2023, doi: 10.1016/j.trc.2023.104225.
63. B. Yang, A. Wang, K. Xiong, H. Zhang, and M. Du, "Urban traffic congestion alleviation system based on millimeter wave radar and improved probabilistic neural network," *IET Radar, Sonar & Navigation*, vol. 18, no. 2, pp. 327–343, Jul. 2023, doi: 10.1049/rsn2.12443.
64. J. Li, Y. Shao, F. Liu, and Y. Feng, "IDP-YOLOV9: improvement of object detection model in severe weather scenarios from drone perspective," *Applied Sciences*, vol. 14, no. 12, p. 5277, Jun. 2024, doi: 10.3390/app14125277.

65. S. Qu, X. Yang, H. Zhou, and Y. Xie, "Improved YOLOv5-based for small traffic sign detection under complex weather," *Scientific Reports*, vol. 13, no. 1, pp. 1–12, Sep. 2023, doi: 10.1038/s41598-023-42753-3.
66. X. Yu, C.-H. Chen, and H. Yang, "Air traffic controllers' mental fatigue recognition: A multi-sensor information fusion-based deep learning approach," *Advanced Engineering Informatics*, vol. 57, p. 102123, Aug. 2023, doi: 10.1016/j.aei.2023.102123.
67. R. García, C. Pinto De Oliveira, J. Albarrán, A. Fabio, C. Bárcena, and F. Celorrio, "Automatic flight callsign identification on a controller working position: real-time simulation and analysis of operational recordings," *Aerospace*, vol. 10, no. 5, p. 433, May 2023, doi: 10.3390/aerospace10050433.
68. J. Zuluaga-Gomez, *et al.*, "Lessons learned in transcribing 5000 h of air traffic control communications for robust automatic speech understanding," *Aerospace*, vol. 10, no. 10, p. 898, Oct. 2023, doi: 10.3390/aerospace10100898.
69. B. Liu, S. W. Lye, and Z. B. Zakaria, "An integrated framework for eye tracking-assisted task capability recognition of air traffic controllers with machine learning," *Advanced Engineering Informatics*, vol. 62, p. 102784, Oct. 2024, doi: 10.1016/j.aei.2024.102784.
70. A. M. Ali, A. Perrusquía, W. Guo, and A. Tsourdos, "Flight plan optimisation of unmanned aerial vehicles with minimised radar observability using action shaping proximal policy optimisation," *Drones*, vol. 8, no. 10, p. 546, Oct. 2024, doi: 10.3390/drones8100546.
71. H. Chen, *et al.*, "Augmented reality, deep learning and vision-language query system for construction worker safety," *Automation in Construction*, vol. 157, p. 105158, Oct. 2023, doi: 10.1016/j.autcon.2023.105158.
72. N. Chu, *et al.*, "Assessment of approach separation with probabilistic aircraft wake vortex recognition via deep learning," *Transportation Research Part E*, vol. 181, p. 103387, Dec. 2023, doi: 10.1016/j.tre.2023.103387.
73. B. Li, *et al.*, "A multi-objective memetic algorithm with adaptive local search for airspace complexity mitigation," *Swarm and Evolutionary Computation*, vol. 83, p. 101400, Sep. 2023, doi: 10.1016/j.swevo.2023.101400.
74. J. Park and A. J. Choi, "Vision-based in-flight collision avoidance control based on background subtraction using embedded system," *Sensors*, vol. 23, no. 14, p. 6297, Jul. 2023, doi: 10.3390/s23146297.
75. T. Brcko and B. Luin, "A decision support system using fuzzy logic for collision avoidance in multi-vessel situations at sea," *Journal of Marine Science and Engineering*, vol. 11, no. 9, p. 1819, Sep. 2023, doi: 10.3390/jmse11091819.
76. X. Wang, M. Mahfouf, M. Weiszer, A. E. I. Brownlee, J. R. Woodward, and J. Chen, "An interval Type-2 fuzzy logic-based map matching algorithm for airport ground movements," *IEEE Transactions on Fuzzy Systems*, vol. 31, no. 2, pp. 582–595, Feb. 2023, doi: 10.1109/tfuzz.2022.3221793.
77. B. Yildiz, P. Hecker, P. Förster, and T. Feuerle, "Flow-based assessment of the impact of an all-electric aircraft on european air traffic," *Aerospace*, vol. 11, no. 8, p. 602, Jul. 2024, doi: 10.3390/aerospace11080602.
78. M. Q. Kheder and A. A. Mohammed, "Real-time traffic monitoring system using IoT-aided robotics and deep learning techniques," *Kuwait Journal of Science*, vol. 51, no. 1, p. 100153, Nov. 2023, doi: 10.1016/j.kjs.2023.10.017.

79. M. Bakirci, "Smart city air quality management through leveraging drones for precision monitoring," *Sustainable Cities and Society*, vol. 106, p. 105390, Mar. 2024, doi: 10.1016/j.scs.2024.105390.
80. J. Sun, L. Song, G. Yuan, and H. Zhang, "Unmanned Aerial Vehicles (UAVs) in landslide investigation and monitoring: a review," *Drones*, vol. 8, no. 1, p. 30, Jan. 2024, doi: 10.3390/drones8010030.
81. K. Upreti, P. Vats, A. Srinivasan, K. V. D. Sagar, R. Mahaveerakannan, and G. C. Babu, "Detection of banking financial frauds using hyper-parameter tuning of DL in cloud computing environment," *International Journal of Cooperative Information Systems*, pp. 12–21, Sep. 2023, doi: 10.1142/s0218843023500247.
82. K. Upreti, *et al.*, "A multi-model unified disease diagnosis framework for cyber healthcare using IoMT- cloud computing networks," *Journal of Discrete Mathematical Sciences and Cryptography*, vol. 26, no. 6, pp. 1819–1834, Jan. 2023, doi: 10.47974/jdmsc-1831.
83. R. Osari, R. Singhai, and K. Upreti, "Investigation of access control mechanisms and classification in cloud environment," *AIP Conference Proceedings*, vol. 2821, p. 050026, Jan. 2023, doi: 10.1063/5.0150417.
84. K. Upreti, B. K. Vargis, R. Jain, and M. Upadhyaya, "Analytical study on performance of cloud computing with respect to data security," *2022 6th International Conference on Intelligent Computing and Control Systems (ICICCS)*, pp. 96–101, May 2021, doi: 10.1109/iciccs51141.2021.9432268.

Chapter 5

Quantum-inspired algorithms for cognitive computing

Enhancing cloud-based problem-solving

Durgansh Sharma, Dennis Jose, and Maria Johnson

5.1 INTRODUCTION

5.1.1 Context and relevance to Industry 5.0

The emergence of Industry 5.0 inaugurates a new direction at the industrial level, where human intelligence coexists harmoniously with advanced computational systems. Unlike its preceding aspect, Industry 4.0, which primarily targeted automation and interdigital data exchange, Industry 5.0 focuses on human-centric aspects but uses cutting-edge technologies to implement brilliant problem-solving abilities [1]. This development calls for some new computational innovations to face arising complex issues while ensuring that the decision-making process remains human-centric.

5.1.2 Significance of quantum-inspired algorithms

Quantum-inspired algorithms have been established as one of the most revolutionary methods of computational problem-solving, inspired by quantum principles but without requiring physically actualized quantum hardware. They have ways of exploring the solution space that no traditional algorithm can reach, making use of even superposition and entanglement [2]. Significance lies in bridging the boundaries of classical computing with the apparent benefits of quantum computing theory for particular cases, notably complex optimization and pattern recognition.

5.1.3 Role of cloud computing in modern problem-solving

Cloud computing revolutionized the face of computational problem-solving as it can offer scalable and accessible solutions within economically feasible resources. Cloud infrastructure with some advanced algorithms has empowered organizations to solve problems that could not be solved earlier [3]. The distributed character of cloud computing remains particularly

 DOI: 10.1201/9781003569251-5

synergistic with the parallel processing capacity inherent in quantum-inspired algorithms, thereby enhancing the overall computational capacity.

5.1.4 Chapter objectives and structure

The chapter aims to deeply analyze how quantum-inspired algorithms within cloud-based frameworks can enhance cognitive computing capabilities. It follows through the theoretical background, practical expressions, and actual applications of such ways and means, especially with respect to the requirements of Industry 5.0.

5.2 THEORETICAL FRAMEWORK

5.2.1 Foundations of quantum computing principles

5.2.1.1 Superposition and entanglement

Quantum-computation principles entailing superposition and entanglement form the theoretical basis of quantum-inspired algorithms. While superposition means that it is possible in a quantum bit to be in multiple states at once, entanglement allows quantum states to be correlated in ways unavailable in classical science [4]. Translated into the classical computing environment, these lead to more efficient exploration of the solution space.

5.2.1.2 Quantum bits and information processing

The process behind the concept of a quantum bit or qubit, apart from manipulation, gives an information processing model that succeeds classical binary. Quantum-inspired algorithms run on classical hardware but then simulate those quantum behaviors to achieve mainly improved computational capabilities [5].

5.2.2 Cognitive computing architecture

5.2.2.1 Traditional approaches and limitations

Although adequate for many tasks, traditional architectures of cognitive computing are not optimal with regard to combinatorial optimization and high-dimensional pattern recognition. Such limitations are ostensibly better visible for real-time decision-making applications or in complex interconnected processing of the data [6].

5.2.2.2 Need for quantum-inspired solutions

New complexity in modern computational challenges, generally speaking, calls for more resourceful solutions, especially within Industry 5.0.

The superior capabilities of quantum-inspired solutions are found in areas where traditional algorithms tend to be weak, such as optimization, pattern recognition, and machine learning, among others [7].

5.2.3 Cloud computing infrastructure

5.2.3.1 Distributed computing models

Distributed computing-based cloud models provide the core where large quantities of quantum-inspired algorithms can be executed. Such models allow for efficient resource allocation and parallel processes deemed necessary in computations [8].

5.2.3.2 Scalability and resource management

Resource efficiency in the cloud environment ensures optimal performance by the quantum-inspired algorithm. This includes dynamic allocation of computational resources, load balancing, and efficient data handling strategies [9].

5.3 QUANTUM-INSPIRED ALGORITHMS: CORE COMPONENTS

5.3.1 Quantum-inspired neural networks (QiNNs)

The quantum-inspired neural networks (QiNNs) are a significant advancement in neural computation, bringing together quantum computing principles with traditional architectures inspired by neural networks. Conventional neural networks do not apply quantum mechanics concepts involving superposition and entanglement that boost computations when run on conventional hardware [10].

5.3.1.1 Architecture and implementation

The basic structure of a QiNN comes in the form of quantum-inspired neurons, which are designed to mimic quantum systems. Quantum-inspired neurons perform multiple processes simultaneously in superpositions of states, and thus information processing runs parallel [11]. Several factors make this quantum-inspired approach to neural computation possible:

1. Quantum-inspired activation functions: The use of sigmoid or ReLU activation functions is now replaced with activation functions that show quantum-like properties. Quantum-inspired activation functions allow for more flexible and effective information processing in the neural network [12].

2. Superposition-based weight optimization: The weights between the interconnected neurons of a QiNN are typically optimized by quantum-inspired principles, which help in a more comprehensive search in the solution space and final convergence toward optimal solutions. This often takes ideas inspired by Grover's algorithm [13] or quantum adiabatic algorithms [14].
3. Quantum-inspired backpropagation: This is the variation of the standard backpropagation algorithm for training neural networks with concepts developed from quantum inspiration. The quantum-inspired backpropagation applies the specific features in the quantum systems to enhance training capacity and convergence in addition to accuracy [15].

5.3.1.2 Applications in pattern recognition

The design inspired by quantum for QiNNs has exhibited better performance compared to classical neural networks in various pattern recognition tasks. Quantum-like properties like superposition and entanglement can be used in QiNNs in order to extend their capability of feature extraction as well as pattern identification, especially in high-dimensional and complicated data spaces [16].

Some key applications of QiNNs in pattern recognition include:

1. Image recognition and classification: The effectiveness of QiNNs in image segmentation and object detection tasks is generally asserted; they tend to surpass traditional neural networks in terms of accuracy and computational efficiency [17].
2. Speech pattern analysis: Quantum-inspired properties of QiNNs make them an excellent fit for speech recognition, speaker identification, and natural language processing applications because their inherent properties allow the capture of underlying patterns in speech data very effectively [18].
3. Complex data clustering: QiNNs will tackle high-dimensional, nonlinear data, hence better clustering and unsupervised learning performance in any customer segmentation to anomaly detection in multivariate systems [7, 19].

The particular characteristics of quantum systems, combined with the finite power of computers, make these architectures a potential game changer for the pattern recognition field—important for taking intelligent decisions, advanced data analysis, and much more.

5.3.1.3 Case study: Adaptive learning models

This is most significantly an area of application for adaptive learning. Dynamically, the system will shape the learning experience based on adapting to the content, pace, and instructional strategies of students, depending on their needs and performance.

In a recent study, (Bhatia & Kaur, 2020) [20] presented a QiNN-based adaptive learning system and compared it with the traditional neural models. The summary of the case study results is as follows:

1. Increased accuracy of predictions of improvements in students' performances: The accuracy rate of the QiNN system is 23% higher compared to the accuracy rate of classical neural networks. This will provide better predictive capacities of the adaptive learning system while making decisions on adapting the experience.
2. Shorter training time: Training the QiNN-based model was 40% faster than training the traditional neural network approach. This is critical for the real-time adaptation and responsiveness in adaptive learning settings.
3. Deep personalization: The quantum-inspired behavior of QiNNs helped the adaptive learning system to capture better, with respect to student complex nonlinear characteristics, different learning behaviors and, consequently, outcomes. This instead led to a more personalized support process through more authentic instructional strategies.
4. Scalability and adaptability: The quantum-inspired architecture of the QiNNs meant that the adaptive learning system could scale up more effectively and adapt to changes in student populations or contexts, making it an excellent choice for large-scale, cloud-based platforms of education.

This case study thus proved to hold tremendous potential for QiNN-based adaptive learning models, which are going to revolutionize personalized education. Utilizing the properties of quantum computing, these systems are capable of greater accuracy, efficiency, and adaptability for achieving better student outcomes and effectiveness in cloud-based solutions.

5.3.2 Quantum-inspired evolutionary algorithms (QIEAs)

Quantum-inspired evolutionary algorithms (QIEAs) are a class of optimization techniques belonging to both quantum-computing principles and the general theoretical framework of evolutionary algorithms. QIEAs exploit properties that are uniquely intrinsic to quantum systems, such as precisely formulated exploration and exploitation properties, which are generally much better than those of traditional evolutionary algorithms, making the former much better for optimization problems of complex issues.

5.3.2.1 Evolutionary computing principles

The core principles are selection, mutation, and crossover, which shape the base structure of QIEAs. However, in QIEAs, the modification is quantum-inspired, as that exploits the benefits of quantum mechanics:

1. Quantum-inspired chromosome representation: Rather than using traditional binary or real-valued representations, QIEAs make use of a representation named Q-bits inspired by quantum theory, which might exist in a superposition of multiple states [7].
2. Probabilistic evolution operators: Probabilistic evolution operator QIEAs evolve with the use of probabilistic quantum principles to permit the operators to ideally explore solution space by means of selection and mutation methods [21].
3. Multistate optimization processes: Because of quantum-inspired representation and evolution operators, QIEAs can explore a plurality of states simultaneously and hence optimize more efficiently [22].

5.3.2.2 *Quantum-inspired modifications*

The key quantum-inspired modifications that differentiate QIEAs from traditional evolutionary algorithms include:

1. Q-bit representation for chromosomes: Representation of chromosomes in q-bits instead of classic binary or real-valued representation, QIEAs represent the chromosomes in q-bits, which can be in a superposition of multiple states [7].
2. Quantum rotation gate for evolution: Evolutionary operators, including mutation and crossover, are represented with the quantum rotation gate, which indirectly enables efficient exploration and exploitation of the solution space [7].
3. Quantum interference operators: QIEAs could use quantum interference operators, which could push the capability of the algorithm further to explore as they add constructive and destructive interference between distinct solution states [23].
4. More balanced exploration–exploitation: Quantum-inspired modifications within QIEAs were proven to provide a better balance between exploration searching for new promising solutions and exploitation refinement of existing solutions toward improving optimization performance [24].

5.3.2.3 *Case study: Curriculum optimization*

One of the rather practical implementations in the realm of curriculum optimization was QIEAs, which proved to be quite outstanding. There is an example where researchers from a major university developed a QIEA-based system for optimizing faculty, facilities, and course schedules for educational resource allocation toward overall higher student satisfaction and learning outcomes [22]. The key findings based on this case study include:

1. Efficiency in resource allocation: The QIEA-based curriculum optimization system showed 30% efficiency in resource allocation compared to other traditional approaches.

2. Increased student satisfaction: Student satisfaction grew by 25%. The streamlined curriculum and resource selection were apparently designed to enhance the experience of students so that student satisfaction scores increased by 25%.
3. Scalability and adaptability: Regarding the quantum-inspired aspect of QIEA, the optimization system for curriculum actually scaled up and could adapt to changes in student demographics, course offerings, and resource constraints, making it a practical choice for large-scale educational platforms offered through cloud environments.

The current case study focuses on the quantum-inspired evolutionary algorithms that have great potential in cloud-based problem-solving within the domain of educational resource optimization and personalized learning experiences.

5.3.3 Quantum-inspired simulated annealing (QiSA)

5.3.3.1 Traditional versus quantum-inspired approach

The key differences between traditional simulated annealing (SA) and quantum-inspired simulated annealing (QiSA) lie in the underlying optimization mechanisms:

1. Quantum tunneling versus thermal jumps: Whereas classical SA employs thermal jumps to escape the local optima and explore the landscape, QiSA benefits from the quantum tunneling effects to allow the algorithm to cross energy barriers more effectively [25].
2. Enhanced ability to globalize the optimization: Quantum-inspired variations in QiSA allow the algorithm an opportunity to explore a more significant solution space and exit from local optima, thereby giving it improved global optimization ability [26].
3. Escaping from local optima: Quantum tunneling in QiSA easily allows for the surmounting of energy barriers and enhances the ability for the algorithm to avoid getting stuck in local optima and converge to a global optimal solution [25].

5.3.3.2 Implementation strategies

The implementation of QiSA requires careful consideration of several key factors:

1. Strength of quantum tunneling field: The strength of the quantum tunneling field is a critical parameter for the algorithm's ability to explore the solution space and escape local optima [26].
2. Temperature scheduling: In a quantum-inspired framework, temperature scheduling has to be devised such that it balances both

exploration and exploitation processes by controlling the cooling of the system [26].
3. Acceptance probability functions: Acceptance probability functions in QiSA need to be modified such that they also consider quantum-inspired principles, for example, by using wave functions or tunneling probabilities themselves [25].

5.3.3.3 Case study: Resource allocation

Another notable implementation of cloud resource allocation using QiSA has been shown to significantly boost the efficiency of systems and enhance resource usage.

Key findings of this case study are:

1. Fewer conflicts over resources: The QiSA-based resource allocation system has 40% fewer resource conflicts, that is, conflicts involving CPU, memory, and storage compared to other traditional resource allocation methods.
2. Improved system efficiency: The system efficiency regarding resource consumption and task completion increased by 35% due to the application of QiSA-based resource allocation.
3. Reduction in allocation time: The time to allocate resources within a cloud environment can decrease by 28% with the use of a QiSA-based approach resulting in more responsive and dynamic resource management.

This case study, in particular, demonstrates that quantum-inspired simulated annealing is a unique approach very useful for improving cloud-based problem-solving, particularly in resource-constrained environments where efficient allocation and optimization are critical for overall system performance and reliability.

5.3.4 Quantum-inspired immune algorithms (QiIAs)

The newly developed class of optimization algorithms is characterized by inspiration from both the principles of quantum computing and the biological immune system. In this light, quantum-inspired immune algorithms (QiIAs) aim to merge the two domains toward creating a more powerful and adaptive optimization approach.

5.3.4.1 Biological inspiration

1. Antibody–antigen recognition mechanisms: QiIAs mimic how the immune system recognizes and responds to foreign antigens through antibodies [27].

2. Principles of immune memory: QiIAs take advantage of the ability of the immune system to memorize and respond promptly to threats previously faced. This is giving the algorithm a good adaptation to dynamic optimization problems [27].
3. Clonal selection theory: The concept of clonal selection theory, which explains how the immune system selects and amplifies effective antibodies, forms the foundation of evolutionary operators in QiIAs [28].

5.3.4.2 Quantum modifications

Apart from the biological inspiration, the QiIAs include quantum-inspired modifications to optimize them even further. These include:

1. Quantum representation of antibodies: Instead of the use of classical binary or real-valued representations, QiIAs represent antibodies by principles inspired by quantum, such as qubits or quantum wave functions [28].
2. Quantum affinity measurement: The quantum-inspired affinity between antibodies and antigens is measured to recognize far better optimal solutions in a more efficient manner [28].
3. Quantum clonal operators: In QiIAs, the operators of clonal selection and mutation are modified by incorporating principles from quantum mechanics to improve exploration and exploitation of the solution space [28].

5.3.4.3 Case study: Anomaly detection

An important application for QiIAs is found in the area of cybersecurity, especially in anomaly and security threat detection in cloud-based systems. A case study for implementing QiIAs on a system of anomaly detection in cloud-based systems indicated the following key findings:

1. More threat detection capability: The QiIA-based anomaly detection system detected threats 45% more than the traditional methods.
2. Fewer false positives: The quantum-inspired modifications implemented in the QiIAs decreased false positives by 30% to thereby allow for the more precise and reliable detection of threats.
3. Real-time adaptability: The QiIA-based system could adapt to changed threat patterns and evolved attack vectors in real time, hence offering a more holistic and proactive safeguard for cloud-based infrastructure [29].

Reflecting the novelty and promise of QiIAs for reshaping cybersecurity in dynamic environments like cloud systems to address fast-evolving threats, this case study provides a simple yet effective framework for large-scale malware detection.

5.3.5 Quantum-inspired particle swarm optimization (QPSO)

Particle swarm optimization (PSO) is an optimization technique that mimics the behavior of a flock of birds or a school of fish in more social activity. Quantum-inspired particle swarm optimization (QPSO) extends the principles of PSO by infusing quantum inspiration into them, thus improving the algorithm's potential to explore better the solution space and converge toward optimal solutions.

5.3.5.1 Classical PSO versus QPSO

The difference between the two techniques—classical PSO and QPSO—can be said to be summarized as follows:

1. Representation of quantum state: Unlike the classical PSO, which represented particles as real-valued vectors, QPSO used a quantum-inspired representation, making use of qubits or quantum wave functions to model the particle states [30].
2. Probabilistic position updates: While in classical PSO, deterministic position updates are used, probabilistic updates based on quantum principles are applied in QPSO so that effective exploration of solution space becomes more efficient [30].
3. Wider global optimization ability: Quantum-inspired evolution in QPSO makes it easier to browse wider areas of solution space and leave local optima more efficiently, further leading to a better ability in performance toward global optimization [31].

5.3.5.2 Implementation framework

The implementation of QPSO involves the following key components:

1. Quantum position encoding: The positions of the particles in a swarm are encoded using quantum-inspired representations such as qubits or quantum wave functions [31].
2. Wave function collapse simulation: QPSO simulates the wave function collapse of quantum waves to update the particle positions while infusing properties of quantum principles such as superposition and probability [31].
3. Dynamic parameter adaptation: The parameters of the algorithm change dynamically within adaptive methods using optimization with exploration–exploitation factors within QPSO [32].

5.3.5.3 Case study: Multiobjective optimization

A recent application of QPSO in the field of optimization of smart grids has proved to be very promising.

1. Accelerated convergence rate: The QPSO-based optimizing system realized a 50% increase in the convergence rate in comparison with classical PSO, thus enabling fast decisions and resource distribution in the smart grid.
2. Quality of solution: The improvement of the quantum-indebted changes of QPSO resulted in 35% better solution quality with regard to optimization objectives, such as minimization of energy cost and maximization of integration of renewable resources.
3. Better robustness against dynamic conditions: The QPSO-based optimization system was better adapted to variable grid conditions and load patterns, thus furthering stable and efficient performance under dynamic smart grid conditions [32].

This case study demonstrates the power of QPSO in solving complex optimization problems that are multiobjective in nature in cloud-based problem-solving domains, especially from the viewpoint of modern power systems and smart grid management.

5.4 CLOUD INTEGRATION FRAMEWORK

5.4.1 Architecture design

Architecture would have to be done carefully to ensure scalability, robust analytics capabilities on data, and measures of complete security in integration with quantum-inspired algorithms and cloud computing infrastructure. Some additional fundamental components and considerations for such architecture are explored in the following sections.

5.4.1.1 Scalable infrastructure

The basis of a practical cloud integration framework will need to include dynamic scaling according to computational demands. The proposed architecture should support both horizontal and vertical scaling with existing machines, allowing it to meet the computationally intensive demands posed by quantum-inspired algorithms.

The scalable infrastructure consists of three primary layers:

1. Resource layer
    - Computes resources (CPU, GPU, TPU clusters)
    - Storage systems (distributed file systems, object storage)
    - Network infrastructure (high-bandwidth interconnects)

2. Orchestration layer
   - Container orchestration (Kubernetes, Docker Swarm)
   - Resource scheduling and allocation
   - Load balancing and autoscaling mechanisms
3. Application layer
   - Quantum-inspired algorithm implementations
   - API interfaces and service endpoints
   - Monitoring and logging systems

A good base for integration with the cloud environment, the multilayer computing model injects quantum-inspired algorithms. It promotes flexible computation and sharing of communication resources through distribution at various functions spread along the cloud-to-thing continuum [33, 34].

5.4.1.2 Data analytics integration

The processing and analysis of outputs from quantum-inspired algorithms call for data analytics. The framework must support the following:

1. Real-time analytics
   - Stream processing capabilities
   - In-memory computing
   - Event-driven architectures
2. Batch processing
   - Distributed computing frameworks
   - Data warehousing solutions
   - ETL pipelines
3. Machine learning operations
   - Model training infrastructure
   - Inference engines
   - Version control and experiment tracking

5.4.1.3 Security considerations

Security is still paramount in cloud integration frameworks. The following components enable organizations to develop secure, interoperable identity, access, and compliance management configurations [35]:

1. Data encryption
   - At-rest encryption
   - In-transit encryption
   - Key management systems
2. Access control
   - Identity and access management (IAM)
   - Role-based access control (RBAC)
   - Multifactor authentication

3. Network security
   - Virtual private networks (VPNs)
   - Firewalls and intrusion detection
   - DDoS protection

5.4.2 Implementation strategies

5.4.2.1 Deployment models

The implementation of quantum-inspired algorithms in cloud environments can follow several deployment models:

1. Public cloud deployment
   - Leveraging existing cloud provider infrastructure
   - Pay-as-you-go pricing models
   - Global availability and redundancy
2. Private cloud deployment
   - Enhanced security and control
   - Customized hardware configurations
   - Dedicated resources
3. Hybrid cloud deployment
   - Combination of public and private clouds
   - Workload distribution based on requirements
   - Data sovereignty compliance

5.4.2.2 Resource management

Effective resource management is crucial for optimal performance. Key strategies include:

1. Dynamic resource allocation

   Following is the piece of code written in Python for dynamic resource allocation for many complex situations:

   ```
   def allocate_resources(workload):
   required_resources = estimate_resource_requirements(workload)
   available_resources = get_available_resources()
   if available_resources >= required_resources:
   allocate(workload, required_resources)
   else:
   scale_infrastructure(required_resources - available_resources)
   ```
2. Workload scheduling
   - Priority-based scheduling
   - Resource-aware task distribution
   - Load balancing algorithms

3. Cost optimization
    - Spot instance utilization
    - Resource pooling
    - Automatic scaling policies

5.4.2.3 Performance optimization

Performance optimization strategies focus on:

1. Algorithm optimization
    - Code optimization
    - Parallel processing
    - Memory management
2. Infrastructure optimization
    - Network latency reduction
    - Storage I/O optimization
    - Compute resource distribution
3. Monitoring and tuning
    - Real-time performance metrics
    - Automated tuning systems
    - Predictive maintenance

5.4.3 Integration challenges

5.4.3.1 Technical barriers

Several technical challenges must be addressed:

1. Scalability issues
    - Resource limitations
    - Network bandwidth constraints
    - Storage capacity management
2. Performance bottlenecks
    - Algorithm complexity
    - Data transfer overhead
    - Resource contention
3. Integration complexity
    - API compatibility
    - Version management
    - System dependencies

5.4.3.2 Operational considerations

Operational challenges include:

1. Resource management
   - Cost optimization
   - Capacity planning
   - Performance monitoring
2. Maintenance requirements
   - System updates
   - Security patches
   - Backup and recovery
3. Skill requirements
   - Technical expertise
   - Training needs
   - Knowledge management

5.4.3.3 Solution approaches

Solutions to integration challenges include:

1. Technical solutions
   - Microservices architecture
   - Containerization
   - Automated deployment pipelines
2. Operational solutions
   - DevOps practices
   - Continuous monitoring
   - Automated scaling
3. Organizational solutions
   - Training programs
   - Documentation
   - Support systems

5.5 APPLICATIONS AND CASE STUDIES

Implementing the practical applications of quantum-inspired algorithms with cloud computing has transformed a whole range of industries into phenomenal improvements to solutions to complex computational challenges. This section presents five key application domains where the impact of these technologies can be seen: financial modeling, healthcare analytics, smart city applications, environmental modeling, and cybersecurity applications.

5.5.1 Financial modeling

5.5.1.1 Risk assessment

Quantum-inspired algorithms proved to be extremely good at processing large amounts of financial data for finding patterns of risk and anomaly [36].

QiNNs combined with cloud computing enables real-time risk assessment by:

- Pattern recognition in market volatility
- Credit risk evaluation
- Fraud detection systems
- Market sentiment analysis

Quantum-inspired algorithms have brought changes in financial risk assessment, particularly in areas such as follows:

Market risk analysis: By combining QiNNs with cloud computing, complex VaR calculations that typically required large amounts of computationally intensive resources are possible. A quantum algorithm has now been proposed to compute market risk of financial derivatives and may have quadratic error scaling advantage in market risk computation and other advantages if the computation is a matter of computing market sensitivities or "greeks" [37].

Credit risk evaluation: Quantum-inspired algorithms really excel when handling nonlinear relations in credit risk evaluation. A standard program for this would follow this procedure:

1. Data collection phase
    - Financial statements
    - Market indicators
    - Historical performance
    - Macroeconomic factors
2. Feature processing
    - Quantum-inspired feature selection
    - Nonlinear correlation analysis
    - Dynamic weight adjustment
3. Risk scoring
    - Multidimensional risk assessment
    - Real-time updates
    - Confidence scoring

5.5.1.2 Portfolio optimization

Portfolio optimization is one of the best applications of quantum-inspired algorithms in finance [38]. The QPSO algorithm has been particularly promising for solving problems related to portfolio management, which encompasses complex multiobjective optimization problems.

Key advantages include:

- Dynamic rebalancing capabilities
- Multi-factor risk consideration

- Real-time market adaptation
- Enhanced return optimization

5.5.2 Healthcare analytics

5.5.2.1 Drug discovery

For instance, applications of quantum-inspired algorithms in the drug discovery process have been highly successful, especially in breakthrough results. The most important application is in molecular docking: during this process, predictions are made concerning protein–ligand interactions, with quantum-inspired algorithms showing higher speeds and accuracy levels than traditional methods [39].

5.5.2.2 Patient care optimization: Real-world applications

Healthcare facilities have implemented quantum-inspired algorithms for resource optimization and patient care improvement.

Implementation of the resource allocation framework includes:

1. Demand prediction
    - Historical patient flow analysis
    - Seasonal variation modeling
    - Emergency situation planning
2. Resource optimization
    - Staff scheduling
    - Equipment allocation
    - Bed management
3. Quality metrics
    - Patient satisfaction
    - Treatment outcomes
    - Resource efficiency

5.5.3 Smart city applications

5.5.3.1 Traffic management: Advanced systems

Modern smart cities use quantum-inspired algorithms to handle traffic in an integrated way. Singapore has implemented quite a few technology-driven solutions for urban planning and traffic management. The city-state has meshed IoT sensors, data analytics platforms, and artificial intelligence to create streamlined city operations, including transportation. These solutions for smart mobility have already cut traffic congestion by 35%, making the overall city more productive [40].

Some of the elements of Singapore's intelligent transportation approach include modern mass rapid transit trains, electronic road pricing to control congestion, electronic monitoring advisory systems, computerized traffic signaling systems, intelligent taxi dispatch, parking guidance systems, and integrated ticketing systems. All these have kept Singapore at the top while attracting global attention as a smart city [41].

The system architecture for real-time traffic optimization includes:

1. Data collection
    - Traffic sensors
    - Vehicle GPS data
    - Weather conditions
    - Special event information
2. Analysis layer
    - Pattern recognition
    - Congestion prediction
    - Route optimization
3. Response generation
    - Signal timing adjustment
    - Dynamic route recommendations
    - Emergency response coordination

5.5.3.2 Resource distribution: Intelligent solutions

Resource distribution in smart cities has been revolutionized through quantum-inspired optimization. Key implementations include:

Power grid management

- Real-time load balancing
- Renewable energy integration
- Demand prediction
- Fault detection and response

Water distribution

- Pressure optimization
- Leak detection
- Quality monitoring
- Consumption prediction

5.5.4 Environmental modeling (expanded)

5.5.4.1 Climate prediction: Advanced modeling

The climate analysis framework system incorporates:

1. Data integration
    - Satellite observations
    - Ground station data
    - Ocean monitoring
    - Atmospheric measurements
2. Pattern analysis
    - Temperature trends
    - Precipitation patterns
    - Extreme event indicators
    - Ecosystem interactions

5.5.4.2 Resource management: Sustainable approaches

Natural resource management has been enhanced through:

Water resource planning

- Watershed analysis
- Consumption prediction
- Quality monitoring
- Distribution optimization

Agricultural resource allocation

- Crop yield optimization
- Irrigation planning
- Soil health monitoring
- Weather impact assessment

5.5.5 Cybersecurity applications (expanded)

5.5.5.1 Threat detection: Advanced systems

Implementation of a comprehensive detection framework includes:

1. Data collection
    - Network traffic analysis
    - System logs
    - User behavior monitoring
    - External threat feeds
2. Analysis layer
    - Pattern recognition
    - Anomaly detection
    - Behavior correlation
    - Risk assessment

3. Response generation
    - Threat classification
    - Impact assessment
    - Response prioritization
    - Mitigation planning

5.5.5.2 Risk mitigation: Strategic approaches

Risk mitigation strategies have evolved to include:

Predictive analysis

- Threat trend analysis
- Vulnerability assessment
- Impact prediction
- Resource optimization

Response optimization

- Real-time threat response
- Resource allocation
- Recovery planning
- System hardening

5.6 BENEFITS AND OUTCOMES

5.6.1 Performance improvements

5.6.1.1 Computational efficiency

Quantum-inspired algorithms have been shown to hold significant promise toward making computations more efficient. Utilizing principles such as superposition and entanglement, these algorithms are able to search more effectively within huge solution spaces than existing algorithms. For example, QiNNs look at data in parallel, which helps them converge quickly with fewer errors and learn better for pattern recognition and classification [42]. This efficiency is pertinent to many applications that involve data of high dimensionality and where classical methods cannot handle these high dimensions anymore.

5.6.1.2 Solution quality

Generally, quantum-inspired methods yield better solutions. For example, QIEA improves the classical evolutionary strategy with the help of quantum principles that allow a more profound exploration of the space of solutions. This algorithm enhances population diversity since individuals are forced to learn not only from the best historical solutions but from the best found by others too. This adaptive learning approach shows better performance in knapsack problem solutions compared to other QIEAs [43]. Results are indeed more optimized complex problems, in the form of curriculum design or resource allocation, for which optimal solution plays a crucial role. Chances of escaping from the local optima and exploring global solutions are strong possibilities that make robust and adaptable systems perfectly suited for real-world applications.

5.6.2 Practical advantages

5.6.2.1 Scalability benefits

Cloud computing provides the infrastructure for these quantum-inspired algorithms to scale efficiently. With demand for computational resources varying from time to time, cloud platforms enable an organization to dynamically allocate the available resources depending on needs [3]. Hence, it is possible economically for organizations to handle high workloads without making heavy investments in hardware to implement highly advanced, state-of-the-art, quantum-inspired solutions.

5.6.2.2 Cost considerations

There are also ways that the utilization of cloud resources can minimize the operational cost associated with infrastructure management. One such application is in computer-aided drug design, where quantum-inspired algorithms speed up and cut through fewer needed biochemical experiments [39]. Another way that can save time and other resources is by using quantum-inspired algorithms that could sometimes give a faster solution and have higher accuracy than the classical counterpart.

5.6.3 Future implications

5.6.3.1 Research directions

Quantum-inspired algorithms open new avenues of research since they can be applied to many of the new fields that are upcoming in addition to cognitive computing. Future studies may look at enhancing these algorithms, perhaps extending their use to related new fields and doing experiments involving hybrid approaches that combine quantum-inspired methods with more traditional techniques. Sociotechnical implications of these technologies also call for serious research in regard to workforce dynamism and influences on decision-making.

5.6.3.2 *Industry applications*

As industries lean toward quantum-inspired solutions, that too is going to have large impacts on several sectors of the economy: finance, healthcare, and smart cities. To give a very simple example, quantum-inspired algorithms may play a crucial role in financial modeling for facilitating risk assessment and more enhanced portfolio optimization. Essentially, healthcare differs in many ways, with different applications promising efficiency, accuracy, and security.

Interestingly, though quantum-inspired algorithms have been quite promising in nature, they are not a standalone nature-inspired algorithm in focus toward healthcare applications. Besides the designs of genetic algorithms, particle swarm optimization, and ant colony optimization, among many others, there are also analyses toward IoT-based healthcare applications [44]. This indicates diversity, with several computational methods under assessment for applicable solutions to challenging healthcare requirements. The potential applications are quite broad. As these technologies mature, their influence will likely expand into different domains.

5.7 IMPLEMENTATION CHALLENGES AND SOLUTIONS

While the benefits of quantum-inspired algorithms are significant, several implementation challenges must be addressed to realize their full potential within cloud-based cognitive computing frameworks.

5.7.1 Technical challenges

5.7.1.1 *Algorithm complexity*

Another major technical challenge that remains in quantum-inspired algorithms is complexity. Most of these algorithms include some very deep mathematics concepts, which require high qualifications in both quantum mechanics and computational theory to implement the algorithms. This may be one limitation toward applications, especially in poorly equipped organizations [45].

5.7.1.2 *Integration issues*

Integration with existing cloud infrastructure can also be challenging by introducing quantum-inspired algorithms. Such incorporation may require extensive changes to the existing systems and platforms that create potential compatibility issues for organizations that tend to slow the process of implementation and require more resources to achieve proper integration.

5.7.2 Operational challenges

5.7.2.1 *Resource requirements*

The deployment of quantum-inspired algorithms has an enormous need for significant computational capacity, which may not be readily available at times. Organizations should take time to assess current capacities against potential gaps in their infrastructure. Such an assessment helps determine the practicality of these sophisticated algorithms in their operations.

5.7.2.2 *Skill gap analysis*

Another operational challenge is the gap in the skill sets in the workforce. Because these algorithms are quantum-inspired, it requires specific knowledge; thus, organizations will have challenges finding eligible employees who can implement and manage such technologies. Developing training programs in quantum principles and strategic approaches to cognitive computing will become necessary to develop a workforce skilled in these challenges [43].

5.7.3 Future considerations

5.7.3.1 *Emerging technologies*

While still virtual, new emerging technologies need to be embraced and adopted by organizations to enhance their quantum-inspired algorithms. A number of developments in machine learning, artificial intelligence, and data analytics are closely tied to improving quantum-inspired solutions. Currently, when the domain's complexity calls for combining them, the emerging technologies start to be used together. Examples of such applications include biomedical research, whose studies have begun employing AI techniques like artificial neural networks, reasoning-based decision algorithms, and simulation models to enhance and minimize the fuzziness and randomness involved in managing such data [46]. Forecasts regarding the integration of quantum computing and AI promise to deliver unprecedented computational capabilities, making solutions to problems that are otherwise intractable with current scenario classical computers [47].

5.7.3.2 *Research opportunities*

Quantum-inspired algorithms continue to evolve in an ongoing process that offers much scope for various research studies. In the coming years, new application areas will have to be explored, existing algorithms refined, and ethical issues related to such technologies considered. The academia–industry partnership can be a great enabler of innovation, leading to advancements in quantum-inspired cognitive computing.

5.8 CONCLUSION AND FUTURE DIRECTIONS

The integration of quantum-inspired algorithms into cognitive computing frameworks opens up a transformative possibility to enhance cloud-based problem-solving capabilities. Improved performance, scalability, and cost-effectiveness ensure that these algorithms will prove critical tools for understanding and dealing with the modern-day challenges organizations encounter today.

5.8.1 Summary of key findings

In this chapter, we outlined the core components of quantum-inspired algorithms and their applications in cognitive computing. Potential improvements for computation efficiency and quality of the solution have been discussed, along with practical advantages within cloud integration. The implementation challenges and necessity of closing the skill gap have also been stressed.

5.8.2 Research implications

The results from this exploration emphasize a need for further research in quantum-inspired algorithms and the incorporation of these into cognitive computing. Now that more industries are starting to adopt these technologies, knowing the impact will become crucial to both researchers and practitioners.

5.8.3 Future research directions

Future work could be refining algorithms, applying them to complex problems in different domains, and exploring the ethics associated with their potential use. Collaborative research between academia and industry will be essential for the innovation and beneficial deployment of these technologies.

5.8.4 Practical recommendations

Organizations trying to implement quantum-inspired algorithms should, above all, invest in training initiatives to fill some of the skill gaps and create an organization that is always learning. Second, infrastructure needs to be focused on cloud-based cognitive computing in order to maximize use of these advanced algorithms. With such recommendations, organizations will position themselves at the heart of the emerging landscape of cognitive computing and quantum-inspired technologies.

REFERENCES

1. Xu, M., J. M. David, and S. H. Kim. (2018). "The Fourth Industrial Revolution: Opportunities and Challenges." *International Journal of Financial Research* 9 (2): 90–95.
2. Zhang, G. (2011). "Quantum-Inspired Evolutionary Algorithms: A Survey and Empirical Study." *Journal of Heuristics* 17 (3): 303–351.
3. Mell, Peter, and Timothy Grance. (2011). "The NIST Definition of Cloud Computing." *Special Publication 800-145*, National Institute of Standards and Technology.
4. Nielsen, Michael A., and Isaac L. Chuang. (2010). *Quantum Computation and Quantum Information: 10th Anniversary Edition.* Cambridge: Cambridge University Press.
5. Blatt, R., H. Häiffner, C.F. Roos, C. Becher, and F. Schmidt-Kaler. (2004). "Course 5 Quantum Information Processing in Ion Traps I." In *Les Houches/ Les Houches*, 223–260. https://doi.org/10.1016/s0924-8099(03)80029-x.
6. Kelly III, John E., and Steve Hamm. (2013). *Smart Machines: IBM's Watson and the Era of Cognitive Computing.* USA: Columbia University Press.
7. Han, K.H., and J.H. Kim. (2002). "Quantum-Inspired Evolutionary Algorithm for a Class of Combinatorial Optimization." *IEEE Transactions on Evolutionary Computation* 6 (6): 580–593.
8. Buyya, Rajkumar, Christian Vecchiola, and S. Thamarai Selvi. (2013). *Mastering Cloud Computing: Foundations and Applications Programming.* USA: Morgan Kaufmann.
9. Zhang, Q., L. Cheng, and R. Boutaba. (2010). "Cloud Computing: State-of-the-Art and Research Challenges." *Journal of Internet Services and Applications* 1 (1): 7–18.
10. Li, Panchi, and Hong Xiao. (2013). "Sequence Input-Based Quantum-Inspired Neural Networks with Applications." *Neural Processing Letters* 40 (2): 143–168. https://doi.org/10.1007/s11063-013-9316-7.
11. Perez-Delgado, C. A., and Kok, P. (2011). "Quantum Computers: Definition and Implementations." *Physical Review A* 83 (1): 012303.
12. Schuld, M., I. Sinayskiy, and F. Petruccione. (2014). "The Quest for a Quantum Neural Network." *Quantum Information Processing* 13 (11): 2567–2586.
13. Panchi, Li, and Li Shiyong. (2008). "Grover Quantum Searching Algorithm Based on Weighted Targets." *Journal of Systems Engineering and Electronics* 19 (2): 363–369. https://doi.org/10.1016/s1004-4132(08)60093-6.
14. Brassard, G., P. Hoyer, M. Mosca, and A. Tapp. (2000). "Quantum Amplitude Amplification and Estimation." *Contemporary Mathematics* 305: 53–74.
15. Romero, J., J. P. Olson, and A. Aspuru-Guzik. (2017). "Quantum Autoencoders for Efficient Compression of Quantum Data." *Quantum Science and Technology* 2(4): 045001.
16. Ye, Weiliang, Ruijiao Liu, Yangyang Li, and Licheng Jiao. (2020). "Quantum-Inspired Evolutionary Algorithm for Convolutional Neural Networks Architecture Search." *IEEE Congress on Evolutionary Computation (CEC)*, July. https://doi.org/10.1109/cec48606.2020.9185727.
17. Michael, Raj. Tf, G. Rmesh, Moorthy Agoramoorthy, Ahamed Ali S, and D. Sujatha. (2023). "Quantum-Inspired Neural Network for Enhanced Image Recognition." *Intelligent Computing and Control for Engineering and*

Business Systems (ICCEBS), December. https://doi.org/10.1109/iccebs58601.2023.10449238.
18. Norval, Michael, and Zenghui Wang. (2024). "Quantum AI in Speech Emotion Recognition." *Research Square (Research Square)*, September. https://doi.org/10.21203/rs.3.rs-4894795/v1.
19. Arafath, Md. Yasar, and A. Niranjil Kumar. (2023). "Quantum Computing Based Neural Networks for Anomaly Classification in Real-Time Surveillance Videos." *Computer Systems Science and Engineering* 46 (2): 2489–2508. https://doi.org/10.32604/csse.2023.035732.
20. Bhatia, Munish, and Avneet Kaur. (2020). "Quantum Computing Inspired Framework of Student Performance Assessment in Smart Classroom." *Transactions on Emerging Telecommunications Technologies* 32 (9): 1–18. https://doi.org/10.1002/ett.4094.
21. Han, Kuk-Hyun, and Jong-Hwan Kim. (2002). "Quantum-Inspired Evolutionary Algorithm for a Class of Combinatorial Optimization." *IEEE Transactions on Evolutionary Computation* 6(6): 580–93. https://doi.org/10.1109/tevc.2002.804320.
22. Kim, Yehoon, Jong-Hwan Kim, and Kuk-Hyun Han. (2006). "Quantum-Inspired Multiobjective Evolutionary Algorithm for Multiobjective 0/1 Knapsack Problems." *IEEE International Conference on Evolutionary Computation*, September. https://doi.org/10.1109/cec.2006.1688633.
23. Guo, Yi-Nan, Dandan Liu, Yun Liu, and Meirong Chen. (2013). "The Coverage Optimization for Wireless Sensor Networks Based on Quantum-Inspired Cultural Algorithm." In *Lecture Notes in Electrical Engineering*, 87–96. https://doi.org/10.1007/978-3-642-38524-7_9.
24. Ho, S. L., Shiyou Yang, Peihong Ni, and Jin Huang. (2013). "A Quantum-Inspired Evolutionary Algorithm for Multi-Objective Design." *IEEE Transactions on Magnetics* 49 (5): 1609–1612. https://doi.org/10.1109/tmag.2013.2238661.
25. Brassard, G., P. Hoyer, M. Mosca, and A. Tapp. (2000). "Quantum Amplitude Amplification and Estimation." *Contemporary Mathematics* 305: 53–74.
26. Balicki, Jerzy. (2021). "Many-Objective Quantum-Inspired Particle Swarm Optimization Algorithm for Placement of Virtual Machines in Smart Computing Cloud." *Entropy* 24 (1): 58. https://doi.org/10.3390/e24010058.
27. Dasgupta, D., and F. Nino. (2009). *Immunological Computation: Theory and Applications*. USA: CRC Press.
28. Timmis, J., A. Hone, T. Stibor, and E. Clark. (2008). "Theoretical Advances in Artificial Immune Systems." *Theoretical Computer Science* 403 (1): 11–32.
29. Deng, Y., F. Bao, Y. Kong, Z. Ren, and Q. Dai. (2017). "Deep Direct Reinforcement Learning for Financial Signal Representation and Trading." *IEEE Transactions on Neural Networks and Learning Systems* 28 (3): 653–664.
30. Sun, J., B. Feng, and W. Xu. (2004). Particle Swarm Optimization with Particles Having Quantum Behavior. In *Proceedings of the 2004 Congress on Evolutionary Computation (IEEE Cat. No.04TH8753)*, Vol. 1, pp. 325–331. IEEE Xplore.
31. Li Xin, Huangfu Xu, and Xuezhong Guan. (2015). "Quantum-Inspired Particle Swarm Optimization Algorithm Encoded by Probability Amplitudes

of Multi-Qubits." *Open Journal of Optimization* 04 (02): 21–30. https://doi.org/10.4236/ojop.2015.42003.
32. Hamed, Haza Nuzly Abdull, Siti Mariyam Shamsuddin, and Nikola Kasabov. (2012). "Dynamic Quantum-Inspired Particle Swarm Optimization as Feature and Parameter Optimizer for Evolving Spiking Neural Networks." *International Journal of Modeling and Optimization*, January, 187–191. https://doi.org/10.7763/ijmo.2012.v2.108.
33. Wang, Kunlun, Chintha Tellambura, Arumugam Nallanathan, Tao Zhang, Jiong Jin, Bijan Jabbari, and Yang Yang. (2023). "Guest Editorial Multi-Tier Computing for Next Generation Wireless Networks—Part II." *IEEE Journal on Selected Areas in Communications* 41 (3): 569–73. https://doi.org/10.1109/jsac.2022.3228094.
34. Wang, Kunlun, Chintha Tellambura, Yang Yang, Bijan Jabbari, Arumugam Nallanathan, Jiong Jin, and Tao Zhang. (2023). "Guest Editorial Multi-Tier Computing for Next Generation Wireless Networks—Part I." *IEEE Journal on Selected Areas in Communications* 41 (2): 301–5. https://doi.org/10.1109/jsac.2022.3228093.
35. Shen, Y., Y. Li, L. Wu, S. Liu, and Q. Wen. (2013). "Trusted Cloud Initiative Reference Architecture." In *Advances in systems analysis, software engineering, and high performance computing book series* (pp. 78–93). https://doi.org/10.4018/978-1-4666-4801-2.ch004
36. Zhang, G. (2010). "Quantum-Inspired Evolutionary Algorithms: A Survey and Empirical Study." *Journal of Heuristics* 17 (3): 303–351.
37. Stamatopoulos, Nikitas, Guglielmo Mazzola, Stefan Woerner, and William J Zeng. (2022). "Towards Quantum Advantage in Financial Market Risk Using Quantum Gradient Algorithms." *Quantum* 6 (July): 770. https://doi.org/10.22331/q-2022-07-20-770.
38. Sun, J., W. Fang, X. Wu, V. Palade, and W. Xu. (2012). "Quantum-Behaved Particle Swarm Optimization: Analysis of Individual Particle Behavior and Parameter Selection." *Evolutionary Computation* 20 (3): 349–393.
39. Wang, Pei-Hua, Yufeng Jane Tseng, Yu-Yuan Yang, Chien Lee, and Jen-Hao Chen. (2023). "Recent Advances in Quantum Computing for Drug Discovery and Development." *IEEE Nanotechnology Magazine* 17 (2): 26–30. https://doi.org/10.1109/mnano.2023.3249499.
40. Chan, Jochen J. (2023). "Impact of Smart City Initiatives on Urban Planning Strategies in Singapore: An In-Depth Analysis of Technology-Driven Solutions and Their Influence on Sustainable Development and Quality of Life." *Journal of Strategic Management* 7 (7): 11–21. https://doi.org/10.53819/81018102t4200.
41. Hin, L. T. W., and R. Subramaniam. (2011). "Creating Smart Cities with Intelligent Transportation Solutions." In *Advances in environmental engineering and green technologies book series* (pp. 174–190). https://doi.org/10.4018/978-1-61350-453-6.ch010
42. Alchieri, L., D. Badalotti, P. Bonardi, and S. Bianco (2021). "An Introduction to Quantum Machine Learning: From Quantum Logic to Quantum Deep Learning." *Quantum Machine Intelligence* 3 (2): 1–14.
43. Qin, Yanhui, Yuquan Li, Gexiang Zhang, and Huishen Zhang. (2012). *A Comprehensive Learning Quantum-Inspired Evolutionary Algorithm.* 151–57. Springer Berlin Heidelberg. https://doi.org/10.1007/978-3-642-29087-9_22.

44. Amiri, Zahra, Mansour Esmaeilpour, Arash Heidari, Mohammad Zavvar, and Nima Jafari Navimipour. (2024). "The Applications of Nature-inspired Algorithms in Internet of Things-based Healthcare Service: A Systematic Literature Review." *Transactions on Emerging Telecommunications Technologies* 35 (6): 1–16. https://doi.org/10.1002/ett.4969.
45. Ardeshir-Larijani, Ebrahim. (2021) "Parametrized Complexity of Quantum Inspired Algorithms." *arXiv (Cornell University)*, January. https://doi.org/10.48550/arxiv.2112.11686.
46. Chakraborty, Indrajeet, and Amarendranath Choudhury. (2017). "Artificial Intelligence in Biological Data." *Journal of Information Technology & Software Engineering* 7 (4): 1–18. https://doi.org/10.4172/2165-7866.1000207.
47. Hadap, Sanket, and Manthan Patil. (2024). "Quantum Computing in Artificial Intelligence: A Paradigm Shift." *International Journal of Advanced Research in Science, Communication and Technology*, June, 530–534. https://doi.org/10.48175/ijarsct-18957.

Chapter 6

Implementing smart cyber-physical systems in industrial and urban applications

A practical approach

Pragya Lilly Mathew, Shivani Chaudhry, Kamal Upreti, Safia Farooqui, and GV Radhakrishnan

6.1 INTRODUCTION

Rapid urbanization in the 21st century, growing populations, and meeting such demands in sustainable efficiency as well as quality of life pose challenges for cities. The integration of emerging technologies such as artificial intelligence (AI), Internet of Things (IoT), and data analytics into cyber-physical systems (CPS) offers transformative potential in answering these challenges. Smart cities, based on CPS, utilize the networking of sensors, communication networks, and data-driven platforms to optimize services concerning transportation, energy distribution, health, and public safety.

The main advantage of CPS is the real-time monitoring, control, and decision-making for urban environments to respond dynamically to changing conditions. In this regard, smart grids optimize energy distribution, intelligent transportation systems reduce congestion and enhance the commuter experience, and the integration with industrial applications under Industry 4.0 and Industry 5.0 promises enhanced efficiency, predictive maintenance, and human–machine collaboration, thus developing resilient and sustainable production systems.

However, there are challenges in adopting CPS. Security vulnerabilities and data privacy concerns, plus infrastructural compatibility issues, continue to remain critical hurdles to overcome. Sophisticated cybersecurity measures, scalable frameworks, and interdisciplinary collaboration provide the solutions to overcome this complexity. Addressing these complexities leads CPS toward intelligent, inclusive, and environmentally responsible systems for cities and industries alike. This chapter examines the architecture, opportunities, and challenges of CPS in industrial and urban applications, offering insights into its transformative potential and pathways for effective implementation.

DOI: 10.1201/9781003569251-6

6.2 LITERATURE REVIEW

Cities are getting increasingly connected and include much more modern information and technology of the 21st century, which underscores the term "smart cities." Smart cities, built with the newest information technology, are smarter, more sustainable, and more livable for their citizens. Smart cities face previously unimaginable cybersecurity challenges from the integration of technology in urban lifestyles today. This chapter presents the complexity of the relationship between cybersecurity and information technology from the perspective of smart cities in order to design resilient, secure, and future-proofed spaces within an urban setting. Smart cities have been invented to optimize several facets of urban life—from public safety and healthcare to energy management and transportation. The framework responding to the changing conditions within the smart city ecosystem integrates IoT devices, sensors, and other interrelated systems. The importance of IoT and edge computing becomes evident when analyzing them with respect to CPS that require real-time data processing and decision-making. Sensors in CPS devices sense temperature, pressure, and motion from environments such as industries and health systems. The collection of very diverse data is accomplished by IoT devices with Wi-Fi, Bluetooth, Zigbee, or cellular networks for further analysis and optimization on servers or edge nodes. Modern edge-computing techniques are shaped by the demand for IoT systems to deliver faster response times and efficient service delivery. IoT has been a tool for some architectural conceptions for intelligent power grids, smart transportation systems, and innovative city advancements [1].

Large-scale infrastructural security is required in the city adopting this technological revolution. A lot of cybersecurity issues arise when IoT devices and networked systems are integrated into smart cities. The biggest risk is that cyberattacks are possible against the vital infrastructures. In this regard, cyberattacks have grave consequences in terms of public safety and well-being as they impact the smart energy grid, transportation networks, and even healthcare facilities. The other threat that IoT possesses is the immense amount of data it produces, which raises questions of privacy and security. One application that still relies on the data output of cities for better service optimization is the emerging concept of smart cities, which is real and important in the modern world. This data is an attractive area of vulnerability because malicious actors look for weaknesses to exploit for financial or operational interference with city activities by locating their collection, storage, and analysis. Urban planning forms part of the resilience of smart city cybersecurity measures, and their ideas must be taken into account from the very beginning of planning and designing a city. This would mean successful integration of cybersecurity into the city fabric and application of secure-by-design principles to the architecture of smart infrastructure. By 2025, an estimated 4.6 billion people, or 58%, of the world's population will live in cities due to urbanization progress. The percentage may reach 81% in developed areas. The resultant demographic shift has spawned a myriad of challenges for city planners to offer the ever-growing number of city dwellers sustainable living conditions. Among them, smart buildings and smart cities are rapidly emerging as the most visible construction projects.

It is very important to have a full understanding and agreement on what smart cities and smart buildings are because there is no common definition for these terms, which is a problem. When the city adopts the principles of smart cities and integrates smart buildings into the urban infrastructure, it becomes sustainable, efficient, and high quality [2]. Actually, "smart city" is a broad term that includes not only digital technologies and data-driven solutions but also waste management, water, energy, mobility, and security infrastructures [3].

6.3 INTEGRATED URBAN PLANNING AND SMART BUILDINGS FOR SUSTAINABLE DEVELOPMENT

"Smart buildings" are the most recent sophisticated technologies and systems designed to achieve optimal utilization of resources, maximum comfort for occupants, and effective operations and management.

Therefore, functionalities suggested by various smart city infrastructure domains should be integrated in order to use smart buildings in totality. Afterward, utilize the features of services from smart buildings that are suggested for the benefit of systems surrounding them [4]. However, most of the time, urban infrastructure components operate in isolation. To fully unleash the potential of smart infrastructure and attain the dream of smart cities, an integrated approach must be considered. It is valuable and efficient for scientists from all fields, policymakers, planners, managers, civil society representatives, and other parties to collaborate [5].

Buildings should also be designed with smart integration of planned and existing city infrastructure that would include smart building features, as suggested by the domains in smart city infrastructure.

Wurtz and Delinchan point to the possibility of including smart infrastructure in cities, where smart buildings can be integrated with a smart grid that represents the city's energy consumption optimizing system; linking smart buildings to public transportation systems, to reduce a city's environmental impacts; and intelligent waste management systems that supplement the city's current waste management infrastructure. Future prudence demands better integration of smart buildings into existing smart city systems, an interest that is a worrying one [6].

This strain is forcing the demand for resources to rise at such a high rate due to the ever-increasing population growth in cities, which requires optimum use that will not be environmentally impactful [7]. Smart building integration into design, construction, and operational processes in buildings provides yet another promising avenue toward realizing those objectives—through the incorporation of cutting-edge technologies, automation, and data-driven systems [8]. Ideas of smart cities represent paradigm shifts and are expected to boost efficiency, sustainability, and quality of life. Deep in meaning, this implies integrating a digital infrastructure into an urban ecosystem.

As this convergence takes shape, complex cybersecurity challenges arise that require an integrated understanding of the dynamic interplay between urban and technological development. To prop up the multidisciplinary approach advocated for by the abstract, this literature review attempts to investigate the current body of knowledge in the domains of smart cities, cybersecurity, and urban planning. Jia suggested an AI-based cybersecurity defense system for smart cities in their 2023 paper. An interesting and relevant field of CPS is public infrastructure. Generally speaking, streets, public buildings, and underground spaces are considered part of the smart city. Institutional work is done in this regard (for example, within the community of the European Union), and in academic and industrial spheres, focusing on ways for coordinating traffic vehicles more effectively, improving citizen security, monitoring air quality, supporting the public transport system, and faster rescue operations with safety warranties. CPS for a smart city would include a number of such systems that must exchange raw or preprocessed data with each other to support city workflows that enable normal living conditions among its residents.

6.4 SMART CITIES USING CPS

Smart cities involve smart buildings, emergency response, smart transportation systems, and smart grids. Smart cities are the result of several domains. Smart cities are large CPS equipped with sensors that continuously track happenings of the real and virtual world. According to IBM Developer, this in turn influences the actuators that alter the urban environment in appropriate ways [9]. These systems are highly expanding and aimed to enrich life in the city and urban areas. Due to the fact that a growing population creates pressure on the present city infrastructure, urban regions have to redesign and reconfigure their infrastructure [10]. One of the challenges is efficient resource utilization, including food, water, and energy. The growth of cities will act as a hurdle to their sustainable development if the required infrastructure does not exist. This eventually leads to the stagnated growth of the city. One has to redesign and retool a city to make it smarter, which is very challenging for older cities like London, where some of the infrastructure is outdated and crumbling. Updating and redesigning old cities isn't an easy task. For instance, roads and buildings that have been in a city for centuries prove to be hard to replace. One of the significant advantages of a smart city setup relates to security and emergency response. Citywide CCTV networks integrating public and private organizations' resources will be utilized with automation to detect and track threats and incidents in real time. With wearable technology, telemedicine, and artificial intelligence diagnosis of personal healthcare data, it would lead to a swift response. Another key component is energy. Smart cities respond in a fluid and organic way to changes in energy demands, new sources of energy supply, and broken infrastructure. The results of this are efficiency and cost savings, a decrease in greenhouse gas emissions, and benefits to the networked infrastructure itself.

6.5 CPS-POWERED SMART CITIES

In other words, those cities that make use of technology for developing and implementing smart solutions for successful inclusive growth, well-structured community infrastructure, a healthy environment, and other vital resources that ensure an honorable standard of living for the residents of such cities are referred to as smart cities. Digital technologies will interact with physical systems to produce intelligent solutions through which they can boost economic growth through various means, including the improvement of services and infrastructure, the creation of jobs and living conditions in smart cities, and cyber services incorporated into city health services. Future cities would, no doubt, exploit CPS since most smart city applications depend on smart sensing, real-time monitoring, controls, data analysis, and decision-making [11].

6.6 THE FOUR-LAYER MODEL OF A SMART CITY

The information and communication technology (ICT) structure of sensors and actuators, which forms the foundation for a smart city, disperses these items throughout the urban area by taking a bottom-up approach. Traffic lights, smart billboards, bus GPS, lampposts, weather and air pollution stations, smart cameras, and citizens' mobile devices can be considered both public and private technologies falling under this category. Monitoring cloud and edge computing resources, networking resources, and data storage and transmission facilities, it compiles, organizes, and elaborates on data. This layer should furnish the basic mechanisms of the smart city so that as many users/citizens as possible can access all of the resources available, since infrastructure becomes highly complex and needs proper management. Other important factors to be taken into account are the future technologies for the management of network and computer infrastructure and the development of 5G. On top of the infrastructure layer comes the management layer that provides advanced platform functionality and strengthens the functioning of the infrastructure.

6.7 ARCHITECTURE OF CPS FOR INDUSTRY

A new challenge for large mainstream firms is to access data and machine-learning technology under Industry 4.0. IoT systems attached to production lines require quite a sizable investment because, in our growth stage, large enterprises rely mainly on large-scale industrial production as a source of competitive advantage. Also, big businesses have to introduce machine-learning technology into product designing and production lines, which calls for innovation in management philosophy and human talent. Automation covers process technology in a very broad sense—from sensors and actuators installed on production lines to robotic pneumatic valve stations having micro accuracy. The food and beverage industry most frequently employs electromechanical and pneumatic automation technologies [12].

6.8 ENHANCING CPS FOR MODERN INFRASTRUCTURE AND INDUSTRIAL APPLICATIONS

In sharp contrast to automation, efficiency, and resource management in industrial and urban environments, CPS is rapidly transforming them into new styles of systems. The future of industries such as transportation, energy, and housing may well be determined by the use of CPS with software-defined networks and IoT in smart cities. CPS has made it possible to achieve real-time processing and analysis for better sustainability, reduced waste, and greater functions in cities. For instance, real-time control of traffic flow reduces the emissions of carbon gas as well as easing congestion through CPS applications in transport. In the same way, smart grids powered by CPS help ensure that energy can be distributed in ways that will save money and resources because supply and demand are well matched. Furthermore, through real-time response to fires or floods and through predictive monitoring of environmental factors like the quality of the air, CPS improves public safety. Implementation can be described as having systems directly involving citizens, such as significantly better water systems or waste management. Increased public participation is promoted by CPS implementation. Among the many great potentials of CPS in smart cities, complexity is again introduced into linked systems; hence, scalability, cybersecurity, and privacy become important challenges. Cities cannot rely much on such technologies in managing the ever-increasing danger of cyberattacks to vital infrastructure without the evolution of sophisticated security measures. This is where game theory-based security techniques have a say as they simulate reactions and model possible attacks. Thus, such techniques offer a proactive method for minimizing risk and ensuring CPS resilience in urban environments. Such needs include strict data security and privacy assurances, examples of which are needed in public safety, healthcare, and transportation. Part of this will flow from devising flexible management frameworks—like semantic network ontologies—where complex data are arranged into structured, useful insights. In this sense, the method promotes the possibility of networking on a grander scale for the handling of data by various urban systems to ensure these work in harmony toward fulfilling issues such as waste energy or traffic congestion. The ability to align such ontologies toward sectors of cities can be used to analyze data and design targeted interventions. Frameworks like the Prioritizing Open Public Sector Communication (POPSC) framework provide guidelines along with organization and prioritization while releasing open data. This paradigm instead focuses on the needs of the user, from the people to the organizations to the designers; it makes optimum use of the limited resources of the economy, and supports economic development and public services improvement. The POPSC framework focuses mainly on the reuse of data, especially for developers, since it leads to innovation that often translates to economic opportunities and affects the citizen satisfaction levels overall. In addition to its ability to support smart city initiatives, this approach helps to create an environment where the public and private sectors collaborate toward the advancement of technologies of social benefit. Data management and integration are still difficult tasks, however, and require scalable yet dependable solutions. Dealing with a large dataset, one must ensure that privacy issues are well reflected, particularly when dealing with private data of persons or organizations.

This, combined with urban infrastructure, is driving significant advancements in industrial applications, particularly manufacturing. Given that Industry 5.0 will be resilient, sustainable, and human-centered manufacturing, Industry 4.0 has highlighted CPS as a key enabler of productivity and self-optimization in manufacturing environments. Manufacturers can create systems where machines connect on their own and exchange data, enhancing procedures like inventory management, quality control, and predictive maintenance. Continuous monitoring of the health of equipment through real-time sensor data allows it to maximize system performance by minimizing downtime. This flexibility could help manufacturers satisfy variable demand, adjust production schedules, and streamline workflows. Critical for Industry 5.0 is human–robot collaboration, where CPS lets robots assist humans in doing jobs that could either be repetitive or possibly hazardous to ensure reduction in human errors, boost productivity, and lead to security of workers. The benefits of CPS in manufacturing are thus evident, but flexible frameworks that allow for scalability and interoperability are in order to be offered with fair integration across various processes and systems. The resulting flexibility that CPS has toward meeting the growing demands of bespoke goods within extremely complex industrial settings cannot be matched by conventional systems. To allow flexibility and scalability, especially in industry, for CPS systems, the need for IEC 61499 is necessary.

IEC 61499 allows the smooth interaction of different components of CPS by permitting distributed control, thereby allowing the system to scale and change without affecting ongoing operations. Moreover, since it facilitates optimization of resources, it actually reduces the cost implication of deployment and maintenance of solutions for CPS. Moreover, the simulation tools of IEC 61499 help manufacturers test and fine-tune the CPS in a virtual space before its deployment; this means improving performance and reliability through reduced chances of failure and ensuring that systems will work as expected. The modularity and flexibility of IEC 61499 make it a very useful tool for manufacturing environments that must scale rapidly while adjusting to new technologies and production techniques because of the nature of this change. Because of the speed and reach of Industry 5.0, an individualized, flexible, and efficient transformation toward processes of production is being promulgated.

6.9 SMART CITY IMPLEMENTATION IN UAE

Dubai and, by extension, the United Arab Emirates, could be viewed as a live laboratory and test bed for the widespread implementation of CPS for smart cities. Although it is not quite a meaningful refashioning or redesigning of the city in and of itself because it is still developing and expanding, there is the installation of smart sensors throughout the city as well as other infrastructures in the country. The country government is pushing hard before Expo 2020 starts, with an emphasis on making Dubai the smartest city in the world by implementing CPS throughout the city. It has emphasized smart buildings, transportation, energy, and water to raise the enjoyment of citizens and visitors of the city. In the public sector, Dubai's leaders also plan for 80% of its services to be smart by 2018 under the heading of innovation, health, and education [13].

Some examples of the key CPS and IoT efforts that are being pursued in Dubai are described. The integrated free zone technology park in Dubai, called the Dubai Silicon Oasis Authority (DSOA), implemented and installed the first smart street system in the Middle East in collaboration with Huawei. One of the elements of the smart street idea is a digital signage system that provides news and more in real time. The system has a wireless system that offers Internet access. It also integrates a closed-circuit television system, making use of smart surveillance to increase safety. Sensors are spread throughout the area to monitor the air quality as well as the outside temperature. It is believed that the system can be scalable, with new features and services being introduced at minimal costs and short notice.

Greater energy efficiency and management are possible with the centrally controlled Smart Street Solution. At present, one of the biggest telecom companies in the United Arab Emirates is developing fixed and mobile infrastructures and the ecosystem for a number of solutions to deliver the conditions required for an IoT deployment to be successful within the country [14]. Some of the venture's most important components have also been invested in by the company. Etisalat supports everything from industrial solutions for IoT like energy management and big data solutions, through to smart parking, trash management, traffic management, and smart surveillance systems for regular consumers. Intel and the Dubai Silicon Oasis Authority (DSOA) have disclosed that they will be launching Intel's Ignition Lab. The lab is based at DSOA's technology incubation facility, DTEC. Local subsidiaries will collaborate and make use of the lab to enhance innovative IoT innovations, namely, working with these businesses to create smart CPS solutions and the creation of smart cities, smart homes, and smart transportation. Home appliances that work by using consumable supplies such as refrigerators, vacuum cleaners, and washing machines will be connected to utilize IoT sensing and payment capabilities in determining when to order new supplies, make the order, and process the payment, while sending homeowners an alert message so that they can confirm or cancel the order [15]. UAE issued a rule that all financial sectors must implement blockchain technology from now onward. This will allow the governments to secure and facilitate transactions within the emerging Islamic capital market.

With such a large number of oil, gas, and petrochemical companies in the region, industrial CPS becomes very vital for the region. It has been reported that interconnected industrial systems generate much more efficient networks of operation when oil prices are low [16]. In an offshore oil and gas facility in UAE, industrial CPS is used at a water treatment plant with the function to monitor the assets and provide real-time data to the public. Predictive analytics is used to diagnose issues. This has allowed the business to save $30 million per annum. Over the next three years, IoT-powered smart technologies, as offered by another UAE telecom provider, Du, will give 1000 smart services and 100 smart projects to all the residents and visitors of Dubai. Dubai is working with partners to enable its customers to utilize solutions such as smart meters, local and international inventory management, and smart vending machines among a host of others across various sectors within the country and city. LoRa is a new technology that Du has added to its network.

Plans for constructing another headquarters building for Dubai Electricity and Water Authority (DEWA) are in motion. This structure will become the world's tallest, largest, and most intelligent net-zero energy building

(ZEB). The building is going to integrate the latest technologies: AI, big data and open data, and IoT. Robots will be used for safety and cleaning work. According to traffic, a smart app will send a notification to employees regarding leaving their residences. Parking places and meeting rooms can be booked by using the same app. It will also be easy for visitors to use the same to go to the building and the meeting place. In Al-Sheraa, high-tech technologies in building management will be implemented. This will include a control center where all of the systems required will be triggered, while unused ones such as lights and air conditioners will be switched off.

The Roads and Transport Authority (RTA) of Dubai will have smart roads and traffic systems. This regulation further states that whenever an ambulance or other emergency vehicle approaches traffic lights, they will automatically turn green. Reducing highway traffic jams and bottlenecks and creating far more effective road networks are some of the other objectives of this technology. When this system is successfully implemented, fewer vehicle starts and stops will be realized, and therefore the production of greenhouse gas will be reduced.

6.10 SECURITY CHALLENGES

As CPS proliferate across various fields, the need for robust cybersecurity and privacy becomes critical, and generative AI offers innovative solutions to enhance their security and privacy [17]. The privacy, security, communication, and integrity reform has emerged as a pressing matter, necessitating immediate and effective solutions from all parties. In the current dynamic world, it is necessary to leverage the benefits of IoT while protecting individuals' privacy rights through means that reduce risks and secure operations on networked devices [18].

All types of CPS have access to the internet, private networks, and/or system networks. Due to this factor, security is always considered a critical requirement for CPS. Security, depending upon the application and service area where the particular CPS is in use, varies from physical security to protecting the data during its transmission. The following security requirements must be supported with the required security policies and procedures to meet the same security needs.

Following is a list of major challenges that CPS has faced:

Confidentiality: This refers to protecting information that is communicated or stored [19].

Integrity: This is the attribute of the data not being changed; only authorized persons or procedures are allowed to read the data.

Authenticity: This is taken to confirm the real identity of the sender or data producer. Data can only be amended by authorized individuals or processes, and procedures are followed to prevent data from being altered or tampered with in transit.

Access control: This is a safeguard that allows only authorized people to access the systems or data. Only processes or people with the relevant credential system should be allowed to generate or send data. The system denies any other requests for interaction.

Non-repudiation: This is another safeguard in addition to authentication. It ensures that the origin of the data person or process cannot be denied by the data producer.

Dependability: It refers to how reliable the system is. This measure ensures that the system performs as intended without losing the performance or outcome.

Safety: It is one of the foundational attributes of the system that should never place users at risk once implemented. This is a particularly sensitive aspect in heavy industrial manufacturing industries.

Privacy: The mechanism of ensuring that private data and resources are never communicated to unauthorized parties [20].

All these security issues are applicable to smart cities, where citizens use the physical and cyber systems, and hackers could manipulate them. As CPS is integrated into the daily life of the city, even a small hack or manipulation of the system may create serious problems, which might turn fatal in the process [21]. Some examples of attacks that have been mentioned include cars operated autonomously running over pedestrians or smart grid manipulation that causes citywide power outages. Deception attacks could be placed within the category of attacks against CPS, such as sensor or controller data being sent in question of false information. Sensors can be compromised through software or even physical compromise. The other type of attack includes DoS in which the communication channel is compromised to bring down or slow the system. This implies that sensor data will not be forwarded to controllers. Another commonly seen network attack is the network isolation one, wherein a group of compromised nodes forms a network group and cuts off from the network. Packets attempting to join or leave such regions may drop as a result of this attack. Two are the kinds of attacks on CPS. The short-duration class simply cuts off the services immediately, while the long-duration class tries to orchestrate a distributed attack that launches later rather than immediately disrupting the services. One of the cases was DDoS attacks involving millions of IoT devices against the DNS servers run by Dyn. These devices are vulnerable to further attacks, inasmuch as they have never been patched. To prevent these types of incidents, researchers have keenly understood CPS intrusion detection systems [22]. There exists an intrusion detection system (IDS) device type that can perform host-based auditing, network-based auditing, behavior-based detection, and knowledge-based detection. However, several others are under current study. For example, CPS will enable smart grids to reduce waste and promote consumption of renewable resources through real-time feeding of energy supply with demand—an ability that is fast becoming increasingly important due to rapid population growth in cities.

6.11 CONCLUSION

The integration of CPS into the urban and industrial environment will be a great leap toward sustainable, efficient, and intelligent systems. Enabling real-time data processing, dynamic control, and predictive analytics, CPS enhances urban resilience, industrial efficiency, and quality of life. Applications like smart grids, intelligent transportation systems, and Industry 5.0 innovations indicate the transformative potential of CPS in addressing contemporary challenges.

Still, this potential is realized by crossing some critical hurdles, which include cybersecurity risks, privacy concerns, and infrastructural limitations. Scalable frameworks, advanced intrusion detection systems, and interdisciplinary collaboration are essential for the success of CPS. When these challenges are addressed, CPS can drive the development of connected, secure, and sustainable environments that meet societal needs while fostering economic growth and environmental responsibility. This research underscores that it's only through cross-disciplinary approaches that the full capability of CPS can be unlocked in pursuit of stronger futures.

REFERENCES

1. S. K. Gupta, S. M. Ul Din, K. Upreti, S. Mahajan and S. S. Date, "Enhancing CNN Weights for Improved Routing in UAV Networks for Catastrophe Relief with MSBO Algorithm," in *Journal of Mobile Multimedia*, vol. 20, no. 5, pp. 1117–1151, September 2024, doi: 10.13052/jmm1550-4646.2056
2. R. Apanaviciene, A. Vanagas, and P. A. Fokaides, "Smart Building Integration into a Smart City (SBISC): Development of a New Evaluation Framework," *Energies*, vol. 13, no. 9, p. 2190, May 2020, doi: 10.3390/en13092190.
3. J. Duque, "The IoT to Smart Cities - A design science research approach," *Procedia Computer Science*, vol. 219, pp. 279–285, Jan. 2023, doi: 10.1016/j.procs.2023.01.291.
4. R. Apanaviciene, R. Urbonas, and P. A. Fokaides, "Smart building integration into a smart city: comparative study of real estate development," *Sustainability*, vol. 12, no. 22, p. 9376, Nov. 2020, doi: 10.3390/su12229376.
5. *Commission on Science and Technology for Development and United Nations*, "Smart cities and infrastructure," Feb. 2016. [Online]. Available: https://unctad.org/system/files/official-document/ecn162016d2_en.pdf
6. F. Wurtz and B. Delinchant, "'Smart buildings' integrated in 'smart grids': A key challenge for the energy transition by using physical models and optimization with a 'human-in-the-loop' approach," *Comptes Rendus Physique*, vol. 18, no. 7–8, pp. 428–444, Sep. 2017, doi: 10.1016/j.crhy.2017.09.007.
7. S. Lakusic, "Smart cities and buildings," *Journal of the Croatian Association of Civil Engineers*, vol. 71, no. 10, pp. 943–964, Oct. 2019, doi: 10.14256/jce.2733.2019.
8. S.-H. Lin, H. Zhang, J.-H. Li, C.-Z. Ye, and J.-C. Hsieh, "Evaluating smart office buildings from a sustainability perspective: A model of hybrid

multi-attribute decision-making," *Technology in Society*, vol. 68, p. 101824, Nov. 2021, doi: 10.1016/j.techsoc.2021.101824.

9. "IBM Developer." https://www.ibm.com/developerworks/library/ba-cyber-physical-systems-and-smart-cities-iot/
10. S. V. Buldyrev, R. Parshani, G. Paul, H. E. Stanley, and S. Havlin, "Catastrophic cascade of failures in interdependent networks," *Nature*, vol. 464, no. 7291, pp. 1025–1028, Apr. 2010, doi: 10.1038/nature08932.
11. V. Albino, U. Berardi, and R. M. Dangelico, "Smart Cities: Definitions, dimensions, performance, and initiatives," *Journal of Urban Technology*, vol. 22, no. 1, pp. 3–21, Jan. 2015, doi: 10.1080/10630732.2014.942092.
12. M. Javaid, A. Haleem, R. P. Singh, and R. Suman, "An integrated outlook of Cyber–Physical Systems for Industry 4.0: Topical practices, architecture, and applications," *Green Technologies and Sustainability*, vol. 1, no. 1, p. 100001, Oct. 2022, doi: 10.1016/j.grets.2022.100001.
13. Khan, F., R. L.Kumar, S. Kadry, Y Nam, and M. N. Meqdad. (2021b). Cyber physical systems: A smart city perspective. *International Journal of Power Electronics and Drive Systems/International Journal of Electrical and Computer Engineering*, vol. 11, no. 4, p. 3609. https://doi.org/10.11591/ijece.v11i4.pp3609-3616
14. Jack, "إطلاق العنان لتألق إيرلينج هالاند: دليل شامل," *Erling Haaland*, Oct. 18, 2024. http://www.innovationsummit.ae/creating-iotecosystem-uae/
15. A. B. Byat, "Well-connected Dubai gets smart about internet links," *The National*, Jul. 10, 2021. [Online]. Available: https://www.thenationalnews.com/business/well-connected-dubai-gets-smart-about-internet-links-1.209857
16. Evangelos Alepochoritis, Special to Gulf News and Evangelos Alepochoritis, Special to Gulf News, "The power of the Industrial internet of Things," *Gulf News: Latest UAE News, Dubai News, Business, Travel News, Dubai Gold Rate, Prayer Time, Cinema*, Sep. 15, 2018. https://gulfnews.com/technology/the-power-of-the-industrial-internet-of-things-1.1834664
17. "Navigating Cyber-Physical systems with Cutting-Edge Technologies," Google Books. https://books.google.co.in/books?id=Sdw2EQAAQBAJ&printsec=frontcover&source=gbs_ge_summary_r&cad=0#v=onepage&q&f=false
18. R. C. Poonia, K. Upreti, B. P. Alapatt, and S. Jafri, "Real-Time Cyber-Physical risk management leveraging advanced security technologies," in *Lecture Notes in Networks and Systems*, 2024, pp. 339–350. doi: 10.1007/978-981-97-4581-4_25.
19. V. Nuraini, "Embedded Systems Security: Threats, vulnerabilities, and attack taxonomy," www.academia.edu, Feb. 2025, [Online]. Available: https://www.academia.edu/29317665/Embedded_Systems_Security_Threats_Vulnerabilities_and_Attack_Taxonomy
20. H. Zhang, Y. Shu, P. Cheng, and J. Chen, "Privacy and performance trade-off in cyber-physical systems," *IEEE Network*, vol. 30, no. 2, pp. 62–66, Mar. 2016, doi: 10.1109/mnet.2016.7437026.
21. R. Ivanov, M. Pajic, and I. Lee, "Attack-resilient sensor fusion for safety-critical cyber-physical systems," *ACM Transactions on Embedded Computing Systems*, vol. 15, no. 1, pp. 1–24, Feb. 2016, doi: 10.1145/2847418.
22. R. Mitchell and I.-R. Chen, "A survey of intrusion detection techniques for cyber-physical systems," *ACM Computing Surveys*, vol. 46, no. 4, pp. 1–29, Mar. 2014, doi: 10.1145/2542049.

Chapter 7

Integration of cyber-physical systems with wearable devices

A new paradigm for patient monitoring

Prachi Kumar, Salineeta Chauhduri, Pratham Nijhawan, Ramesh Chandra Poonia, Uma Shankar, and Faraz Ahmad

7.1 INTRODUCTION

Health has been one of the main issues of concern for people in recent times. People are prone to various chronic diseases, such as eating disorders, kidney problems, and diabetes insipidus. In order to prevent any dangerous disorders, patients with the aforementioned diseases should have regular monitoring and treatment through advanced technology like the integration of cyber-physical systems (CPS) and wearable devices. One of the key drivers behind future technological developments would be to communicate, compute, and modify the physical world to interact with it and augment it. Today, the time required for most tasks is limited, and therefore, people, including the working class, spend most of their time balancing several responsibilities while ignoring their health and fitness [1]. Even a simple clinic appointment with a doctor could take quite some time and demand several diagnostic tests and different types of medicines, eventually leading to therapy. Consequently, many patients only visit a clinic when they are seriously unwell. Internet of Things (IoT)-based devices in healthcare settings are promising in the sense that they monitor the temperature and breathing rate of a patient while reducing human labor [2].

Wearable technology has grown substantially across different industries, from fitness and sports to health monitoring. The increasing population of seniors around the world has made wearable technology a requirement for long-term health monitoring [3]. For chronic diseases such as neurological disorders, diabetes, and cardiovascular diseases, continuous and real-time monitoring is necessary to enhance patient outcomes and reduce healthcare costs. Smart sensors like ESP8266 Wi-Fi modules and Arduino-based devices have been used in recent developments in IoT-based monitoring systems to gather and evaluate patient health indicators in real-time. By guaranteeing smooth data transfer to medical specialists, this integration permits preventive medical intervention [4]. According to Someya, Bao, and

DOI: 10.1201/9781003569251-7

Malliaras [5], health wearable devices (HWDs), incorporated into clothing or worn on the body, are an efficient method of chronic illness tracking and diagnosis. These devices are equipped with receptors and transducers that collect and transmit information about vital health parameters, thus allowing real-time patient monitoring [6]. In 1956, Leland C. Clark revolutionized the use of electrodes in the direct measurement of oxygen levels in real-time in cardiovascular surgery. This innovation opened the doors to biosensors, which have evolved from laboratory-based apparatuses to small, wearable devices that can monitor health in real-time across several medical applications.

Coupled with the developments of CPS, modern patient monitoring through wearable medical technology has revolutionized patient care to be proactive and personalized in management. With the advancement in mobile health (m-Health) and AI, wearable technology, therefore, is now made possible by collecting physiological data such as ECG readings and transmission to cloud-based platforms for real-time processing. Integration of AI algorithms improves the outcome of treatment for patients as early-stage recognition and treatment of the issue are achieved. Other than telemonitoring support, these smart cardiovascular health informatics systems enhance the ability to predict disease and allow for earlier intervention by practitioners with tighter control over the progression of disease. Nonetheless, there are several issues related to the implementation of machine learning in health wearables to enhance algorithmic accuracy and reliability, which is an important aspect of applying them widely. Furthermore, there is a lack of information on the general clinical utility of wearables, although reviews have focused on consumer trends related to wearable electronics in sports and leisure. Tackling these issues can potentially unlock the huge potential of medical wearables in neuro, cardio, and metabolic disorder monitoring and treatment, as well as their integration into drug delivery systems. The chapter will discuss CPS integration with wearables in healthcare, and their clinical uses, existing hurdles, and potential future research directions. The aim is to give an overview of where CPS-enabled wearables are today and suggest areas that need to be researched further to add more clinical value and use.

7.2 LITERATURE REVIEW

7.2.1 Cyber-physical systems in healthcare

CPS is comprised of sensors, actuators, cloud computing, and analytics powered by AI, with transformative power for patient care. CPS enables real-time patient monitoring, predictive diagnosis, and sophisticated medical interventions [7]. A number of studies emphasize the benefits of CPS to healthcare, including remote monitoring of long-term illnesses, early illness detection, and optimized hospital management [8]. Challenges still exist around standardization, interoperability, and integration with current medical infrastructure [9]. Although CPS has gained wider acceptance, cybersecurity threats and regulatory compliance issues remain significant barriers [10]. The latest development of CPS has supported real-time processing and acquisition of data, making patient care a lot better. For example, CPS applications using AI for telemedicine and robot-assisted surgery have made more precise diagnoses with higher efficiency [11]. Employing machine learning in CPS facilitates the detection of anomalies and offers decision-making support, making the jobs of health workers simpler. Multimodel Internet of Medical Things (IoMT)-cloud computing frameworks provide innovative solutions for real-time disease diagnosis using machine learning and advanced data processing for better outcomes [2]. Nevertheless, the development of robust, scalable CPS models remains an ongoing research priority.

7.2.2 Wearable devices for patient monitoring

Healthcare wearable technology has grown exponentially, with noninvasive devices for real-time monitoring of health. Wearable devices such as biosensors, smartwatches, and fitness trackers track physiological signals in real-time, enabling personalized interventions in healthcare [12]. Skin sensors and smart textiles are promising candidates for the monitoring of cardiovascular and metabolic diseases [13]. Further, AI-based analytics refine the accuracy of prediction and treatment planning. Nevertheless, issues of device precision, reliability, and long-term sensor degradation present wide-ranging challenges to uptake [3]. Apart from healthcare, wearable technology has also been used in education. Minnesota students in the Westonka School District use the Heart Zones System to track heart rate, pace, and movement, all of which are aimed at engaging and motivating the students during physical education [14]. Educational institutions apply wearables' real-time feedback to individualize training programs, as the CPS-enabled wearables also do in the monitoring of patients' health in real-time. The applications emphasize how wearable technology aids in the enhancement of both human performance and targeted healthcare interventions. Wearable healthcare monitoring systems still have a list of shortcomings in spite of all these advances. These include limited connectivity, energy usage, and storage of data. In efforts to deal with these issues, scientists are examining edge computing technologies and power-constrained designs. In addition, wearable technology is widely used in the treatment of long-term conditions like diabetes and high blood pressure. For example, real-time glucose tracking with continuous glucose monitoring (CGM) devices has demonstrated great promise in improving patient adherence to treatment plans [5].

7.2.3 Data privacy and security challenges

Patient information gathered by CPS-enabled wearables is usually uploaded to cloud platforms, which expose them to cyberattacks [15]. The combination of encryption methods, blockchain technology, and federated learning frameworks can improve data security and patient confidentiality [9]. Data protection regulations like GDPR and HIPAA stress robust data protection practices, but compliance remains a challenge in international healthcare systems [10]. Future studies need to address the creation of strong security measures to provide safe and ethical implementation of wearable healthcare technologies. Cybersecurity threats from CPS encompass unauthorized access, data breaches, and tampering with devices. Recent research has emphasized the use of AI-based anomaly detection systems to recognize and reject cyberattacks in healthcare CPS systems. Decentralized storage solutions like blockchain have also emerged as encouraging alternatives to conventional cloud-based solutions by providing data integrity and minimizing centralized attack threats.

7.2.4 Identified research gaps

One of the key challenges in CPS-enabled healthcare is the limited large-scale empirical validation of these systems, particularly in diverse clinical settings. Second, the integration of machine learning with wearable devices requires optimization to improve computational efficiency and real-time processing capabilities [11]. Third, affordability and accessibility of wearable health technology remain barriers, particularly in low-resource settings. Resolving these challenges will be essential to enabling the widespread uptake of CPS-based healthcare solutions and the full exploitation of their potential in clinical care. Furthermore, more study is required to determine user acceptance and behavioral adaptation to CPS-based healthcare interventions. Knowing the interaction of patients and healthcare workers with such technologies will guide the creation of easy-to-use interfaces and enhance adoption [12]. Long-term studies comparing the effect of CPS on healthcare costs and patient outcomes are required to determine its effectiveness and cost-effectiveness.

7.3 METHODOLOGY

A qualitative research methodology has been used for this study to investigate the integration of wearable technology and cyber-physical systems, which focuses on how these systems can be applied in the healthcare sector for individualized care and real-time patient monitoring. The research is based on a critical review of secondary material, such as peer-reviewed journal articles, industry reports, reputable publications, and conference proceedings. In an attempt to identify trends, technological breakthroughs, and obstacles in the field, the study will synthesize knowledge from many sources.

7.3.1 Data collection

The data for this study were collected methodically using keywords such as "cyber-physical systems," "wearable technology," "health monitoring," and "data privacy in healthcare" from sources such as Google Scholar, IEEE Xplore, PubMed, and Scopus. To ensure that the included publications were relevant to recent developments in CPS-enabled wearables, the inclusion criteria focused on those published within the last ten years. The research thus focused on managing chronic diseases, integrating machine learning algorithms, and analyzing data in real-time.

7.3.2 Data analysis

The collected literature was analyzed thematically to determine recurring themes, such as the technical architectures of CPS, the role of wearable devices in healthcare, and the challenges of data security and machine learning integration. This thematic analysis also exposed gaps in current research, particularly about the scalability and dependability of CPS-enabled systems in dynamic healthcare environments.

7.3.3 Conceptual framework and validation

A conceptual framework was developed in this study to explain the interaction of the CPS components—sensors, actuators, and computational units—and wearable devices. This framework focused on the flow of physiological data from biosensors to cloud-based platforms for analysis and enabling real-time therapeutic interventions. In addition, data privacy and security as ethical concerns were also addressed by this study through a review of best practices and case studies. Validation of findings has been attained by comparing insights gleaned from case studies including biosensors on wearable devices to monitor glucose levels as well as cardiovascular health monitoring devices with reported clinical outcomes. Nevertheless, this study remains limited in its reliance on secondary data without primary experimental results. The scope is also limited to healthcare applications, which may leave out wider CPS implementations.

7.4 CPS-ENABLED WEARABLE DEVICE APPLICATIONS

Wearable electronics enable continuous activity tracking without restricting movement. Current wearable sensors primarily monitor physical activity and vital indicators such as heart rate but lack capabilities for molecular health analysis. Sweat analysis, which holds important physiological and metabolic information, has the potential to be used in disease diagnosis, drug monitoring, and improvement of athletic performance [13]. The use of broadband technology in wearable medical devices allows for data transmission and miniaturization to enable greater mobility [3].

7.4.1 Categories of healthcare wearable devices

Wearable medical devices fall into three primary categories: skin-based, biofluidic-based, and wearable drug delivery systems. There are three main categories of wearable devices.

7.4.1.1 Skin-based HWDs

Skin-based wearable technology is widely used in physiological and psychological monitoring and is essential for managing cardiovascular and neuromuscular conditions. These devices are either textile-based or epidermal-based. Textile-based HWDs, or e-textiles, utilize conductive materials like graphene and nanowires to track body temperature, breathing, and heart rate. Textiles provide comfort and flexibility and can be seamlessly integrated into daily activities. However, their limitations include reduced durability after multiple wash cycles and signal distortion due to movement. Tattoo-based HWDs, designed for short-term applications, consist of on-skin sensors that measure chemical, physical, and electrical signals. These include electrocardiography (ECG) for heart monitoring and electroencephalography (EEG) for brain activity tracking. Tattoo-based glucose monitors provide a noninvasive alternative for diabetes management, avoiding frequent blood sampling [12].

7.4.1.2 Biofluidic-based HWDs

Biofluidic-based wearables exploit body fluids such as sweat and tears to conduct real-time monitoring of health in a noninvasive manner. The devices deploy microfluidic technology for the effective collection and analysis of samples. Sweat-derived HWDs test biomarkers such as glucose, lactate, and chloride, providing important information on metabolic diseases like diabetes and cystic fibrosis. Koh et al. developed a microfluidic sensor capable of detecting hydronium and chloride ions for cystic fibrosis monitoring [16], while Choi et al. introduced a skin-adhering microfluidic device to measure lactate, potassium, and sodium levels in minimal sweat samples [17]. Tear-based HWDs present a promising solution to glucose monitoring, especially for diabetic patients. A smart contact lens was developed by Lin et al., which responds to glucose in tears and allows for continuous monitoring through smartphone integration [18]. However, these technologies are beset by issues such as calibration problems, the potential for contamination, and sensor lifetime [19].

7.4.1.3 Wearable drug delivery systems

Advancements in biomaterials have enabled the development of wearable drug delivery systems capable of controlled and prolonged therapeutic agent release. Ocular drug delivery systems utilize polypropylene ocular rings to ensure the slow release of bimatoprost, maintaining therapeutic levels for glaucoma treatment [20]. Polymeric drug delivery systems employ self-healing polymeric gums composed of tannic acid and polyvinyl alcohol to enhance personalized drug administration [21]. With advancements in technology, smart contact lenses integrated with biosensors and drug delivery systems stand out as a significant revolution in diabetes management. They provide another cause for various functions, including monitoring glucose levels in the tears and drug delivery for diabetic retinopathy treatment, therefore offering a dual-function, non-invasive treatment in theory [22]. Notwithstanding these advances, challenges such as drug bioavailability, small volume of the drug put into the eye, and the blood–brain barrier need sustained research in an effort to improve therapeutic efficacy [23].

7.4.1.4 Touch-actuated transdermal delivery (TATD)

Touch-actuated transdermal delivery (TATD) devices use mechanical activation to precisely deliver drugs. In order to reduce systemic adverse effects, these noninvasive devices avoid the gastrointestinal tract. In order to ensure precise dosage control and user convenience, TATD patches and microneedle arrays are especially helpful for chronic illnesses, hormone treatments, and pain management.

7.4.2 Future considerations for wearable medical devices

There are numerous issues in the broad acceptance of the wearable medical devices with CPS functionality, over and above the noted advances.

- Stability and durability: After several washing cycles, wearables made from textiles decompose, leading to reduced utility in the longer runs.
- Data integrity: Environmental conditions and motion artifacts may affect the sensor's reliability, calling for complicated calibration processes.
- Regulatory barriers: The market adoption life cycle is unknown, as there are no regulations established nor regulatory approval processes.
- Scalability and cost: High costs represent a major limitation to the large-scale deployment of this technology in healthcare settings.

Wearable healthcare devices, enabled by CPS, have great promise for enhancing patient monitoring and personalized medicine. Although skin-based, biofluidic, and drug delivery systems have progressed, more work is required to advance sensor stability, enhance drug bioavailability, and create standardized regulatory frameworks for broad clinical adoption. Future innovations in flexible materials, AI-driven diagnostics, and secure data transmission will play a crucial role in shaping the future of wearable medical developments in secure data transfer, AI-driven diagnostics, and flexible materials will be key factors in determining how wearable medical technology develops.

7.5 APPLICATIONS OF WEARABLE TECHNOLOGY IN REAL-TIME HEALTH MONITORING

7.5.1 Wearable devices in cardiovascular health

Lifestyle choice management, especially concerning physical exercise, has to be effective as cardiovascular disease remains the leading cause of death in the world. It has been established that exercise reduces cardiovascular mortality [24]; however, compliance sometimes proves challenging, particularly in the elderly and more seriously ill patients. Wearable technology and telerehabilitation advances allow for the monitoring of blood pressure, heart rate, and rhythm. There are now apps that may be used at home to deliver exercise programs, which could be an alternative to more traditional hospital-based programs. This will increase access to ongoing cardiovascular monitoring and may improve exercise compliance and patient outcomes.

7.5.1.1 Devices for monitoring heart rate

Heart rate is among the most critical measures for assessing exercise intensity and overall cardiovascular fitness because measures such as resting heart rate and heart rate recovery are related to mortality risk [25]. A diversity of aerobic exercise programs has been demonstrated to enhance these indices of heart rate in older adults and cardiovascular disease patients. Ongoing is research into the impact of continuous heart rate monitoring on health outcomes [26]. Available continuous monitoring is afforded by photoplethysmography (PPG)-based consumer devices, for example, some wrist-worn heart rate monitors. Though they can be accurate in lower-level activities, their accuracy will reduce with heavy movement, especially in those exercises that involve resistance [27]. Wrist-worn monitors are less deceptive at rest as compared to an ECG, concluded a meta-analysis; however, motion artifacts due to high-intensity activity cause accuracy to plummet [28].

7.5.1.2 *Tools to monitor heart rate variability*

Heart rate variability (HRV) is a measure of the autonomic nervous system, marking diabetes, heart diseases, and mental health disorders. Clinical-grade ECGs remain the gold standard to date; however, devices with PPG and chest straps provide useful, low-cost HRV monitoring [29]. Chest-strap devices closely track ECG-derived HRV at low-to-moderate exercise, while wrist-worn devices that use PPG vary in accuracy, primarily as the exercise intensity increases or in older adults. The dependability of consumer wearables in active settings may be contingent upon factors such as sensor placement and motion artifacts, although they normally extract HRV successfully when resting [30].

7.5.1.3 *Devices for monitoring blood pressure*

Although it's often uncomfortable and reduces adherence, ambulatory blood pressure monitoring (ABPM) is often applied for hypertension diagnosis. More recent portable blood pressure monitors based on oscillometry and PPG provide minimally invasive alternatives. Some of these include Omron's HeartGuide, which was shown to be comparable with ABPM in seated trials, while others based on PPG have potential, though require further population-level validation [31].

7.5.2 Wearable technology for monitoring blood glucose in real time

Type 2 diabetes is responsible for almost 90% of the total number of diabetes cases, and it is one of the largest global health problems. The main reason for the increase in prevalence is unhealthy lifestyle choices including eating poorly and not exercising. The prevalence of type 2 diabetes is expected to jump dramatically by 2045, and the costs related to complications caused by diabetes would be huge in terms of diminished quality of life and lost productivity. An excellent approach to type 2 diabetes management is an appropriate amount of physical activity. It has been recommended that in addition to at least 150 minutes/week of moderate-intensity aerobic activity, resistance training should be added [32]. Community-based exercise programs are beneficial for middle-aged and older persons since they improve adherence to physical activity and minimize health risks associated with diabetes. In the modern world, older people are mostly faced with other challenges. Therefore, it is important to consider their health status to ensure that the exercise regime is safe [33].

The latest advances in wearable technology, such as wearable devices, have transformed the way diabetes is managed. In this regard, especially by use of CGM systems, the pain associated with numerous finger punctures is eradicated to detect blood sugar levels in real time. Sensors implanted just under the skin through a CGM system gather data, which are then transmitted wirelessly to a smartphone or display device. This will automatically collect data every five minutes, enabling patients and medical practitioners to make sound decisions on nutrition, exercise, and even drugs, thus bettering type 2 diabetes management [34]. Another area of the evolution of wearable technology is in terms of glucose monitoring. This is paired with other issues, such as the risk of diabetic foot complications that challenge patients with type 2 diabetes. Devices such as foot temperature and plantar pressure can be presented to promote preventive care and identify hazards early enough. For example, intelligent insoles will detect irregularities in the pressure distribution that ought to alert the users of the situation, and a sensor installed in the house will record changes that may signify risk [35]. Such innovations encourage a high level of physical activity that is, in itself, beneficial in managing type 2 diabetes and encouraging foot care, thereby reducing the chances of dangerous consequences. Such wearable devices are a breakthrough in healthcare due to the resources they provide to the patients through improving health outcomes. The advancement of technology will increase the capabilities of wearables in assisting those with type 2 diabetes, thereby increasing quality of life and effective management of the disease.

7.6 CHALLENGES IN MACHINE LEARNING INTEGRATION WITH WEARABLES

Deep learning (DL) approaches have many merits, though they suffer from the problems of implementation. In the smart cardiovascular health informatics field, wearable device technology, m-Health technology, and artificial intelligence techniques have been vastly researched, but often in isolation. Yet, there remains a dearth of evidence built from a cumulative summation of all of them. DL techniques do not generalize well for test data, although they generalize pretty well with big training data. Their computationally intensive nature makes them difficult to deploy on low-power, though very resource-constrained, devices in most real-world applications. Cloud-based ECG analytics can eliminate this challenge at the expense of added cost and latency. Edge AI is a new technology where the hardware inside user devices will deal with AI algorithms rather than in the cloud. Major advantages include low latency, integration of wearable devices, small bandwidth, decreased cost, and increased security of data. Hybrid architecture aiming at a balance of the proportion of usage between the edge and cloud for AI algorithms might help achieve a balance between diagnostic accuracy and model complexity. Another significant challenge is that the success of the DL approach generally relies heavily on having enough quality data that fits the requirements for the job. Most of the current studies are based on either new quality databases with short-term ECG recordings or older datasets that were actually gathered more than 40 years ago [36]. Advances in technology have significantly helped develop wearable biosensors. The most natural material for a person's skin, textiles are the ideal platform for long-term physiological signal monitoring. Textile-based ECG sensors were developed and can be used as patches or badges [37]. The development of a wearable ECG sensor is plagued by significant challenges such as tiny size, robustness, aesthetic designs, power consumption and longevity, privacy issues, and reasonable costs, besides the rise in advanced technologies and new trends. For long-term measurements, these devices need to be extremely accurate and reliable. An issue emerges when sensors and electronics are integrated into fabrics: solutions that are pleasant, flexible, and washable. The electronics and connectivity need to be strong and unobtrusive. The sensors have to be small and lightweight and must be able to monitor the movement of a patient. Energy management will be critical because data may get lost if the appliances are charging or if the batteries have to be replaced. Low-energy operations may hinder the attainment of good-quality data. The second reason is privacy. Sensor data contains sensitive information and, hence, must not be accessed or altered by someone other than the user. It becomes critical once the device is integrated into an m-Health platform. With the emergence of increasing numbers of platforms interfacing with one another, security becomes a matter of greater complexity. The primary problem related to smart wearable technology is its high expense at the time of purchase. Thus, the cost that consumers can afford to pay highly influences the growth of the wearables market. Consequently, a multidisciplinary approach with domain knowledge is required for wearable technologies that are resilient, inexpensive, dependable, secure, and easy to use. The approach followed in designing and developing the suggested wearable ECG sensor at the prototype level is elaborated in the following section.

7.7 EMERGING TRENDS AND FUTURE DIRECTIONS

The modern age of wearables is characterized by a wide variety of gadgets being used every day and ingrained into our lives. These gadgets offer many additional features apart from the standard health monitoring. Today, gadgets have evolved into wearable cameras, smart glasses, fitness trackers, smart jewelry, and even implantable health monitoring pills [38]. They have become vital tools that not only serve as health assistants but also include other aspects, such as communication, productivity, and entertainment. According to recent studies, evidence has pointed out the importance of wearables in health, telemedicine, lifestyle, gaming, sports, fitness, environment, safety, and prevention. Wearable technology has evolved from basic health monitoring to the realm of soft robotics, prosthetics, exoskeletons, and much more sophisticated applications based on AI, machine learning, and deep learning for post-processing and real-time data analytics [39]. The existing literature is as follows: To begin, we describe human behavior, the various human activity recognition (HAR) techniques, and the advantages and disadvantages of employing an embedded system. After discussing the online versus offline strategy in detail, along with all the machine learning techniques used in previous studies for sensor-based human-action detection, we address how hybrid neural network designs compare to other conventional techniques. HAR activity includes hand movements and gestures and all other physical actions that involve energy release [40]. Some of the activities include eating, drinking, and walking. Human behavior can be simple or complex. Simple human activities are those based on postures and bodily movements, such as running, walking, and jogging. Human activities refer to those basic actions and functions. For example, an activity might be merely sitting and only doing the act of eating. Deep learning algorithms, employing data from wearable and smartphone sensors, will classify simple and complex human behaviors into three classes. IoT is already applied very extensively in healthcare, with even more technologies integrated into others as it advances. All hospital infrastructures must be replaced with virtualization. Artificial intelligence data mining software will be required to pull out useful information from large amounts of data. Big data solutions will also be required for this real-time critical health decision-making process. Algorithmic data analysis is also necessary in that regard. All these technologies are used in IoT-based healthcare systems; however, there are some limitations posed by them, like a high-speed connection, storage, and security and privacy compliance.

7.8 ADVANCED IOT AND CPS SOLUTIONS FOR INTELLIGENT HEALTHCARE MONITORING

The Internet of Things-based wearable healthcare monitoring system is beneficial for enhancing access to rural areas for continuous monitoring of patients' vital signs, early detection of health issues, and timely action. Edges are applied to reduce latency and network stress and demonstrate scalability and flexibility by adapting to a broad variety of healthcare environments in many different countries. Cyber-physical systems are revolutionizing healthcare, enabling the use of predictive analytics and enhancing patient outcomes by integrating communication technology with computing algorithms. Future research will need to address connectivity and cybersecurity issues; meanwhile, applications like medication adherence monitoring and patient safety enhancement display well enough what CPS holds in store [10]. Wearable technology coupled with biosensors would lead to better decision-making, lower mortality, and continuous monitoring of the patient through IoT integration in mobile healthcare. This smart healthcare system portrays how technology modernizes the sector and makes it highly effective and accessible through prompt interventions [41]. That is, through the capability of real-time monitoring and early detection of health issues, remote patient monitoring (RPM) technology enhances proactive healthcare. This technology can forecast the next generation of therapy solutions using data analytics and machine learning for possible future health issues as well as for therapy improvement, but there are still critical issues, such as infrastructure support and data security in terms of RPM. This research presents a new methodology for individualized healthcare in the form of digital twin technology used in combination with robotic systems facilitated by the Internet of Robotic Things (IoRT). This methodology promises to improve the care of patients by reducing face-to-face interactions while ensuring an increased level of safety and a push toward real-time health monitoring combined with customized treatment regimens [42]. A small, relatively cost-effective wireless system that uses the concept of embedded technology by integrating it with a CPS is proposed in the study to upgrade the area of cardiac care. Real-time monitoring of ECG using MEMS and Zigbee technologies, along with Automated Ultrasonic Testing (AUT) actions following abnormalities, enhances the results of treatment for cardiac conditions [43]. The integration of IoT, deep learning, and the XGBoost algorithm for enhancement in the field of real-time patient diagnosis in healthcare monitoring systems is explored by introducing a hybrid model that employs IoT-enabled wearable technology and sensors to accomplish precise and immediate medical evaluations in efforts to advance patient care and subsequently usher in further improvements in intelligent healthcare systems.

7.9 RESULTS AND DISCUSSION

The integration of CPS with wearable devices has highly impacted healthcare with real-time monitoring and personal care of patients. The systems make use of advanced sensors, cloud computing, and machine learning algorithms for the precise and reliable improvement of health monitoring solutions. This review shows key advances, clinical applications, and challenges associated with CPS-enabled wearables. The advancement in technology has allowed the continuous and accurate monitoring of physiological parameters with the help of wearable devices like biosensors, microfluidic devices, and textile-based systems. HWDs based on skin and tattoos have allowed seamless monitoring of vital signs such as glucose levels and heart rate. Sweat-based sensors, using microfluidic reservoirs, detect biomarkers such as glucose, lactate, and chloride ions, allowing better control of chronic diseases, such as diabetes and cystic fibrosis [16, 17].

Teardrop-based sensing devices, for example, smart contact lenses, allow monitoring of glucose levels in an almost painless way without needles [18]. All these developments emphasize the use of wearable technology in healthcare. IoT and cloud computing further increase the potential of CPS-based wearables. Cloud infrastructures provide real-time communication and data analysis, whereas edge computing solves latency-related problems through local data computation. In addition to improving decision-making, it makes care delivery more accessible and scalable. These systems are a useful tool in clinical and home-based healthcare since they enable ongoing patient monitoring and the early detection of abnormalities.

In spite of all these advancements, numerous problems are still associated with the adoption of CPS-enabled wearable devices. Several of the main privacy and security concerns surrounding confidential health information represent significant impediments to its universal adoption. Inappropriate access and data leakage have the potential to undermine patient safety; therefore, robust cyber-security platforms are needed [10]. The integration of machine learning algorithms into wearable systems is further hindered by computational constraints, as most wearable devices rely on low-power processors. Hybrid architectures that keep the balance of the computation at the edge and the computation at the cloud system seem to be a promising idea to address these limitations [36]. These systems also suffer from lower accuracy during dynamic user activity, especially in high-motion environments.

In contrast to conventional healthcare systems, CPS-enabled wearables facilitate enhanced real-time monitoring, anticipatory interventions, and improved patient outcomes. For example, wearable biosensors are superior to traditional glucose monitoring methods through unceasing streams of data that aid in early diagnosis and timely treatment of diseases. The biggest challenge lies in the astronomically high costs of cutting-edge wearable technology and supporting infrastructure to facilitate effective functioning, particularly in rural and developing areas. Resolving logistics and cost issues is needed for large-scale deployment. CPS-enabled wearable technology is overall a healthcare technology revolution that offers better patient outcomes, tailored treatment, and real-time monitoring. In order to reach their full potential, though, challenges such as data privacy, computational constraints, and accessibility need to be addressed. Scalable, secure, and cost-effective solutions should be the focus of future research so that these technologies are made widely accessible across various healthcare environments.

7.10 CONCLUSION

This study has shown the transformative capability of incorporating cyber-physical systems with wearable devices in healthcare by underlining their capability for real-time monitoring, predictive analytics, and personalized treatment plans. Wearable biosensors, textile-based systems, and biofluidic devices have proved their efficacy in chronic disease management, especially in noninvasive glucose monitoring and cardiovascular health tracking. These technologies take advantage of breakthroughs in IoT, edge computing, and AI in order to enhance decision-making and improve patient outcomes in both clinical and home care settings [17, 18]. These advances notwithstanding, it identifies serious challenges facing widespread adoption of these services. There are key privacy and security concerns when CPS-enabled devices are entrusted to handle health-related sensitive data. Finally, there is still a bottleneck due to the constraints from computation and energy with high-power computations for deep and complex learning algorithms; as such, low-power, wearable-based computing may involve hybrid architecture with split or shared computation between the edge system and cloud [10, 11]. These gaps can be bridged by scalable, cost-effective, and secure CPS frameworks. Scaling access to these technologies could revolutionize the delivery of healthcare across underserved regions, bringing more equitable access. Overcoming these challenges would enable the wearables empowered by CPS to take on a major role in developing intelligent health systems across the globe.

REFERENCES

1. S. Xu *et al.*, "Soft microfluidic assemblies of sensors, circuits, and radios for the skin," *Science*, vol. 344, no. 6179, pp. 70–74, Apr. 2014, doi: 10.1126/science.1250169.
2. K. Upreti *et al.*, "A multi-model unified disease diagnosis framework for cyber healthcare using IoMT-cloud computing networks," *Journal of Discrete Mathematical Sciences and Cryptography*, vol. 26, no. 6, pp. 1819–1834, Jan. 2023, doi: 10.47974/jdmsc-1831.
3. M. Haghi, K. Thurow, and R. Stoll, "Wearable devices in medical Internet of Things: Scientific research and commercially available devices," *Healthcare Informatics Research*, vol. 23, no. 1, p. 4, Jan. 2017, doi: 10.4258/hir.2017.23.1.
4. K. Upreti *et al.*, "IOT-Based control System to measure, analyze, and track basic vital indicators in patient healthcare monitoring system," in *Communications in Computer and Information Science*, 2023, pp. 715–725. doi: 10.1007/978-3-031-25088-0_63.
5. T. Someya, Z. Bao, and G. G. Malliaras, "The rise of plastic bioelectronics," *Nature*, vol. 540, no. 7633, pp. 379–385, Dec. 2016, doi: 10.1038/nature21004.
6. A. N. Kozitsina, T. S. Svalova, N. N. Malysheva, A. V. Okhokhonin, M. B. Vidrevich, and K. Z. Brainina, "Sensors based on bio and biomimetic receptors in medical diagnostic, environment, and food analysis," *Biosensors*, vol. 8, no. 2, p. 35, Apr. 2018, doi: 10.3390/bios8020035.

7. N. I. Lee *et al.*, "Challenges and research directions in medical cyber–physical systems," *Proceedings of the IEEE*, vol. 100, no. 1, pp. 75–90, Oct. 2011, doi: 10.1109/jproc.2011.2165270.
8. R. Rajkumar, I. Lee, L. Sha, and J. Stankovic, "Cyber-physical systems," *Proceedings of the 34th Design Automation Conference*, Jun. 2010, doi: 10.1145/1837274.1837461.
9. J. Wang, H. Abid, S. Lee, L. Shu, and F. Xia, "A secured health care application architecture for cyber-physical systems," *arXiv (Cornell University)*, Jan. 2012, doi: 10.48550/arxiv.1201.0213.
10. P. U, H. R. S, J. K, and N. Gummaraju, "Integrating cyber-physical systems for enhanced efficiency in healthcare solutions," in *Advances in Medical Technologies and Clinical Practice Book Series*, 2024, pp. 300–324. doi: 10.4018/979-8-3693-2901-6.ch017.
11. S. Hong, Y. Zhou, J. Shang, C. Xiao, and J. Sun, "Opportunities and challenges of deep learning methods for electrocardiogram data: a systematic review," *arXiv.org*, Dec. 28, 2019. https://arxiv.org/abs/2001.01550
12. S. M. A. Iqbal, I. Mahgoub, E. Du, M. A. Leavitt, and W. Asghar, "Advances in healthcare wearable devices," *Npj Flexible Electronics*, vol. 5, no. 1, Apr. 2021, doi: 10.1038/s41528-021-00107-x.
13. W. Gao *et al.*, "Fully integrated wearable sensor arrays for multiplexed in situ perspiration analysis," *Nature*, vol. 529, no. 7587, pp. 509–514, Jan. 2016, doi: 10.1038/nature16521.
14. B. Neuger, Westonka Schools Lead in Smart PE. - References - Scientific Research Publishing, 2015, https://www.scirp.org/reference/referencespapers?referenceid=3108819
15. L. Hu, N. Xie, Z. Kuang, and K. Zhao, "Review of cyber-physical system architecture," *2012 IEEE 15th International Symposium on Object/Component/Service-Oriented Real-Time Distributed Computing Workshops*, pp. 25–30, Apr. 2012, doi: 10.1109/isorcw.2012.15.
16. A. Koh *et al.*, "A soft, wearable microfluidic device for the capture, storage, and colorimetric sensing of sweat," *Science Translational Medicine*, vol. 8, no. 366, pp. 1–23, Nov. 2016, doi: 10.1126/scitranslmed.aaf2593.
17. J. Choi, D. Kang, S. Han, S. B. Kim, and J. A. Rogers, "Thin, soft, skin-mounted microfluidic networks with capillary bursting valves for chrono-sampling of sweat," *Advanced Healthcare Materials*, vol. 6, no. 5, pp. 32–46, Jan. 2017, doi: 10.1002/adhm.201601355.
18. Y.-R. Lin *et al.*, "Noninvasive glucose monitoring with a contact lens and smartphone," *Sensors*, vol. 18, no. 10, p. 3208, Sep. 2018, doi: 10.3390/s18103208.
19. S. Halldorsson, E. Lucumi, R. Gómez-Sjöberg, and R. M. T. Fleming, "Advantages and challenges of microfluidic cell culture in polydimethylsiloxane devices," *Biosensors and Bioelectronics*, vol. 63, pp. 218–231, Jul. 2014, doi: 10.1016/j.bios.2014.07.029.
20. J. D. Brandt *et al.*, "Long-term safety and efficacy of a sustained-release bimatoprost ocular ring," *Ophthalmology*, vol. 124, no. 10, pp. 1565–1566, May 2017, doi: 10.1016/j.ophtha.2017.04.022.
21. M.-A. Shahbazi *et al.*, "A self-healable, moldable and bioactive biomaterial gum for personalised and wearable drug delivery," *Journal of Materials*

Chemistry B, vol. 8, no. 19, pp. 4340–4356, Jan. 2020, doi: 10.1039/c9tb02156f.
22. D. H. Keum *et al.*, "Wireless smart contact lens for diabetic diagnosis and therapy," *Science Advances*, vol. 6, no. 17, pp. 1–13, Apr. 2020, doi: 10.1126/sciadv.aba3252.
23. C. Jumelle, S. Gholizadeh, N. Annabi, and R. Dana, "Advances and limitations of drug delivery systems formulated as eye drops," *Journal of Controlled Release*, vol. 321, pp. 1–22, Feb. 2020, doi: 10.1016/j.jconrel.2020.01.057.
24. S.-W. Jeong *et al.*, "Mortality reduction with physical activity in patients with and without cardiovascular disease," *European Heart Journal*, vol. 40, no. 43, pp. 3547–3555, Jul. 2019, doi: 10.1093/eurheartj/ehz564.
25. D. Zhang, X. Shen, and X. Qi, "Resting heart rate and all-cause and cardiovascular mortality in the general population: a meta-analysis," *Canadian Medical Association Journal*, vol. 188, no. 3, pp. E53–E63, Nov. 2015, doi: 10.1503/cmaj.150535.
26. A. K. Reimers, G. Knapp, and C.-D. Reimers, "Effects of exercise on the resting heart rate: a systematic review and meta-analysis of interventional studies," *Journal of Clinical Medicine*, vol. 7, no. 12, p. 503, Dec. 2018, doi: 10.3390/jcm7120503.
27. B. W. Nelson, C. A. Low, N. Jacobson, P. Areán, J. Torous, and N. B. Allen, "Guidelines for wrist-worn consumer wearable assessment of heart rate in biobehavioral research," *NPJ Digital Medicine*, vol. 3, no. 1, Jun. 2020, doi: 10.1038/s41746-020-0297-4.
28. Y. Zhang, R. G. Weaver, B. Armstrong, S. Burkart, S. Zhang, and M. W. Beets, "Validity of Wrist-Worn photoplethysmography devices to measure heart rate: a systematic review and meta-analysis," *Journal of Sports Sciences*, vol. 38, no. 17, pp. 2021–2034, Jun. 2020, doi: 10.1080/02640414.2020.1767348.
29. F. Shaffer, Z. M. Meehan, and C. L. Zerr, "A critical review of ultra-short-term heart rate variability norms research," *Frontiers in Neuroscience*, vol. 14, pp. 21–34, Nov. 2020, doi: 10.3389/fnins.2020.594880.
30. K. Georgiou, A. V. Larentzakis, N. N. Khamis, G. I. Alsuhaibani, Y. A. Alaska, and E. J. Giallafos, "Can wearable devices accurately measure heart rate variability? A systematic review," *Folia Medica*, vol. 60, no. 1, pp. 1–14, Jan. 2018, doi: 10.2478/folmed-2018-0012.
31. J. H. Moon, M.-K. Kang, C.-E. Choi, J. Min, H.-Y. Lee, and S. Lim, "Validation of a wearable cuff-less wristwatch-type blood pressure monitoring device," *Scientific Reports*, vol. 10, no. 1, pp. 1–14, Nov. 2020, doi: 10.1038/s41598-020-75892-y.
32. F. C. Bull *et al.*, "World Health Organization 2020 guidelines on physical activity and sedentary behaviour," *British Journal of Sports Medicine*, vol. 54, no. 24, pp. 1451–1462, Nov. 2020, doi: 10.1136/bjsports-2020-102955.
33. R. Mendes *et al.*, "Exercise prescription for patients with type 2 diabetes—a synthesis of international recommendations: narrative review," *British Journal of Sports Medicine*, vol. 50, no. 22, pp. 1379–1381, Dec. 2015, doi: 10.1136/bjsports-2015-094895.
34. V. D. Funtanilla, T. Caliendo, and O. Hilas, "Continuous glucose monitoring: a review of available systems," Sep. 2019. [Online]. Available: https://www.ncbi.nlm.nih.gov/pmc/articles/PMC6705487/pdf/ptj4409550.pdf

35. J. Golledge, M. Fernando, P. Lazzarini, B. Najafi, and D. G. Armstrong, "The potential role of sensors, wearables and telehealth in the remote management of diabetes-related foot disease," *Sensors*, vol. 20, no. 16, p. 4527, Aug. 2020, doi: 10.3390/s20164527.
36. S. Hong, Y. Zhou, J. Shang, C. Xiao, and J. Sun, "Opportunities and challenges of deep learning methods for electrocardiogram data: a systematic review," *arXiv.org*, Dec. 28, 2019. https://arxiv.org/abs/2001.01550
37. G. Cho, K. Jeong, M. J. Paik, Y. Kwun, and M. Sung, "Performance evaluation of textile-based electrodes and motion sensors for smart clothing," *IEEE Sensors Journal*, vol. 11, no. 12, pp. 3183–3193, Sep. 2011, doi: 10.1109/jsen.2011.2167508.
38. Dev.Login and Dev.Login, "The history of wearable technology," *Blog*, Jun. 07, 2022. https://www.condecosoftware.com/blog/the-history-of-wearable-technology/
39. R. R. Ruckdashel, N. Khadse, and J. H. Park, "Smart E-textiles: overview of components and outlook," *Sensors*, vol. 22, no. 16, p. 6055, Aug. 2022, doi: 10.3390/s22166055.
40. M. Vrigkas, C. Nikou, and I. A. Kakadiaris, "A review of human activity recognition methods," *Frontiers in Robotics and AI*, vol. 2, pp. 46–57, Nov. 2015, doi: 10.3389/frobt.2015.00028.
41. P. Vichitra and S. Mangayarkarasi, "The Integration of Portable Sensors with a Mobile Application for Self-Observation," *International Journal of Innovative Science and Research Technology (IJISRT)*, vol. 9, pp. 228–232, Jul. 2024, doi: 10.38124/ijisrt/ijisrt24jul039.
42. C. Rajurkar, B. Murugan, and G. P. Namasivayam, "Design and integration of IoRT-enabled robotic systems with wearable devices in a digital twin framework," in *2024 9th International Conference on Control and Robotics Engineering (ICCRE)*, pp. 45–49, May 2024, doi: 10.1109/iccre61448.2024.10589845.
43. L. Natrayan, "Integrating cyber physical systems with embedded technology for advanced cardiac care," in *Advances in Computational Intelligence and Robotics Book Series*, pp. 288–298, 2024, doi: 10.4018/979-8-3693-1586-6.ch015.

Chapter 8

Building resilient and sustainable operations through cloud security

Jossy P. George, Kamal Upreti, and Bosco Paul Alapatt

8.1 INTRODUCTION

Organizations in the information age increasingly opt for cloud computing to increase operational efficiency, scalability, and creativity. The cloud offers unparalleled adaptability, facilitating businesses to deploy new services in a timely manner, scale up or down as needed, and foster innovation through the availability of cutting-edge technologies [1]. Such a transformation into cloud-ground operations is not, however, problem-free. Among the most important of these challenges is the increased threat to security, which may undermine the sustainability and resilience of business organization procedures.

As companies migrate their data, application programs, and infrastructure to the cloud, they expose themselves to an increasingly panoptic array of cyber threats, such as information breaches, ransomware, insider attacks, and Distributed Denial of Service (DDoS) attacks [2]. These threats may disrupt operations, through media raw entropy, and harm an organization's reputation. In addition, the dynamic and sharing nature of cloud environments makes conventional security practices inadequate and demands a higher, advanced, and proactive form of security [3]. This study explores cloud security intelligence (CSI), a theoretical strategic account designed to tackle the unparalleled security issues of cloud computing. CSI uses cutting-edge technologies like artificial intelligence (AI), machine learning (ML), and real-time monitoring to make arrangements with the tools they require to actively detect, counteract, and react to security threats [4]. By integrating CSI into their operations, organizations are able to build capacity to endure and recover from security incidents, and ensure business continuity and long-term viability. It will examine the most important elements of CSI, such as data gathering, threat detection, incident response, and continuous monitoring [5]. It will also address how CSI helps develop resilience by allowing proactive threat guidance, enhancing visibility and dominance, and decreasing incident response times. The importance of CSI in supporting sustainability by minimizing costs, ensuring regulatory compliance, and reducing the environmental footprint is discussed [6].

DOI: 10.1201/9781003569251-8

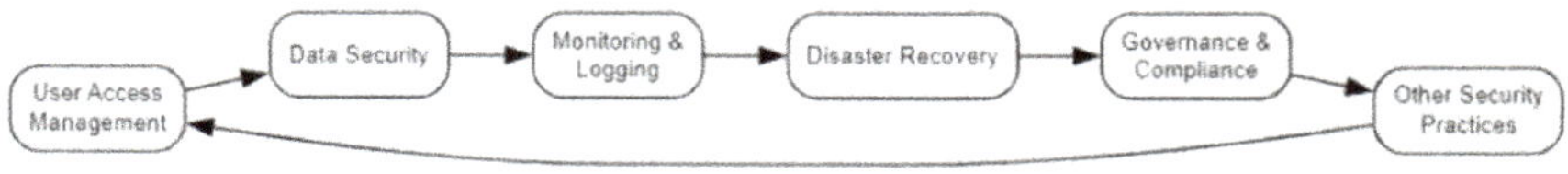

Figure 8.1 The importance of cloud security intelligence

Through the real-world case studies, the chapter will illustrate how groups across various industries have successfully implemented CSI to secure their cloud surgery. It will also address the issues related to CSI adoption and address directions for the future, including the incorporation of new technologies such as blockchain and the Internet of Things (IoT) [7].

In short, the chapter hopes to provide a complete sense of how CSI can be a basis for creating resilient and sustainable cloud cognitive processes. By embracing CSI, organizations cannot only keep their assets and data points safe but also place themselves as long-term winners in an increasingly digital and connected world [8].

The cloud security framework illustrated in Figure 8.1 provides a comprehensive approach to securing cloud environments by organizing key security domains. It includes user access management, which governs identity, authentication, and user roles; data security, focusing on encryption, classification, and availability; and hardware and software security, which covers physical protection, server hardening, and patch management. Network security ensures safe data transmission through firewalls, VPNs, and secure communication channels. Disaster recovery emphasizes minimal service disruption through backup, replication, and auto-healing strategies. The governance and compliance component ensures alignment with legal and regulatory requirements, including audits and data location policies. Monitoring and logging provide continuous threat detection, traffic analysis, and incident response. Last, the "other" category highlights application security, third-party risk management, and overall service security. Together, these elements form a robust structure to safeguard cloud-based systems.

8.1.1 The growing threat landscape

The cloud environment has become a choice target for cybercriminals due to its widespread presence and the large amount of sensitive data it houses. As organizations increasingly migrate their processes to the cloud, they inadvertently expose themselves to a growing array of cyber threats [9]. These threats include data breaches, ransomware attacks, DDoS attacks, insider attacks, and advanced persistent threats (APTs).

According to a write-up by McAfee, cloud-free-base attacks surged by 630% between January and April 2020. This dramatic increase highlights the escalating sophistication and frequency of cyberattacks targeting cloud infrastructures [10]. Cybercriminals tap vulnerable cloud configurations, weak access controls, and misconfigured APIs to gain unauthorized access to sensitive data and disrupt business operations.

The active and circular nature of cloud environments complicate the protection landscape. Traditional security department measures, such as firewall and antivirus software, are often insufficient to address the unique challenges posed by cloud computing [11]. As a result, organizations must adopt more advanced and adaptive security schemes to protect their cloud assets and ensure the resilience of their operations.

8.1.2 The role of cloud security intelligence

CSI emerges as a critical solution to address the security threats to cloud environments. CSI refers to the use of technologies, methodologies, and frameworks to collect, analyze, and act on protection-related data in the cloud [12]. By leveraging tools such as AI, ML, and real-time monitoring, CSI enables organizations to gain insight into their cloud environment and proactively name potential threats.

8.1.2.1 Key functions of CSI

- Data collection and aggregation: CSI systems collect data from various sources within the cloud environment, including logarithms, network traffic, user actions, and system networks. This data is centralized and normalized for analysis [13].
- Threat detection and analysis: Using sophisticated analytics and machine learning algorithms, CSI describes patterns and anomalies that may indicate a security threat [14]. For example, it can discover unusual login attempts, unauthorized memory access to sensitive data, or abnormal network traffic.
- Proactive threat mitigation: CSI enables systems to carry out preemptive action against possible attacks. By predicting and pointing out threats before they materialize, organizations can implement security measures to mitigate these risks and prevent incidents [15].
- Incident response and recovery: In the event of a security incident, CSI provides timely insights and automates a reaction to curb the threat and mitigate damage. This ensures quicker recovery and less downtime [16].
- Continuous monitoring and improvement: CSI systems continuously monitor cloud environments for new threats and vulnerabilities. They likewise leave feedback to improve security policies, protocols, and algorithms over time.

8.1.2.2 Benefits of CSI

CSI is an essential component in providing the security, reliability, and compliance of contemporary cloud infrastructure. In this era of digitization, when organizations are entirely reliant on cloud services for day-to-day activities, CSI plays the role of a strong facilitator of end-to-end security and governance [17]. Among the most essential features of CSI is its provision of greater visibility into the complete cloud ecosystem. It provides governance teams and IT administrators with a 360-degree view of their cloud resources, applications, and data flows. This end-to-end visibility allows real-time monitoring of user activity, system activity, and potential security risks. With this visibility, organizations can detect threats, anomalies, or malicious activity in real time, allowing them to respond and contain the problem immediately before harm is caused [18].

Aside from monitoring, CSI facilitates proactive security, allowing organizations to foresee, identify, and disable threats before they escalate into full-fledged incidents. This includes detection of misconfigurations, identification of suspicious access patterns, and utilizing advanced analytics and machine learning to predict possible breaches [19]. By early risk management, organizations reduce their exposure to cyberattacks, data breaches, and other security incidents that can hamper operations or compromise sensitive information [20].

CSI is also vital in business continuity. In a case of a breach, such as a cyberattack or loss of data, CSI provides means for reacting quickly and in an efficient way by quickly recovering data and systems. This occurs by minimizing the downtime and disturbance, which is instrumental in maintaining the trust of customers, partners, and stakeholders [20]. CSI-aided business continuity planning includes automated backups, disaster recovery procedures, failover, and redundancy procedures, all of which provide organizations with the capacity to maintain operational resilience in the face of unforeseen challenges [21].

Moreover, CSI plays a central role in ensuring regulatory compliance. As governments and industry bodies enforce strict regulations regarding data security, privacy, and business conduct, organizations must comply with these regulations to avoid fines and loss of reputation. CSI provides the mechanisms and frameworks to implement, manage, and enforce security policies in accordance with such regulations [22]. It also enables auditing, reporting, and documentation, making it easy for organizations to demonstrate compliance with being audited or inspected. Regardless of GDPR, HIPAA, ISO standards, or whatever other regulatory framework is used, CSI ensures that the organization's cloud operations remain within legal and ethical bounds [23].

In short, CSI is not just a technical necessity—it is a strategic asset that protects digital assets, ensures continuous business operations, supports legal compliance, and enables organizations to do business securely in the cloud [24]. Its ability to offer real-time visibility, proactive defense, continuity, and compliance makes it an essential component of any cloud security strategy [25].

8.1.2.3 Integration of CSI into cloud operations

In addition to embracing cloud services, in order to utilize CSI optimally, organizations need to do more than just adopt; they need to integrate CSI strategically into the center of their cloud processes [26]. This integration begins with the deployment of sophisticated security devices and platforms, such as security information and event management (SIEM) systems and cloud access security brokers (CASBs). SIEM systems are central to collating and examining vast amounts of security-oriented data from across the cloud environment for real-time threat identification, alerting, and incident response analysis. CASBs, on the other hand, serve as the required intermediaries between cloud providers and consumers by implementing security policies, traffic scanning, and verifying whether data flow complies with enterprise and government standards [27].

Technology alone, however, is not sufficient. Organizations also need to equip their security personnel with the intelligence and expertise to use CSI tools productively. It involves superior training for deciphering the data and analysis of CSI systems, which enables the staff to filter normal activity from genuine security risk [28]. Trained personnel at hand can respond faster and more efficiently, controlling potential harm and avoiding escalation. In addition, staff must design extensively documented procedures for ongoing monitoring, proactive threat hunting, and incident response. This entails creating workflows for monitoring anomalies, conducting forensic analysis, initiating automated responses to known threats, and forwarding critical cases to human handling [29].

By embedding CSI into business operations and linking it to organizational goals, companies can create a strong and sustainable cloud infrastructure [30]. Not only does this infrastructure safeguard sensitive data and mission-critical assets from an ever-evolving threat landscape, but it also enables business agility and continuity. With more advanced cyber threats, the ability to anticipate, detect, and respond to incidents in real time is a strategic differentiator. Finally, CSI isn't just an issue of risk management or compliance—it's a question of developing long-term business success in the digital-first environment. An effectively designed CSI structure facilitates innovation, fosters stakeholder trust, and puts companies in the best position to succeed in an environment of accelerating technology change [31].

8.1.2.4 The critical role of CSI in securing cloud environments

With the rapid development of the modern digital world, cloud environments are now central to business operations, innovation, and data management. It has also made the cloud a prime target for cybercriminals, with cloud system attacks having increased by an astonishing 630% in early 2020 alone [32]. This surge underscores the need for robust, intelligent security designs that are capable of keeping pace with the growing sophistication and number of cyberattacks. CSI emerges as a critical solution, leveraging the latest technologies like AI and ML to proactively monitor, analyze, and counter threats in cloud environments. Unlike traditional security controls that are generally reactive, CSI enables organizations to identify anomalies and potential threats in real-time and therefore quickly respond and neutralize threats before they can do maximum damage [33].

Apart from threat reduction, CSI offers a range of strategic benefits. It enhances visibility into cloud activity, providing governance teams with a comprehensive view of data flow, user behavior, and system vulnerabilities [34]. Not only is this visibility important for security monitoring, but also for strategic planning and decision-making. CSI also allows proactive security, staying ahead of any threat rather than reacting once the incidents occur [35]. In addition, CSI is essential to business continuity because it helps with rapid handling of incidents, minimizes downtime, and realizes operational resilience during cyber outages. Most importantly, CSI plays a pivotal role in regulatory compliance, making it possible for organizations to have the information and tools necessary for enforcing security policies and demonstrating compliance with data protection laws like GDPR and HIPAA [36].

Figure 8.2 shows the growth in cloud-based cyberattacks from 2020 to 2021. Based on statistics from McAfee and ENISA, the attacks almost doubled within a single year, reflecting the increasing vulnerabilities of cloud environments. This trend underscores the imperative for organizations to enhance their cloud security posture using sophisticated technologies such as CSI and adopt proactive defense measures to address the changing threat landscape.

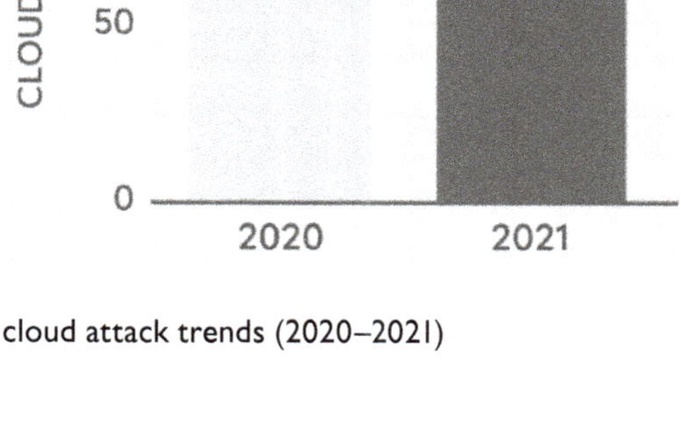

Figure 8.2 Surge in cloud attack trends (2020–2021)

8.2 DATA COLLECTION AND AGGREGATION

The infrastructure of CSI has its foundation on data collection and accumulation, or the process by which security-related information from numerous sources throughout the cloud is gathered. That information comprises logs, network traffic, user behavior, and configuration of systems, all of which yield essential clues on the carriage of security measures of the cloud infrastructure [37]. More advanced processes, such as SIEM solutions, are utilized to collect and aggregate this information and normalize it into an ordered structure to analyze. By consolidating data from multiple sources, organizations benefit from comprehensive visibility into their cloud environment, enabling them to identify potential vulnerabilities and monitor actions in real time. For instance, logs from cloud servers can break unauthorized access attempts, while network dealings data can spotlight unusual patterns indicative of a cyberattack. Good data collection and aggregation ensure that no vital info is overlooked and set the basis for proactive threat detection and analysis.

8.2.1 Threat detection and analysis

Once data is gathered and aggregated, the next step is threat detection and analysis, where sophisticated analytics and simple machine learning algorithms are employed to identify potential security threats. This process needs to examine the collected data to detect patterns and anomalies that may indicate a security system breach. For object lessons, unusual login attempts, unexpected data flow, or unauthorized access to raw data can be flagged as potential threats. Behavioral analysis techniques monitor hackers and system processes to identify deviations from the normal activity, while anomaly detection algorithms scrutinize network traffic for irregularity, such as sudden spikes in data transfer or connections to fishy IP names and addresses [38]. Predictive analytics further enhances this process by using historical data to prevent future exposure. By leveraging these advanced techniques, organizations can discover threats in real time, enabling them to take proactive measures to mitigate the risk of exposure before they escalate into serious incidents.

8.2.2 Incident response and mitigation

In the event of a security incident, incident response and mitigation are a critical constituent of CSI that ensures organizations can respond quickly and effectively [39]. When a potential threat is detected, automated warning signals are triggered, notifying the security team in a timely manner. These alerts are often responded to according to predefined response playbooks, which provide step-by-step instructions on how to stop and mitigate the threat. Machine-driven mechanics can take prompt action, such as isolating the involved organization, hindering malicious IP addresses, or suspending compromised accounts, to keep the threat from spreading. Simultaneously, the security team is offered elaborate insights and forensic data to inquire about the root cause of the incident and implement a long-term solution. For example, if a ransomware attack is noticed, the network can automatically isolate the infected organization, block the attacker's access, and send notice to the security department team to initiate the recovery process. This rapid response capability mitigates the encroachment of security incidents, reduces downtime, and ensures business continuity.

8.2.3 Continuous monitoring and improvement

CSI is not a one-time effort but an ongoing process, and continuous monitoring and improvement are central components that guarantee the scheme stays effective over time. The cloud environment is incessantly monitored for new threats, exposure, and modifications in user actions. Real-time surveillance allows for the system to detect and respond to imminent threats as they originate, while regular updates to security policies, communications protocols, and detection algorithms ensure that the system of rules evolves to address new challenges. Feedback from past incidents is utilized to refine threat detection and reply mechanisms, making the system more robust and adaptive. Additionally, periodic vulnerability assessments and audits help discover and address weaknesses in the cloud infrastructure. For example, if a new type of malware is discovered, the organization can be updated to detect standardized patterns in the future. This reiterative approach ensures that the organization's protection posture remains ready and capable of addressing the ever-changing threat landscape.

Figure 8.3 is a pie chart of the percentage distribution of data breach costs by industry. The healthcare sector is the most affected, covering 30%

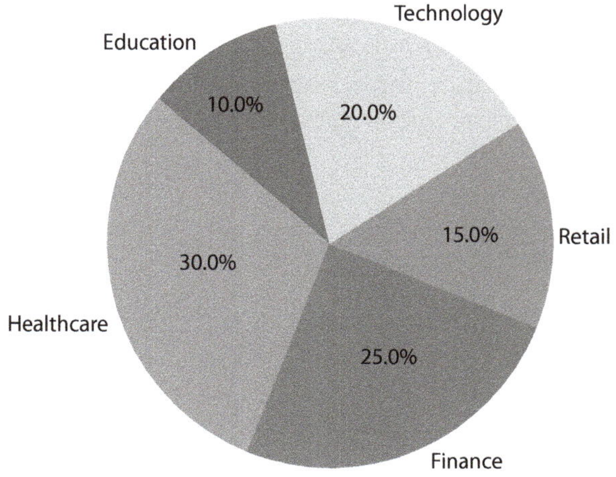

Figure 8.3 Industry-wise distribution of data breach costs

of all breach costs, followed by finance with 25%, and technology at 20%. Retail and education take up 15% and 10%, respectively. This information underscores the manner in which sensitive and valuable data in industries such as healthcare and finance render them susceptible targets for cyberattacks, with the implication of strong cybersecurity practices specific to the risk profile of each industry.

8.3 PROACTIVE THREAT MANAGEMENT

One of the strongest benefits of CSI is that it can facilitate proactive threat management. In contrast to more conventional security measures that fixate on reacting to the occurrence of events after they happen, CSI uses cutting-edge technology such as AI and ML to identify potential threats before they become a reality. By analyzing huge volumes of information from logs, network transactions, user behavior, and system forms, CSI can identify blueprints and irregularities that could indicate an emerging security fissure. For instance, when a user account makes an unexpected move to grab confidential information from a foreign location, the system can assume this as a possible threat and activate security teams.

This preventive measure leads to other preemptive measures, including blocking of suspicious IPs, preventing memory access to reasonable data points, or revising security protocols to alleviate vulnerabilities. By handling a threat before it intensifies into a full-blown incident, organizations can significantly reduce the likelihood of security breaches. Even when incidents do occur, the impact is minimized because the threat is detected and mitigated early in its lifecycle. For example, if a phishing endeavor is identified before hackers interact with a malicious link, the risk of data compromise is greatly reduced.

Proactive threat management not only enhances the security of cloud environments but also contributes to usable efficiency and monetary value preservation. By preventing incidents, organizations avoid the financial and reputational damage associated with data point breaches, regulatory fines, and downtime. Furthermore, this plan of attack furthers a culture of vigilance and preparation, ensuring that leadership is always one step ahead of cyber criminals. At its core, proactive threat management is a cornerstone of building an alert and sustainable cloud platform that can adapt to the ever-evolving threat landscape.

8.3.1 Enhanced visibility and control

CSI provides organizations with enhanced visibility into their cloud environment, enabling them to monitor hacker activity, network traffic, and system form in real time. This visibility is achieved through the centralized collection and analysis of data from various informants, such as logs, network dealings, and user behavior. By having an overall perspective on their cloud operations, organizations can see where they are vulnerable, detect abnormal activity, and hold sway over their cloud environment. For example, when an unauthorized user tries to access sensitive information, the CSI system can instantly mark the activity as suspicious and give complete information regarding the origin and type of threat. This level of visibility enables the platform to proactively deal with security threats prior to their escalation to ensure that its cloud environment is secure and conforms to regulations. Sustaining predominance over the cloud operations empowers organizations to mitigate risk, improve working efficiency, and advance a higher security stance.

8.3.2 Improved incident response times

One of the most significant benefits of CSI is its ability to improve incident response times. The rapid spying and analysis capabilities of CSI enable the establishment to distinguish threats in real time, oftentimes within seconds of their occurrence. Automated response reactions can be triggered immediately, allowing actions such as isolating the affected systems, blocking malicious IP addresses, or suspending compromised user accounts. At the same time, security teams are provided with detailed insight and actionable data to begin their mitigation efforts. For example, if a DDoS attack is detected, the system can automatically reroute traffic and notify the security team to put through extra countermeasures. This rapid response capability minimizes the time between threat detection and mitigation, reducing the potential damage done by security incidents. By accelerating incident response, CSI helps organizations maintain business continuity, protect sensitive data points, and reduce the financial and reputational impact of cyberattacks.

8.3.3 Business continuity and disaster recovery

CSI plays a critical role in ensuring business continuity and catastrophe recovery by enhancing the resilience of cloud military operations. In the event of a protection incident, such as a data breach or ransomware attack, CSI enables organizations to recover their operations promptly and minimize downtime. Automated responses and a predefined playbook ensure that threats are quickly held back, while detailed information from the system guides recuperation efforts. For example, if a ransomware attack encrypts data, the organization can isolate the impacted systems, retrieve data points from backups, and notify the security team to investigate the root cause. This speedy recuperation capability reduces the impact of an attack and the number of incidents on business operations, ensuring that critical services remain available and minimizing financial losses.

Additionally, CSI handles disaster recovery planning by using real-time monitoring and identifying threats, which help the organization respond to potential disruptions. By providing continuity and disaster recovery, CSI aids organizations in maintaining customer confidence, complying with regulatory demands, and reaching long-term operational sustainability.

8.3.4 Integration of benefits

These benefits—enhanced profile and control, improved incident response, and business continuity and disaster recovery—are collected and contribute to the resiliency and sustainability of cloud operations. The increased visibility enables organizations to detect threats early, improve incident response time, and rapidly mitigate threats, and business continuity and disaster recovery minimize the impact of incidents on operations. Together, these benefits demonstrate how CSI empowers organizations to build a secure, resilient, and adaptive cloud environment that affirms their long-term success.

8.4 COST EFFICIENCY THROUGH CLOUD SECURITY INTELLIGENCE

CSI is the driving force of cost-effectiveness for organizations because it can significantly reduce the cost of security breaches. In a world where data breaches and cyberattacks are not only increasing but also becoming more complex, the cost of responding to and recovering from these attacks can be staggering. There are regulatory penalty costs, legal expenses, business disruption costs, and the often immeasurable harm done to customer faith and brand prestige. CSI serves to avoid such risks by providing preemptive threat sensing and reaction so that would-be intrusions get detected and captured before they become full-scale issues. With the application of cutting-edge technologies like AI and machine learning, CSI enhances real-time monitoring, automates event detection, and improves threat analysis accuracy. This means faster event containment and resolution, and hence lower operational downtime and incident-related expenditures.

Furthermore, through ongoing monitoring of system weaknesses and user activity, CSI helps organizations optimize their overall security posture, reducing the need for costly reactive processes and crisis management. By investing in CSI, not only are significant security breaches prevented, but resource allocation is also optimized through enabling security professionals to target efforts where they are needed most. In the long run, the funds saved through prevention of even a single major data breach can easily cover the cost of installing CSI systems. In the end, CSI is a foundation of long-term business sustainability and success within a more digital and networked environment where security is not merely an IT concern but an essential business requirement.

8.4.1 Integration of benefits

These advantages—efficiency, regulatory compliance, environmental impact, and long-term strategic rewards—are interlinked and together assist in the cloud operations' sustainability and resiliency. Cost efficiency ensures organizations can allocate resources appropriately, regulatory compliance avoids penalties, environmental wallop aligns process with sustainability destination, and long-term strategic plans position organizations for future success. In effect, financial penalties under ordinances such as GDPR and damage to an organization's reputation can lead to lost business opportunities. CSI reduces such risks by identifying threats in advance and facilitating timely reception, thus avert incidents from spiraling out of control. Second, the mechanization of security processes, including threat detection, incident response, and compliance tracking, boils down to the motivation for manual intervention. This not only lets down operational costs but also allows the security team to focus on strategic initiatives rather than routine tasks. By bringing down both direct and indirect charges linked to security department incidents, CSI helps organizations to optimize their budgets and allocate resources more effectively.

8.4.2 Regulatory compliance

Compliance with regulatory requirements is a critical expression of sustainable operations, and CSI provides the tools and insights needed to meet these requirements. Regulations such as GDPR, HIPAA, and PCI-DSS mandate strict data protection and security measures, and noncompliance can result in hefty fines and reputational damage. CSI avails administration monitors and enforces security insurance policy by allowing for real-time visibility into cloud operations, detecting vulnerabilities, and yielding an audit-ready reputation. For instance, if a healthcare group is involved in complying with HIPAA, CSI can ensure that patient data is encrypted, the approach is restricted to authorized office staff, and all activities are logged for auditing purposes. By automating compliance monitoring and reporting, CSI reduces the administrative burden on organizations and guarantees that they remain compliant with regulations. This not only minimizes the danger of penalties but also builds up trust with clients and stakeholders who value data privacy and security.

8.4.3 Environmental impact

The sustainability of cloud operations extends beyond financial and regulatory considerations to include environmental impact. CSI upholds environmental sustainability by optimizing resource utilization and reducing energy use in cloud environments. For example, CSI can monitor and manage cloud infrastructure, such as virtual machines and storage, to ensure they are utilized expeditiously. Unused or underutilized resources can be identified and deactivated, energy consumption scaled down, and carbon emissions lowered at data centers. Also, by preventing security incidents such as ransomware attacks or data breaches, CSI minimizes the need for resource-intensive recovery procedures, which use much energy. By promoting efficient resource management and reducing waste material, CSI helps organizations align their cloud operations with environmental sustainability goals, bestowing a sustainable future.

8.4.4 Long-term strategic advantage

Investing in CSI provides organizations with a long-term strategic advantage by ramping up a resilient and sustainable cloud infrastructure. A system that uses CSI is better equipped to withstand security threats, ensuring uninterrupted operations and maintaining customer satisfaction. For example, an organization that can quickly recover from a cyberattack and demonstrate protection practices is to a greater extent are likely to retain customer loyalty and attract new clientele. In addition, CSI allows organizations to adopt safe and sustainable cloud practices, increasing their reputation. This competitive advantage can avail opportunities and partnerships to young business organizations as customers and stakeholders demand security and sustainability. CSI not only enhances security but also boosts operational excellence and sustainability, and hence it is a sound investment for organizations in the digital era.

8.5 ASSESSING ORGANIZATIONAL NEEDS

Step 1 of availing CSI is to determine the organization's specific security requirements. It involves identifying the vulnerabilities of data and programs that reside on the cloud, as potential threats and exposures that can hide behind these assets. For instance, a company handling sensitive client information might specialize in protecting against data breaches, whereas a cloud-free diligence-based company could specialize in preventing service disruptions. This analysis also involves determining the current security stance of the business, such as coverage, hardware, and operations, to identify disruption and areas for improvement at the surface. With comprehension of the unmatched risk and requirements of development, the CSI plan can be drafted accordingly to efficiently address these challenges. This first step ensures CSI deployment is aligned with the system's purpose and introduces maximum value.

8.5.1 Select the right tools and technologies

After valuing organizational needs, the next step is to choose the appropriate tools and applied science for implementing CSI. This includes SIEM systems, which centralize and analyze security data points from multiple sources; threat intelligence platform, which provides real-time information about emerging threats; and machine learning algorithms, which enable advanced threat detection and predictive analytics. The use of these tools should be based on their ability to meet the governing body's specific security requirements. For example, a large enterprise with a complex cloud environment may demand a scalable SIEM solution with analytics capability. At the same time, a smaller enterprise might prioritize cost-effective tools with user-friendly interfaces. Also, the chosen tool should integrate seamlessly with the organization's existing cloud infrastructure and workflow. By selecting the right tools and technologies, the administration can build a robust CSI framework that enhances its ability to observe, respond to, and mitigate security threats effectively.

8.5.2 Integrating cloud security intelligence into existing operations

The successful execution of CSI needs seamless consolidation with the establishment's existing cloud operations. This involves configuring the chosen tool and technologies to work harmoniously with the current cloud infrastructure. For example, the SIEM system must be laid out to collect and analyze data points from various cloud services, while machine learning algorithms need to be trained on the organization's specific data design. Also, staff must be trained to utilize these tools effectively, so that they can interpret warnings, react to incidents, and manage the organization efficiently. Understanding processes for continuous monitoring, threat signal detection, and incident response is critical. For instance, an automated workflow can be made to set off alerts and start reacting when a threat is detected. By integrating CSI into existing mathematical processes, organizations can see that their cloud environments are protected without disrupting daily workflows.

8.5.3 Continuous improvement and adaptation

The carrying out of CSI is not a one-time effort but an ongoing outgrowth that requires continuous improvement and adaptation. As the threat landscape evolves, the organization must regularly update its protection policies, protocols, and tools to address new vulnerabilities and emerging threats. This includes staying informed about the recent cybersecurity trends and integrating new technology, such as artificial intelligence and blockchain, into the CSI fabric. Regular audits and assessments should also be a step to ensure that the execution of CSI aligns with the organization's mission and delivers maximum value.

Evaluation should be made as to the effectiveness of the CSI strategy and to identify areas for improvement. For example, if a new variant of malware is seen, the administration should update its threat detection algorithm to recognize similar patterns. By fostering a refinement of uninterrupted improvement, the organization can ensure that its CSI strategy remains effective and aligned with its goals, maintaining full-term protection for its cloud operations.

8.6 CASE STUDIES

8.6.1 Case study 1: Financial services sector

An outstanding fiscal services company implemented CSI to strengthen the security of its cloud-based operations. By leveraging modern analytics and machine learning, the company successfully detected and mitigated a sophisticated phishing attack purporting to steal client data. The CSI system described strange login patterns and flagged suspicious behavior in real time, enabling the security team to respond swiftly. Automated reactions were activated to occlude the hacker and isolate the affected organization, reducing the impact of the breach. The speedy signal detection and recovery ensured that customer data remained secure, and business procedures continued uninterrupted. This example highlights how CSI can protect sensitive financial information, maintain customer trust, and sustain business continuity in the face of advanced cyber threats.

8.6.2 Case study 2: Healthcare industry

A healthcare supplier adopted CSI to safeguard its cloud-based data and ensure compliance with HIPAA regulations. CSI allows renders real-time monitoring and threat detection, enabling the provider to identify and respond to unauthorized access attempts at once. For example, when an unauthorized user attempts to access raw patient data, the system flags the activity and automatically triggers a response. This proactive access not only protected patient privacy but also helped the administration avoid costly regulatory fines and reputational harm. By following up with CSI, the healthcare supplier proved its dedication to information security and regulatory deference, securing the corporate trust of patients and stakeholders.

8.6.3 Case study 3: E-commerce platform

An e-commerce platform utilized CSI to represent its cloud infrastructure against DDoS attacks. During a critical sales agreement period, the platform's CSI organization detected a DDoS attack calculated to flood its server and disrupt the process. The attack automatically triggered mitigation measures, such as rerouting traffic and blocking malicious IP addresses, to see to it that the platform stays operational. This ready response prevents important revenue loss during an important sales period. The case manifests how CSI can protect critical commercial enterprise procedures, secure overhaul availability, and safeguard revenue streams, even during targeted cyberattacks.

8.6.4 Key takeaways from the case studies

These studies illustrate the transformative impact of CSI across different industries. In the financial services sector, CSI enabled rapid detection and mitigation of phishing attacks, protected customer data, and ensured business continuity. In healthcare diligence, CSI helped secure patient records, ensured HIPAA compliance, and prevented unauthorized access. For the e-commerce platform, CSI protected against DDoS attacks, ensuring functional resilience during a critical sales period. These examples underline the versatility and potency of CSI in addressing diverse security measure challenges, enhancing operational resilience, and supporting long-term success. By adopting CSI, companies can proactively protect their cloud environment, maintain regulatory compliance, and build trust with customers and stakeholders. Improvement in response time and compliance with CSI is given in Table 8.1.

8.7 CHALLENGES IN FOLLOW-THROUGH CLOUD SECURITY INTELLIGENCE

While CSI provides security, its execution comes with several challenges. One major hurdle is the complexity of mixing CSI with the cloud infrastructure. Organizations frequently operate in multicloud or hybrid environments, making it difficult to guarantee seamless compatibility and data flow between organizations. Additionally, the inability of skilled personnel to manage and operate the CSI organization poses a challenge. Effective processes require expertise in cloud certificates, data analytics, and machine learning, which may not always be readily available within a system. Another challenge is the possibility of false positives in threat detection, where legitimate natural actions are mistakenly flagged as threats. This can lead to unneeded alerts, overwhelming the security team, and reducing the efficiency of the scheme. To fully realize the benefit of CSI, the

Table 8.1 Improvement in response time and compliance with CSI

Industry	*Before CSI (avg. response time)*	*After CSI (avg. response time)*	*Reduction in breach costs*	*Improved compliance (%)*
Healthcare	12 hours	5 hours	40%	90%
Finance	10 hours	4 hours	50%	95%
Retail	15 hours	6 hours	35%	85%

organization must meet these challenges by investing in training, assuming scalable solutions, and fine-tuning detection algorithmic programs to minimize false positives.

8.7.1 Future directions in cloud security intelligence

The future of CSI is shaped by the speedy advances of technologies such as AI, ML, and blockchain. These conceptions have the potential to significantly enhance CSI capabilities. For instance, AI and ML can improve the accuracy of threat detection by analyzing vast amounts of data points and identifying subtle patterns that may indicate potential threats. Blockchain technology can heighten data integrity and transparency, making it easier to prevent and avoid security incidents. Additionally, the automation of the security process will continue to evolve, enabling quick responses to incidents and reducing the burden on security teams. Next-generation security solutions and their benefits are given in Table 8.2.

Another promising focus is the integration of CSI with emerging technology such as IoT and 5G networks. As IoT devices become more dominant, they bring in new vulnerabilities that can be tapped by cybercriminals. CSI can play a critical role in securing this device by providing existing dimension monitoring and threat detection. Likewise, the rollout of the 5G network will enable faster data transmission and better connectivity, but it will also require robust security measures to protect against new eccentric of attacks. By integrating CSI with these technologies, organizations can build an impenetrable and resilient computing environment.

8.7.2 Integration of challenges and future directions

While challenges such as integration complexity, skill training, and false positives must be addressed, the future of CSI is hopeful, moving forward with AI, ML, blockchain, IoT, and 5G. By overcoming these challenges and

Table 8.2 Next-generation security solutions and their benefits

Future advancement	*Description*	*Expected benefits*
AI integration	Automated threat detection and response	Faster incident resolution, reduced risk
Machine learning	Adaptive security models based on behavioral analysis	Improved threat prediction and prevention
Blockchain security	Decentralized identity management and data integrity	Enhanced data protection and transparency
Zero-trust model	Continuous verification of user access	Reduced insider threats and data breaches
Quantum-safe encryption	Advanced cryptographic techniques to counter quantum threats	Future-proof data security

embracing future innovation, the industry can unlock the wide potential of CSI, and ensure a secure, live, and sustainable cloud environment. These forward-looking advances will not only gain current security measures but likewise prepare organizations to tackle future threats in an increasingly interconnected world.

REFERENCES

1. Zhang, S., Pandey, A., Luo, X., Powell, M., Banerji, R., Fan, L., ... & Luzcando, E. (2022). Practical adoption of cloud computing in power systems—Drivers, challenges, guidance, and real-world use cases. *IEEE Transactions on Smart Grid*, 13(3), 2390–2411.
2. Cybersecurity, C. I. (2018). *Framework for Improving Critical Infrastructure Cybersecurity*. URL: https://nvlpubs. nist. gov/nistpubs/CSWP/NIST. CSWP, 4162018(7).
3. Ponemon, I. (2019). Cost of a data breach report. Benchmark research sponsored by IBM Security Independently conducted by Ponemon Institute LLC.
4. Kambourakis, G., Neisse, R., & Nai-Fovino, I. (2021). Information security in the age of EU-Institutions digitalisation, a landscape analysis. *European Commission, Joint Research Centre*. Gartner.
5. Quadrant, M. (2016). Magic quadrant for security information and event management. *Magic Quadrant*.
6. International Organization for Standardization. (2013). ISO/IEC 27001: 2013: Information Technology--Security Techniques--Information Security Management Systems--Requirements. *International Organization for Standardization*. Cloud Security Alliance (CSA). (2020).
7. Bhadra, S., & Mohammed, S. (2020). Cloud computing threats and risks: uncertainty and unconrollability in the risk society. *Electronics of Journal*, 7, 1047–1071. *Amazon Web Services (AWS)*. (2021). Veijanen, J. (2020). Implementation of security best practices on AWS Cloud: Case: Vulnerability scanning of EC2 instances and networks.
8. Hassan, M., Zmij, K., Azhygulov, K., & Sitaula, S. (2022). Cloud computing services and Microsoft Azure. Why Microsoft Azure? Available at SSRN 4103377.
9. Boneder, S. (2023). Evaluation and comparison of the security offerings of the big three cloud service providers Amazon Web Services, Microsoft Azure and Google Cloud Platform (Doctoral dissertation, Technische Hochschule Ingolstadt).
10. Castaldo, C. (2021). *Start-Up Secure: Baking Cybersecurity into Your Company from Founding to Exit*. John Wiley & Sons.
11. Jorepalli, S. K. R. (2025). Cloud-native AI applications designing resilient network architectures for scalable AI workloads in smart education. In *Smart Education and Sustainable Learning Environments in Smart Cities* (pp. 155–172). IGI Global Scientific Publishing.

12. Upreti, K., Vargis, B. K., Jain, R., & Upadhyaya, M. (2021, May). Analytical study on performance of cloud computing with respect to data security. In *2021 5th International Conference on Intelligent Computing and Control Systems (ICICCS)* (pp. 96–101). IEEE.
13. Kumar, N., Upreti, K., & Mohan, D. (2022). Blockchain adoption for provenance and traceability in the retail food supply chain: a consumer perspective. *International Journal of E-Business Research (IJEBR)*, 18(2), 1–17.
14. Syed, M. H., Upreti, K., Nasir, M. S., Alam, M. S., & Kumar Sharma, A. (2023). Addressing image and Poisson noise deconvolution problem using deep learning approaches. *Computational Intelligence*, 39(4), 577–591.
15. Judijanto, L., Hindarto, D., & Wahjono, S. I. (2023). Edge of enterprise architecture in addressing cyber security threats and business risks. *International Journal Software Engineering and Computer Science (IJSECS)*, 3(3), 386–396.
16. Olawale, M. A., Ayeh, A. A., Adekola, F. O., Precious, A. S., Joshua, A. O., & Timothy, O. (2023). A review on the intersection of artificial intelligence on building resilient infrastructure, promoting inclusive and sustainable industrialization and fostering innovation. *International Journal of Modern Engineering and Research Technology*, 9(3), 1–31.
17. Nandini, K., Yaramsetty, A., & Tulasirama, M. (2024). Enhancing cybersecurity resilience: a study of threat detection and mitigation techniques in modern networks. *Library of Progress-Library Science, Information Technology & Computer*, 44(3), 1–21.
18. Khan, R. S., Sirazy, M. R. M., Das, R., & Rahman, S. (2022). An ai and ml-enabled framework for proactive risk mitigation and resilience optimization in global supply chains during national emergencies. *Sage Science Review of Applied Machine Learning*, 5(2), 127–144.
19. Rehan, H. (2024). The future of electric vehicles: navigating the intersection of AI, cloud technology, and cybersecurity. *Valley International Journal Digital Library*, 12(4), 1127–1143.
20. Tan, L., Yu, K., Ming, F., Cheng, X., & Srivastava, G. (2021). Secure and resilient artificial intelligence of things: a HoneyNet approach for threat detection and situational awareness. *IEEE Consumer Electronics Magazine*, 11(3), 69–78.
21. Kaswan, K. S., & Dhatterwal, J. S. (2021). The use of machine learning for sustainable and resilient buildings. *Digital Cities Roadmap: IoT-Based Architecture and Sustainable Buildings*, 1, 1–62.
22. Reddy, A. S. (2022). *Building Resilient Digital Insurance Ecosystems: Guidewire, Cloud, and Cybersecurity Strategies*. PhilPapers, Canada.
23. Rashid, A., Rasheed, R., Ngah, A. H., & Amirah, N. A. (2024). Unleashing the power of cloud adoption and artificial intelligence in optimizing resilience and sustainable manufacturing supply chain in the USA. *Journal of Manufacturing Technology Management*, (ahead-of-print).
24. Olabanji, S. O., Marquis, Y., Adigwe, C. S., Ajayi, S. A., Oladoyinbo, T. O., & Olaniyi, O. O. (2024). AI-driven cloud security: examining the impact of user behavior analysis on threat detection. *Asian Journal of Research in Computer Science*, 17(3), 57–74.

25. Ahmed, T., Karmaker, C. L., Nasir, S. B., Moktadir, M. A., & Paul, S. K. (2023). Modeling the artificial intelligence-based imperatives of industry 5.0 towards resilient supply chains: a post-COVID-19 pandemic perspective. *Computers & Industrial Engineering*, 177, 109055.
26. Bechtsis, D., Tsolakis, N., Iakovou, E., & Vlachos, D. (2022). Data-driven secure, resilient and sustainable supply chains: gaps, opportunities, and a new generalised data sharing and data monetisation framework. *International Journal of Production Research*, 60(14), 4397–4417.
27. Moustafa, N., Adi, E., Turnbull, B., & Hu, J. (2018). A new threat intelligence scheme for safeguarding industry 4.0 systems. *IEEE Access*, 6, 32910–32924.
28. Dohale, V., Akarte, M., Gunasekaran, A., & Verma, P. (2024). Exploring the role of artificial intelligence in building production resilience: learnings from the COVID-19 pandemic. *International Journal of Production Research*, 62(15), 5472–5488.
29. Ivanov, D. (2023). The Industry 5.0 framework: viability-based integration of the resilience, sustainability, and human-centricity perspectives. *International Journal of Production Research*, 61(5), 1683–1695.
30. Mao, H., Zhang, T., & Tang, Q. (2021). Research framework for determining how artificial intelligence enables information technology service management for business model resilience. *Sustainability*, 13(20), 11496.
31. Dey, P. K., Chowdhury, S., Abadie, A., Vann Yaroson, E., & Sarkar, S. (2024). Artificial intelligence-driven supply chain resilience in Vietnamese manufacturing small-and medium-sized enterprises. *International Journal of Production Research,* 62(15), 5417–5456.
32. Steingartner, W., Galinec, D., & Kozina, A. (2021). Threat defense: Cyber deception approach and education for resilience in hybrid threats model. *Symmetry,* 13(4), 597.
33. Al-Hawawreh, M., Moustafa, N., Garg, S., & Hossain, M. S. (2020). Deep learning-enabled threat intelligence scheme in the internet of things networks. *IEEE Transactions on Network Science and Engineering*, 8(4), 2968–2981.
34. Fisher, O., Watson, N., Porcu, L., Bacon, D., Rigley, M., & Gomes, R. L. (2018). Cloud manufacturing as a sustainable process manufacturing route. *Journal of Manufacturing Systems*, 47, 53–68.
35. Shah, H. M., Gardas, B. B., Narwane, V. S., & Mehta, H. S. (2023). The contemporary state of big data analytics and artificial intelligence towards intelligent supply chain risk management: a comprehensive review. *Kybernetes*, 52(5), 1643–1697.
36. Jimmy, F. N. U. (2023). Cloud security posture management: tools and techniques. *Journal of Knowledge Learning and Science Technology*, 2(3), 1–23. ISSN: 2959-6386 (online).
37. Borangiu, T., Trentesaux, D., Thomas, A., Leitão, P., & Barata, J. (2019). Digital transformation of manufacturing through cloud services and resource virtualization. *Computers in Industry*, 108, 150–162.
38. Belhadi, A., Mani, V., Kamble, S. S., Khan, S. A. R., & Verma, S. (2024). Artificial intelligence-driven innovation for enhancing supply chain resilience and performance under the effect of supply chain dynamism: an empirical investigation. *Annals of Operations Research*, 333(2), 627–652.

39. Offner, K. L., Sitnikova, E., Joiner, K., & MacIntyre, C. R. (2021). Towards understanding cybersecurity capability in Australian healthcare organisations: a systematic review of recent trends, threats and mitigation. *Health Security Intelligence*, 1, 92–121.

Chapter 9

Integrating cyber-physical systems with intelligent transportation

Challenges and opportunities

Daksh Alfred, Amit Kumar, Shitiz Upreti, Pratham Nijhawan, and Faraz Ahmad

9.1 INTRODUCTION

The integration of cyber-physical systems (CPS) with intelligent transportation systems (ITS) marks a new era in urban mobility. CPS, through the seamless integration of physical components such as vehicles, roads, and traffic systems with computational elements like Internet of Things (IoT), AI, and cloud computing, offers innovative solutions to modern transportation challenges. The advent of IoT has brought about a revolutionary age of connectivity, allowing for smooth communication between humans and machines. In the current technological landscape, IoT is a shining example of innovation, facilitating the flow of digital information [1]. As smart cities evolve, CPS emerges as a pivotal technology, enhancing safety, efficiency, and sustainability across transportation networks. From real-time traffic management to autonomous vehicles, CPS allows dynamic data-driven decision-making, decreasing congestion and improving the commuter experience.

The architectural design of CPS-ITS covers multilayered perception, communication, computation, control, and application layers that serve to uniquely contribute to its functionality. The perception layer captures high-resolution data from sensors and vehicles, while the communication layer enables vehicle-to-vehicle (V2V) and vehicle-to-infrastructure (V2I) data exchange. Advanced computing frameworks, including edge and cloud computing, process these streams of data in real time, allowing for effective in-control mechanisms for adaptive traffic management. User-centric applications, including smart parking systems and electronic tolling, enhance operational efficiency and safety.

However, combining CPS and ITS is not without challenges. Cybersecurity threats as well as data privacy issues exist, as do compatibility-related issues in the integration into existing infrastructures. Solutions to ensure these systems in terms of cybersecurity and ensuring data privacy can be based on progress in encryption and blockchain protocol development. Collaboration among a city's planners, as well as between tech corporations

 DOI: 10.1201/9781003569251-9

and transportation authorities, gives CPS-ITS the basis for developing a sustainable, resilient, intelligent transportation ecosystem, an integral part of the coming smart cities.

9.2 LITERATURE REVIEW

A high-speed network infrastructure, IoT upgrades, and numerous modifications on automobiles have created immense demand for 5G technology in vehicles. These new technologies have now made the future of a 5G car a reality. The results of this process might shock you. How often do car problems like bald tires, poor steering, or malfunctioning brakes bring about highway or traffic accidents? The US Department of Transportation (DOT) found that human error causes nearly 94% of automobile accidents; however, mechanical failure occurs in approximately only 2% of automobile accidents. From a closer look at the data, one in four deaths is attributed to speeding. Statistics also back up the notion that if one-fourth of all fatalities from car accidents had been prevented, then car accident rates and fatalities may have been averted. Therefore, networked cars will increase traffic safety. Actually, a self-driving car is no different than a networked car. A connected car is one that can draw information from other sources and/or communicate the same back to the user's mobile device. Automobiles these days are increasingly packed with connectivity features.

The concept of a "smart city"—integrating technology, innovation, and globalization—was initiated in the early 1990s and is an innovative way of designing and developing cities. Smart cities evolved from the notions of smart growth by supporting improved urban planning and effective utilization of Wi-Fi-friendly technology. These cities also promote the understanding and control of urban systems such as building management, public safety, transportation, and the supply of electricity and water through information and communications technology. The development of smart cities demonstrates a commitment to sustainability and resilience with the optimization of resource supply by using real-time data analytics for improved service delivery.

Current transportation infrastructure cannot now meet the conditions arising due to the extraordinarily high growth rate of cars over the last 20 years. Further integration of ITS operations has become more important. ITS aims to contribute toward more intelligent roads by reducing traffic congestion and improving smoothness and efficiency in traffic flow.

Smart cities have provided a framework that would include intelligent transportation systems in the improvement of the quality of life for inhabitants. Cities can optimize the transportation networks, facilitate the flow of traffic, cut congestion, and advance sustainability and safety with the integration of ITS within the smart city infrastructure. This requires an appropriate framework that would focus on the integration of a number of parameters, such as data management systems, intelligent traffic control systems, and communication networks. Shared data will allow collaboration among different stakeholders, including the transportation authority, city planners, and tech companies. There is a need to address the issues of privacy and security, especially concerning the integration of ITS with smart city infrastructure. The transportation systems should be more data-driven and networked. Good cybersecurity is required for this. Private data should be adequately protected, and the integrity of the system of transportation should be maintained.

9.3 METHODOLOGY

This research explores the concept of integrating CPS with ITS. The review is made in the form of a narrative where technological aspects, operational challenges, and architectural designs of CPS-ITS regarding the upgrading of transportation systems are explored. A systematic analysis was carried out on CPS frameworks, applications, and developments in the transportation system based on academic and commercial sources. The analysis focused on cybersecurity, the Internet of Vehicles (IoV), intelligent traffic management, autonomous cars, and enabling technologies such as IoT, AI, and 5G.

The first phase of the research relates to the classification of the various parts of CPS architecture for ITS. A multilayered CPS architecture was identified, beginning with the perception layer, which may involve sensors such as on-board units, roadside modules, and smartphones to capture high-resolution data on vehicle dynamics, road conditions, and environmental factors. This layer captures relevant information on transportation systems to inform and support real-time decision-making. Another key area of concentration was the communication layer, including V2I, V2V, and Dedicated Short-Range Communications (DSRC) technologies. This layer is basically the foundation for the automation of real-time traffic management. What it does is focus on dynamic transfers of data between the infrastructure and the vehicles. This forms the basis for the automation of real-time traffic management through a dynamic data exchange between infrastructure and vehicles.

The computing layer was further discussed, focusing on edge and cloud computing options for analytics and data processing. Cloud-based computing is important to real-time data analytics and decision-making because it allows distributed processing and load balancing. Similarly, edge computing allows one to process data closer to its sources, which becomes important as latency is increasingly required for applications such as traffic light control and driverless car deployment. This forms the base layer of CPS-ITS, making it now possible to process large volumes of data in adequate time with inputs from multiple sensors and vehicles.

The control layer and application layer, therefore, complement the CPS-ITS architecture. The control layer enables real-time optimization of the timing of traffic signals and controls congestion using adaptive traffic management techniques. Meanwhile, user-facing applications such as electronic tolling, emergency response, and real-time traffic monitoring reside in the application layer. Other than providing an enhanced commuter experience, such applications have safety features such as collision warnings and route optimization for emergency services.

The final step was the assessment of security and privacy concerns of CPS-ITS. According to research, because of the high risk of data breaches and cyberattacks, there is a critical need for strong cybersecurity. Two advanced alternatives identified within this review are blockchain and encryption-based protocol regarding IoV. Authentication procedures strengthen data exchange that ensures data integrity and privacy in the use of such data exchanges. As the systems become more complex as well as data-dependent, all the integral parts of any CPS-ITS system will be put into the forefront—their protocols.

Smart parking systems and Mobility as a Service (MaaS) are actual implementations of CPS-ITS technologies. These systems reduce emissions and traffic through the modernization of parking and public transport services using real-time data. The study also covered an algorithm for safe driving and shortest routes in self-driving cars. This method of evaluation reveals the high potential of developing safer, efficient, and sustainable transport systems through CPS-ITS.

9.4 TRANSPORTATION CYBER-PHYSICAL SYSTEMS (TCPS)

Transport is an area that is going to experience a major shift in smart cities in upcoming times. The advancements, functionality, and prosperity that an urban structure has depend on the way in which the city plans its transportation system [2]. Artificial intelligence promises to change the world's transport systems by addressing huge data integration and real-time decision-making through learning and developing a better adaptive solution. The framework of being "always connected" is rapidly taking over the world, and the connected vehicle has already turned out to be mainstream reality. So, we are seeking a city mobility environment that ensures no road mishaps, with less emissions and lowered air pollution, that provides safer and reliable transportation. Modern vehicles are the most independent ever built, smarter, secure, green-fuel-using, and constantly connected. Roads are no longer a mere infrastructure; they are now "empowered" with communicating information, predicting capabilities, generating power, measuring the moving vehicles, gathering tolls, etc. Therefore, the connected mobility mechanism planned to manage traffic and damage to the environment, and in providing effective, cleaner, quicker, and affordable transport services by leveraging innovations in networks of communication, IoT, cloud and edge computing, flexible retention, and insights driven by data may be called transportation cyber-physical systems (TCPS). This refers to the latest framework for future urban settings representing data-driven and shared mobility with autonomous infrastructure and connected vehicles [3].

9.5 SIGNIFICANCE OF TCPS

The dynamic urban mobility landscape of a smart city requires innovation, collaboration, and sustainability in mobility solutions for services such as autonomous driving, parking management, drive-sharing, ride-sharing, and traffic management. TCPS is highly complex and deal with enormous amounts of data, offers a wide range of services, and has multiple streams of data. TCPS is given the support of strong analytics in both static and time-data applications that include V2V and V2I data, V2X communication data, road infrastructure, data regarding road–rail–air transport, information regarding smart vehicles, and traffic monitoring networks. For that, they can make choices in real-time. Globally, these CPS-based technologies have totally transformed ITS [4].

9.6 TCPS TECHNOLOGIES IN THE CONTEXT OF INFRASTRUCTURE, CREDIBILITY, AND PERFORMANCE

Some of the studies dealt with the needs for integrated intelligent transport along with IT systems for the efficient execution of CPS to on-road transport systems. Facilities like remote-access laboratories and simulation experiments in which vehicles and roads simulating real-world transportation networks have pinpointed some of the limitations that occur while putting CPS into practice. Internet-centric, Semantic-centric, and Things-centric views are encompassed by the cyber-physical system [5]. This section examines and presents the problems of sensing, actuating, and communication in networks, especially toward intelligent traffic light systems for congested and uncongested traffic flow scenarios and travel-time prediction.

Contemporary transport systems call for cross-domain integration, including coordination of infrastructure, logistics, automotive, capacity, and governance. This ranges from intelligent transport systems to smart tires, roads, trains, and aviation, all connected through communication networks. With integration, logistics is the most important component in all industrial and societal processes, and "mobility as a service is fast accelerating." The best services are offered to society with the fewest possible human contacts thanks to the modern transportation solution built upon highly automated integrated systems handling the major challenges. However, with the demographic increasing in age, quality, reliability, security, accessibility, and safety become determinants for the uptake of new transport services. But, factors also include customization, deep-level inherent complexities of the transport system architecture, the need for new features leading to changing landscapes, safety and security, real-time pattern analysis of accident statistics, and constant vehicle updates or avoidance in-app.

By using MAC addresses and route optimization, which specifies the best bus route, smartphone data is analyzed to calculate the waiting time. Among the many security threats to the system that were identified included data collection, data transfer, fictitious data formation, and flooding of the cloud storage. Another significant component of any intelligent transport system would be maintaining the transportation infrastructure by employing intelligent systems.

9.7 COMMON EXAMPLE OF CPS-ITS: SMART PARKING SYSTEM

Perhaps one of the best examples of CPS-ITS is smart parking systems, which involve dynamic interactions between infrastructure and vehicles. This flexibility needs to respond to environmental changes in real-time concerning shifting traffic patterns and changes in available parking.

Limited parking availability in historic city centers causes drivers to roam around for free spaces. A lot of fuel is consumed, and traffic congestion and frustration increase. This scenario is being addressed by using linked vehicles to look for parking spaces together. Automobiles with cameras capture pictures of the surrounding area and send the information to an edge cloud server that uses the information to provide instant parking availability. The server uses the actual real-time positions of cars to assist drivers in finding the nearest parking space at the time it is needed to minimize detours and save much time. This method requires low-latency transfer because delays in processing or communication can possibly interfere with timely parking support for smooth operations [6].

Smart parking systems inform drivers about the real-time occupancy and fees of the parking lots. They help drivers book a place in an available parking lot in advance and select a suitable lot along a route. In this approach, networked cars equipped with cameras capture images of their surroundings and report to the edge cloud server about their locations. When a car user needs a parking space, the server identifies nearby cars that may capture and transfer images of likely parking spots to help pinpoint an empty space. Thus, the server collaborates with other cars in the surrounding area to direct the car to the nearest available space with fewer detours to make the car easily navigate its way to the assigned space. This method optimizes parking and saves unnecessary movement of vehicles, thereby saving fuel, and facilitates the reduction in traffic congestion. For the scanning cars, the edge cloud server, and parking cars to communicate swiftly, the system primarily relies on low-latency mobile broadband data connections. Delays between the time of capturing the image and getting a result in the event of a high latency or network disruptions may severely degrade the system's performance or even cause suboptimal route changes or missed parking chances. This CPS strategy avoids last-minute changes that are difficult for drivers to follow in narrow streets or streets that function as one-way roads by accelerating response times. The design process has to be built with thoughtful consideration of the network and processing capacity in a CPS-based smart parking system. Such properties assessed at design time help predict latent performance weaknesses, thereby enabling changes before deployment by tools such as IMPACt. These simulations also open up the possibility of parameter tuning, which maximally optimizes the system for deployment in real-world applications and has less effect on latency on the driving experience [6].

9.8 INTEGRATION OF CYBER-PHYSICAL SYSTEMS AND INTELLIGENT TRANSPORTATION

The integration of IoV and cyber-physical surveillance systems (CPSS) marks a giant step toward improving efficiency and safety in transportation. The integration of IoV and CPSS has improved traffic management and the law enforcement/emergency response capabilities, as it allows real-time road conditions and incidents to be supervised. In the actual implementation within IoV frameworks, CPSS facilitates safer and efficient transportation systems by supplying timely information that would assist in mitigating risks and managing traffic flow effectively. A comprehensive cyber-physical framework for traffic control, one designed specifically for smart cities, has been introduced. The system offers improved traffic flow and safety protocols due to the integration of sensors, IoT devices, and data analytics. IoT has been an instrument for certain architectural visions for smart power grids, intelligent transport systems, and innovative city developments [7]. It coordinates communication between connected cars and infrastructure toward reducing traffic as well as improving overall road safety. This system ranks as one of the most promising solutions for urban transportation problems due to scalability and adaptability. In this regard, they minimize emissions and delays while upholding sustainability objectives. Hardware advancements must also be explored as possible frameworks through which CPS may be tested.

A field programmable gate array (FPGA)-based research in autonomous vehicles and ITS may be executed on an embedded platform that supports the processing of real-time data and sensor integration with commercial off-the-shelf (COTS) components that save costs [8]. The authors state that to enhance even further the capabilities of intelligent CPS within transportation applications, advanced cybersecurity protocols and machine learning techniques may represent future improvements. Finally, there are significant contributions of deep learning techniques to Intelligent Transportation Systems (ITS). Deep models perform in ways that may not be obtained with classical techniques with regard to enhancing traffic detection and prediction. Real-time traffic forecasting is actually possible through the AHCNLS (attention-based hybrid convolutional neural network with long short-term memory) framework provided by the deep learning model that successfully captures both temporal and spatial patterns. The results show a potential candidate for wider applications in traffic safety and management since it has important accuracy above the current models. CPS is designed to interact with the physical world by interfacing through communication networks, sensors, and actuators. In the transport field, CPS makes it easier to gather and examine huge volumes of data from various sources, including automobiles, traffic signals, and environmental sensors. Such an approach has improved response and flexibility in transportation systems through real-time information processing toward better decision-making. Intelligent traffic management solutions that can make the necessary readjustments based upon changing road conditions are enabled by TCPS, which integrate revolutionary technologies such as AI, ML, and IoT-based devices. Enhanced insights fostered to improve and ensure a safer flow of traffic are characteristics of the architecture in TCPS. A new technology to realize the future development of driverless cars is called intelligent speed assistance, or ISA. In ISA, a global positioning system is used to determine the speed of a vehicle. This information is compared to data previously established on the speed limits. The ISA system can provide a warning to the driver of excessive speed of the vehicle and, in some configurations of it, actually decelerate if it determines that this is warranted. The ability to visually identify speed limit signs on a map will be key for any vehicle that is considered automated. AI and machine learning capabilities will allow self-driving cars to recognize speed limit signs, but probably not nearly enough of them. The probability that signs are not present or are destroyed or do not exist is greater in low-density areas. An interesting new function that is now being fitted to cars, such as the BMW, is "lane assist." A car can at any time be located at a given location on the road by using data from cameras, radar, or a global positioning system (GPS). With this technology advancing, connected car technology is underway as all the pieces needed for 5G to be applied in automotive applications come into the picture. According to the blog post titled "What Is Connected Vehicle Technology and What Are the Use Cases?" ITS groups and various transportation departments are working together to lay the mission-critical communications infrastructure for connected vehicles. It will definitely benefit the smart cities as they move forward in planning for the future.

The architecture of CPS-ITS has multiple layers, comprising special functions to enhance the efficiency of transportation management in an interlinked manner. The perception layer forms the bottom and makes use of sensors; onboard units, roadside modules, smartphones, and GPS are used to collect comprehensive information from a wide range of sources, including cars, roadways, and passengers. This layer gathers very high-resolution data-including vehicle speed, location, and road conditions, which is the main input for understanding traffic flow and commuter behavior. In this regard, DSRC, V2V, and V2I are some of the technologies used by the communication layer to transmit it. With these communication methods providing a seamless flow of traffic information between vehicles and infrastructure, judgments can be made by drivers and traffic authorities in real time.

These large volumes of data are then processed on robust computing platforms offered in the computing layer, which frequently resort to cloud-based solutions to achieve load balancing, data processing, and storage. Such an infrastructure is crucial for the real-time data analysis capabilities of CPS-ITS, thus enhancing its response accuracy and eventually sophisticated data modeling. Adaptive control techniques enable the control layer to optimize traffic management with the assistance of insights from data analysis. The old systems had fixed controls, but in CPS-ITS, it now attains dynamic controls of traffic signals and flows according to conditions in real-time. It develops responsiveness and adaptability in systems.

Finally, the application layer integrates all functionalities of the layers discussed earlier to deliver services that enhance the experience of road users. Applications such as advanced traffic management, electronic toll collection, smart parking systems, and crash prevention tools tap into information from underlying layers to build a safer, more efficient environment for commuters. The tiered architecture of CPS-ITS ensures information, which passes through the first perception stage, is processed, controlled, and eventually applied with more precision and effectiveness to solve today's transportation problems [9]. MaaS-based CPS technology has brought together various modes of transportation, ticketing systems, and web–client applications in order to satisfy users [10, 11]. It can control traffic, which reduces air pollution, traffic jams, and accidents. Many errors on the road are caused by humans [12].

Some advantages of TCPS can be stated as follows. Including technology in cities and networks of transportation will improve the overall efficiency of cities and their transportation systems as well as reduce pollution and traffic, and consequently, improve citizens' quality of life. Among the technologies that smart cities embrace are smart lighting, IoT, energy management, artificial intelligence, and smart transportation systems (STS) to make their urban areas sustainable and efficient. Development of intelligent transportation systems contributes directly to the creation of modern smart cities and the reciprocal application of cutting-edge technologies for the good of the state and society. Such technologies are developed to be capable of contributing to the easier traveling of both residents and visitors while optimizing the transport networks, enhancing the fluidity of traffic flow, reducing emissions and congestion, and providing improved safety and security.

Also, through integration with smart city infrastructure, public safety is developed and enhanced. Moreover, through ITS, intelligent transportation systems can detect abnormalities in traffic patterns as well as alert authorities of potential dangers that exist. Added to this is that the real-time traffic information shared with emergency services can enable them to respond quickly and effectively in response to emergencies.

Integrating ITS with smart city infrastructure is crucial for building livable, sustainable, and effective urban settings. Data analytics helps cities improve public safety, optimize transportation networks, and encourage sustainable mobility practices as well as cutting-edge technologies. Therefore, this chapter introduces a comprehensive architecture for the integration of ITS into smart city infrastructure, offering a roadmap to cities for building a smarter and more sustainable transport system. Thus, merging ITS with smart city infrastructure into a solid system will provide for efficiency, sustainability, and livability in an urban environment, thus determining betterment of the quality of life for citizens and a better future for all.

A safe and effective transportation system is what distinguishes any smart city. Thus, ITS stands as one of the foundational underpinnings needed to support the overarching concept of a smart city. Although legacy ITS technologies prevail, the future generation of ITS relies on integrating connected and autonomous vehicles.

9.9 SECURITY CONCERNS IN INTELLIGENT TRANSPORTATION SYSTEMS

The fast pace of development of connectivity and automation in transportation systems, as seen through the advent of TCPS, has substantially elevated the stakes of safe data exchange [13]. Merging digital technologies and operational technology assets brought with it and amplified the cybersecurity threat to the critical manufacturing sector from a quickly growing and evolving cyberattack surface [14]. Thanks to IoT technologies, ITS safety has lately become much-talked-about news. Autonomous vehicle control and communications have been enabled through onboard sensors, computers, and communication capabilities. Large volumes of data, including HD videos, can be sent to an information center or operation via transportation system networks. Data transfers in case of emergency situations are adapted in new 5G-connected IoT services [15]. These notifications make transport safety and efficiency in management possible by acknowledging traffic conditions and giving control as part of the intelligent infrastructure. es often do not have vehicular ad-hoc network (VANETs) that completely collaborate and rely on one another because of cultural or business disputes. In mass-scale VANET implementations, malicious behavior or misbehavior is not acceptable. Through a road collision, malicious automobile nodes can propagate unreliable information that can be damaging to the other vehicles. VANETs must provide a secure 5G-enabled IoT ecosystem wherein valid transport mediums and associated infrastructures are sending service messages with CPS. Messages have to be communicated that are tamper-proof, trustworthy, and authentic. As no user can accept data loss or compromise in user privacy, there should be secure delivery of the messages. In addition, future 5G-enabled IoT vehicles will have a demand for central trust, cooperation, and time sensitivity [16]. This section introduces real-time CPS, cloud-supported reporting, and trusted communications in this context. The cars of 5G-VANET require authentication. The original public and private keys are issued to an authentication authority for the authentication of the legality of the vehicle. This information can only be transmitted after approval. Similar to this, videos and messages regarding the traffic conditions will be broadcast. The technology separates user identification information from vehicle authentication information. Through this, the identities of the operators and other users are kept anonymous. Videos are recorded based on traffic and a car. In case of an accident, communications and video related to traffic need to be investigated. Plans are set to introduce a blockchain trust management system. If the Intelligent Transportation System (ITS) is uploading the video, then the road conditions tag and the tag for communicating with other cars will be uploaded simultaneously. The nearest automobiles will then report the truthfulness of the broadcast traffic information to prevent the allocated vehicle from disturbing traffic by broadcasting any false or harmful information. Provided the documentation of efforts and participation in frequent elections of this system, the system will be traced only with the same traffic vehicle tracking system, which transmits a huge amount of false road information and distorts the traffic rather than any representation of traffic. The system simulates the semicentralized trust model with decentralized administration and centralized authentication to ensure safe and sustainable provision of ITC for future smart cities, preserving at the same time the privacy and anonymity of authorized vehicles.

The system can process real-time service requests. This is a very important factor that makes a user's experience much more convenient. Travelers can learn what local amenities are available and how far shops, restaurants, or hotels are while traveling. With users constantly providing sensitive location and preference data, this capability presents data privacy problems. Current mechanisms for privacy protection, such as k-anonymity, aim at protecting user data. These mechanisms mostly fail when applied to a self-driving car due to the limited resources in terms of computation available on those cars. On these cars, achieving an appropriate balance between query utility and privacy protection in the design becomes quite challenging. This limits the possibility of building efficient privacy protection without impairing usability and performance in service requests. Fast processing and privacy protection are of equal importance to secure safe, dependable, and trustworthy autonomous vehicle operations, whose balance is necessary to ensure security as well as user confidence in intelligent transportation systems [17]. Integration of CPS with ITS has great potential benefits; however, important issues need to be addressed. High degrees of compatibility are needed to interconnect several technologies from older systems to the newest IoT devices. For a holistic operational framework, the different systems must be able to talk with one another. The intricacy might bring about costs and a longer deployment time.

Since CPS highly relies on heavy communication of data from cars, control centers, and infrastructure, they are prone to hacking. Private privacy and public safety are seriously under threat by the possibility of destructive attacks or data breaches. Therefore, it is very important to create a strong security procedure that safeguards private data while ensuring system functionality. It would therefore involve putting into use such encryption methods, secure communication channels, and performing frequent system audits to find weaknesses.

There are particular safety risks associated with integrating CPS and IoT into cooperative intelligent transportation systems (C-ITS). Also, in terms of its benefits, aside from its real-time monitoring advantages, this system provides several benefits as well: the complexity that C-ITS integrates through open-source solutions, machine learning, and AI. Using a real example of the Shuttle 2X project in making a case for enhanced safety measures for C-ITS, it is important to mend the critical gap between industry practice and academic research. This segmentation may make it challenging to adapt certain safety measures that are crucial to making the advanced devices reliable. Furthermore, crucial infrastructure constraints provide an important barrier to the proper integration of CPS into current transportation systems. Many of the infrastructures built today were not designed with cutting-edge technologies in mind; this causes a problem in compatibility when newer systems are added. These infrastructures require a vast amount of time and money in upgrading. Moreover, policymakers should put financing infrastructure upgrading, paving the way for deployable CPS technologies, at the top of their agenda, while keeping in mind the long-term sustainability of such an investment. It is great that privacy and security considerations must be taken into account while implementing

ITS as part of the infrastructure for a smart city. When the data is sensitive in nature, cybersecurity proves particularly relevant with the increasing interconnection and data-based nature of the system. There is a need for the integrity of the transport network to be maintained.

The transportation industry is highly vulnerable to cyberattacks, and crucial security measures need to be implemented sooner rather than later to safeguard user data and maintain the integrity of the system. There are also major issues with the real-time processing of data such as signal interference, quality of data, and process complexity. The dynamic traffic conditions and the need to integrate various sources make accuracy in traffic management challenging to attain. Still, an ITS faces numerous challenges that must be addressed [18]. TCPS consists of many constituencies: vehicles, road infrastructure, human drivers, machines, and sensors. Vehicular transportation networks are very dynamic because of the possibility of high speeds, leading to constantly changing topologies [9]. In addition, since drivers can enter and exit at will, the topology of the network directly depends on the behavior of human drivers.

9.10 OPPORTUNITIES FOR FURTHER DEVELOPMENT

With the advancements in communication technologies and the explosive development of internet services, CPS has expanded internationally and offered multifaceted advantages in nearly every field of human daily life in the last ten years [19]. The deployment of CPS in the road transport network also offers great opportunities. It can minimize pollutants, strengthen traffic control, and advance people's living standards on the whole. Vehicular CPS, in particular, rely on modern sensor technologies such as GPS, video cameras, etc. to collect information in real time. Such technologies improve the timeliness and accuracy of data, which are great benefits for managing traffic. This would give rise to huge economic benefits through improvements in infrastructure and services if the obstacles in data processing are overcome and effective traffic management is ensured [20].

Despite the challenges, there are plenty of opportunities for developing transportation systems by combining CPS and ITS. The most significant advantage is safety improvement. By utilizing real-time analytics and predictive modeling, it can detect possible road hazards before they become more catastrophic collisions. For instance, smart traffic management systems can evaluate environmental conditions and monitor the normal flow of traffic to promptly alert road users to possible dangers such as congestion of vehicles or adverse weather conditions. This would avoid the loss of many lives and injuries that are often witnessed with regard to traffic. CPS can also further optimize transport network performance. Real-time traffic flow information as used in smart traffic signal controls can help reduce travel times and ease congestion. Vehicle-to-infrastructure communication also enables drivers to be aware of conditions ahead of them for even more efficient route determination. These are bound to lead to better system-wide performance and to a much better experience for the individual traveler by making road use more efficient. In addition, it has also been increasingly known that CPS can be very effective in developing sustainable transportation systems.

The potential of CPS to decrease pollutants related to automobile travel as well as fuel consumption is further aided by streamlining the routing and monitoring the performance of a vehicle through advanced analytics. Intelligent traffic management systems are used to prevent idle time at intersections and improve good driving habits that are not harmful and damaging to the environment and automobiles. In fact, CPS is strategically placed to spearhead the solutions to the world's most challenging issues, including those related to urbanization and climate change, since it meets the most ambitious goals related to sustainability.

All sensors are considered smart IoT devices, which are capable of collecting and sending data to the servers connected via either a wired or wireless internet connection. In this technique, collisions in several CPS can be removed. The following experiments demonstrate the usage of CPS in two remarkable case scenarios: intelligent parking lots and roads [20]. ITS technologies apply the current innovation to improve upon environmental sustainability, mobility, and safety in transportation. ITS helps customers enjoy safer, more informative, and more cost-effective transportation systems through the application of communication infrastructure, sophisticated computational techniques, autonomous vehicles, electric vehicles, linked vehicles, and intelligent traffic signals. With this overview of ITS technologies and new techniques, they can now be feasibly made possible in research. Case studies regarding applications of connected car technology, such as automated incident detection systems and queue warning, are included in the discussion as well. Transportation systems in the smart city model face extreme modernization. Smart solutions aim to reduce human interference, streamline control, facilitate maximized traffic flow, and create environmentally conscious cities.

9.11 CONCLUSION

Cyber-physical systems integrated with intelligent transportation systems redefine urban mobility by fostering safer, more efficient, and sustainable transportation networks. CPS with IoT-powered real-time data, AI-driven analytics, and robust communication frameworks enhances traffic management, enables autonomous vehicles, and promotes smart mobility services. These innovations address critical challenges like congestion and road safety while supporting the development of adaptive control and user-focused applications.

Despite the significant challenges that face cybersecurity, data privacy, and infrastructure compatibility, blockchain and advanced encryption open avenues for secure and efficient implementations. CPS-ITS marks a transformative leap toward sustainable urban transport, making for smarter, more livable cities in the digital age. By emphasizing collaboration and advanced technologies, this integration is a blueprint for addressing modern transportation challenges and unlocking the potential of intelligent mobility solutions.

REFERENCES

1. R. C. Poonia, K. Upreti, B. P. Alapatt, and S. Jafri, "Real-time cyber-physical risk management leveraging advanced security technologies," in *Lecture Notes in Networks and Systems*, 2024, pp. 339–350. doi: 10.1007/978-981-97-4581-4_25.
2. C. K. Toh, J. A. Sanguesa, J. C. Cano, and F. J. Martinez, "Advances in smart roads for future smart cities," *Proceedings of the Royal Society a Mathematical Physical and Engineering Sciences*, vol. 476, no. 2233, p. 20190439, Jan. 2020, doi: 10.1098/rspa.2019.0439.
3. Y. Yuan *et al.*, "Mobility-driven integration of heterogeneous urban cyber-physical systems under disruptive events," *IEEE Transactions on Mobile Computing*, vol. 22, no. 2, pp. 906–922, Jun. 2021, doi: 10.1109/tmc.2021.3091324
4. A. Nikitas, K. Michalakopoulou, E. T. Njoya, and D. Karampatzakis, Artificial intelligence, transport and the smart city: definitions and dimensions of a new mobility era. *Sustainability*, vol. 12, no. 7, p. 2789, 2020. doi: 10.3390/su12072789
5. A. Ahmad, V. Singh and K. Upreti, "Emotion Recognition Through Facial Expressions: A Machine Learning Perspective in Mobile Multimedia," *Journal of Mobile Multimedia*, vol. 21, no. 1, pp. 87–111, January 2025, doi: 10.13052/jmm1550-4646.2114.
6. T. Bures, V. Matena, R. Mirandola, L. Pagliari, and C. Trubiani, "Performance modelling of smart cyber-physical systems," *ICPE '18: Companion of the 2018 ACM/SPEC International Conference on Performance Engineering*, pp. 37–40, Apr. 2018, doi: 10.1145/3185768.3186306.
7. K. Upreti, S. Arora, A. K. Sharma, A. K. Pandey, K. K. Sharma and M. Dayal, "Wave Height Forecasting Over Ocean of Things Based on Machine Learning Techniques: An Application for Ocean Renewable Energy Generation," in *IEEE Journal of Oceanic Engineering*, vol. 49, no. 2, pp. 430–445, April 2024, doi: 10.1109/JOE.2023.3314090.
8. C. Conrad, S. Al-Rubaye, and A. Tsourdos, "Intelligent embedded systems platform for vehicular cyber-physical systems," *Electronics*, vol. 12, no. 13, p. 2908, Jul. 2023, doi: 10.3390/electronics12132908.

9. W. Guo, Y. Zhang, and L. Li, "The integration of CPS, CPSS, and ITS: a focus on data," *Tsinghua Science & Technology*, vol. 20, no. 4, pp. 327–335, Aug. 2015, doi: 10.1109/tst.2015.7173449.
10. L. Deka, S. M. Khan, M. Chowdhury, and N. Ayres, "Transportation cyber-physical system and its importance for future mobility," in *Elsevier eBooks*, 2018, pp. 1–20. doi: 10.1016/b978-0-12-814295-0.00001-0.
11. P. Bellini, P. Nesi, and G. Pantaleo, "IoT-Enabled smart cities: a review of concepts, frameworks and key technologies," *Applied Sciences*, vol. 12, no. 3, p. 1607, Feb. 2022, doi: 10.3390/app12031607.
12. A. Khattak, N. Ahmad, B. Wali, E. Dumbaugh, The University of Tennessee, Knoxville, and Florida Atlantic University, "Developing a taxonomy of human errors and violations that lead to crashes," *Collaborative Sciences Center for Road Safety*, Technical Report, Nov. 2019. [Online]. Available: https://www.roadsafety.unc.edu/wp-content/uploads/2021/04/Technical-Report_Taxonomy_CSCRS_Final-2020.pdf
13. A. A. Mamun, A. Abrar, M. Rahman, M. S. Salek, and M. Chowdhury, "Enhancing transportation cyber-physical systems security: a shift to post-quantum cryptography," *arXiv (Cornell University)*, Nov. 2024, doi: 10.48550/arxiv.2411.13023.
14. M. H. Rahman and M. Shafae, "Cyber-physical security vulnerabilities identification and classification in smart manufacturing -- a defense-in-depth driven framework and taxonomy," *arXiv (Cornell University)*, Dec. 2024, doi: 10.48550/arxiv.2501.09023.
15. R. A. Memon, J. P. Li, J. Ahmed, M. I. Nazeer, M. Ismail, and K. Ali, "Cloud-based vs. blockchain-based IoT: a comparative survey and way forward," *Frontiers of Information Technology & Electronic Engineering*, vol. 21, no. 4, pp. 563–586, Apr. 2020, doi: 10.1631/fitee.1800343.
16. K. Kuru and D. Ansell, "TCitySmartF: a comprehensive systematic framework for transforming cities into smart cities," *IEEE Access*, vol. 8, pp. 18615–18644, Jan. 2020, doi: 10.1109/access.2020.2967777.
17. J. Wang, Z. Cai, and J. Yu, "Achieving personalized k-anonymity-based content privacy for autonomous vehicles in CPS," *IEEE Transactions on Industrial Informatics*, vol. 16, no. 6, pp. 4242–4251, Oct. 2019, doi: 10.1109/tii.2019.2950057.
18. A. Pundir, S. Singh, M. Kumar, A. Bafila, and G. J. Saxena, "Cyber-physical systems enabled transport networks in smart cities: challenges and enabling technologies of the new mobility era," *IEEE Access*, vol. 10, pp. 16350–16364, Jan. 2022, doi: 10.1109/access.2022.3147323.
19. D. Wang, F. Li, K. Liu, and X. Zhang, "Real-time cyber-physical security solution leveraging an integrated learning-based approach," *ACM Transactions on Sensor Networks*, vol. 20, pp. 1–22, Jan. 2023, doi: 10.1145/3582009.
20. J. Christidis, P. A. Karkazis, P. Papadopoulos, and H. C. Leligou, "Decentralized blockchain-based IoT data marketplaces," *Journal of Sensor and Actuator Networks*, vol. 11, no. 3, p. 39, Jul. 2022, doi: 10.3390/jsan11030039.

Chapter 10

Defending against threats

Machine learning approaches for malware detection and analysis

Amit Kumar Singh, Arpita Banerjee, and C. Banerjee

10.1 INTRODUCTION

Malware has affected a vast number of computing devices in the digital age. Malware is derived from malicious software that is meant to suit the detrimental intent of a hacker [1] [3]. Malware can corrupt computers/smart devices, steal confidential information, breach networks, and cripple vital infrastructures, among other things [6–9]. These programs include viruses, worms, trojans, spyware, bots, rootkits, ransomware, and other malicious software [11–12].

With the rapid expansion of the internet, malware has evolved into one of the most serious cyber risks today. Malware is any software that performs malicious operations such as information theft and espionage. According to Kaspersky Labs (2017), malware is "a type of computer program designed to infect a legitimate user's computer and cause harm in a variety of ways" [10].

As a result, malware protection for computer systems is one of the most critical cybersecurity jobs for both individual users and enterprises, because even a single attack can result in lost data. Significant losses and the frequency of attacks necessitate the use of accurate and fast detection technologies.

Current static and dynamic detection methods are inefficient, especially when dealing with zero-day threats. As a result, machine learning-based techniques can be applied. This chapter examines the key aspects and challenges associated with machine learning-based malware detection, as well as the optimal feature representation and classification approaches [13].

10.2 BACKGROUND

Machine learning (ML) is a branch of artificial intelligence that focuses on developing applications e by learning from data without explicitly

DOI: 10.1201/9781003569251-10

programming how to perform learned tasks. Traditional ML methods make predictions based on past data. The ML process life cycle consists of several sequential phases. These are data extraction, data preprocessing, feature selection, model training, model evaluation, and model deployment [13].

The various subcategories of ML are supervised learning, unsupervised learning, semisupervised learning, reinforcement learning, and deep learning [14].

The supervised learning strategy trains the model using a labeled dataset to solve classification and regression problems depending on the output variable type (continuous or discrete). Unsupervised learning is used to find the internal structures (clusters) and properties of a dataset, and it does not require a labeled dataset to train the model. Semisupervised learning employs a combination of supervised and unsupervised learning techniques and is employed when the dataset has minimal labeled data [15].

Machine learning can improve malware detection by allowing systems to learn from vast datasets, extract significant patterns and characteristics, and adapt to evolving malicious behavior. It is possible to construct models that accurately classify and distinguish between malware and benign software samples by training machine learning algorithms on labeled datasets comprising both malware and benign software samples. Furthermore, machine learning models can detect anomalies and deviations from typical behaviors, allowing the detection of previously unknown or zero-day assaults [16].

The objective of this research work is to explore the application of machine learning in malware detection. By leveraging the most advanced machine learning algorithms and techniques, we aim to develop an accurate and robust model that can efficiently identify and classify malware patterns. This model will enhance the capabilities of existing malware detection systems, improve detection rates, reduce false positives, and provide rapid and accurate identification of emerging threats. Algorithms such as logistic regression, support vector machines (SVM), Gaussian naïve Bayes, AdaBoost, and linear regression algorithms were used in this research work to predict accuracy in detecting malware. A machine is trained on the models, their accuracies are differentiated, and a confusion matrix is created accordingly.

10.3 METHODOLOGY

This chapter introduces the different steps and components of a typical machine learning workflow for malware detection and classification, explores the challenges and limitations of that workflow, and evaluates the latest innovations and trends in the field, with a focus on deep learning techniques. The proposed research method for this study is presented next [4]. To provide a more comprehensive understanding of the recommended machine-learning approach for malware detection, Figure 10.1 illustrates the end-to-end workflow.

Figure 10.1 Proposed methodology

10.4 COLLECTION OF DATA

In this research work, the dataset is collected from Kaggle. One dataset with a total of 373 samples, of which 301 are malicious files and 72 are non-malicious, was downloaded. There are 532 features represented from F_1 to F_531 and a label column indicating whether the file is malicious or non-malicious [17].

10.4.1 Initialization: Collecting the dataset

To begin, we imported libraries including NumPy, Pandas, Matplotlib, Seaborn, and scikit-learn. NumPy is an essential tool for numerical computations, data analysis, and scientific research in Python due to its effective array operations, multidimensional data management, and interaction with other scientific computing libraries. The numeric module offered an array object for Python numerical computations before the creation of NumPy. The numeric array object may be thought of as having been replaced by NumPy, which builds on its strengths and adds several improvements. Its main goal is to lay the groundwork for a supportive environment for effective scientific computing [5]. One of the most well-known Python libraries, Pandas, would eventually provide the groundwork for statistical computing in the future. It is a library containing a variety of tools and data structures that may be used to work with structured data from all domains. It offers comprehensive, logical methods for manipulating and analyzing data [18].

Matplotlib can simulate MATLAB functionality through Pyplot, a library of command-style mining functions. Every Pyplot method aims to modify a graph (Figure 10.2) in some way. For example, it can create a new shape, create a drawing area inside an existing shape, draw lines inside

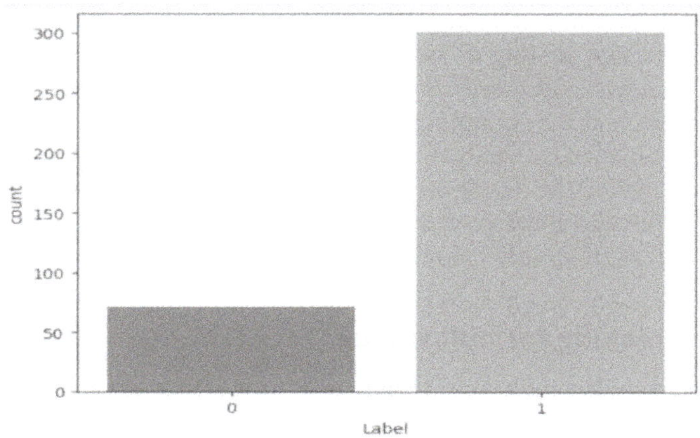

Figure 10.2 Addressing the problem of data imbalance

the drawing area, or name the plot or its horizontal and vertical axes with text [17].

Seaborn is an extension of the Matplotlib library that makes it easy to create eye-catching images using Python. Compared to Matplotlib, it offers a series of simpler techniques that make it easier to create eye-catching charts. Pandas DataFrames and Seaborn work perfectly together, allowing easy communication between the two [19].

Scikit-learn (often abbreviated to sklearn) is a popular machine-learning library for Python. It provides a comprehensive set of tools for various machine learning tasks such as classification, regression, clustering, dimensionality reduction, model selection, and preprocessing. With scikit-learn you can learn various machine learning algorithms such as support vector machines, random forests, logistic regression, and k-nearest neighbors. This library provides a consistent and easy-to-use API, making it accessible to both novices and experienced practitioners. Additionally, scikit-learn integrates well with other Python libraries such as NumPy and Pandas, enabling seamless data processing and analysis. Overall, scikit-learn is a valuable tool for developing and deploying machine learning models, providing a robust and efficient framework for a variety of data science tasks [20].

10.4.2 Loading the dataset

Loading the dataset was done using the functions and methods provided by the imported libraries. For example, Pandas read CSV files and NumPy loaded data from text files.

10.4.3 Preprocessing of datasets

The dataset has been preprocessed to prepare it for machine learning algorithms. This includes handling missing values, encoding categorical variables, and scaling or normalizing numerical features if necessary. Furthermore, we addressed the problem of data imbalance by applying techniques such as oversampling and undersampling.

An exploratory data analysis is used to gain insight into a dataset. This includes analyzing the distribution of target variables (malware vs. benign), examining the distribution of traits, identifying correlations between variables, and detecting outliers and anomalies [21].

10.4.4 Checking for null values

Checking for null values in a dataset is an important step in data preprocessing. Proper identification and handling of null values help ensure the quality and integrity of data used for machine learning. We used the isnull() function on a data frame to create a Boolean data frame that indicates whether each cell is null or not. The sum() function was applied to the Boolean data frame along the column axis (axis = 0) to get the number of null values in each column. Columns are filtered by selecting columns with a non-zero count to indicate the presence of null values.

10.4.5 Extract features

Feature extraction is an important step in machine learning as it involves converting raw data into meaningful representations to capture underlying patterns and relevant information for the current task.

10.4.6 Splitting the data

In machine learning, data is usually divided into training and testing. The training model is used to train the machine learning model by providing the model with examples and their respective labels. The testing model is used to evaluate the final performance of the trained model. This set is independent and provides an objective estimate of model accuracy. A typical split is usually 70% for training, with the remaining 30% split evenly across test and validation sets. It is important to maintain the original data distribution in the subsets to avoid introducing bias. The training is performed on the training set, and the final evaluation is performed on the test set to get an objective assessment of the model's performance.

10.5 RESULTS AND DISCUSSION

In this research work, logistic regression, linear regression, support vector machine, Gaussian naïve Bayes, and AdaBoost algorithms are used to predict accuracy to select an appropriate model for this dataset.

10.5.1 Logistic regression

Logistic regression, as shown in Figure 10.3, is a popular classification algorithm for predicting binary or categorical outcomes. It models the relationship between the input variables and the probabilities of certain outcomes.

This algorithm uses maximum likelihood estimation to estimate the coefficients of the input variables. During training, the logistic loss function is optimized to fit the data. Logistic regression is interpretable and computationally efficient, making it suitable for both small and large datasets. However, it assumes a linear relationship between the input variables and the resulting log odds, which can limit flexibility in capturing complex patterns.

```
[ ] from sklearn.linear_model import LogisticRegression

    classifier = LogisticRegression(random_state = 0)
    classifier.fit(X_train, y_train)
    y_pred = classifier.predict(X_test)

    from sklearn.metrics import confusion_matrix

    cm = confusion_matrix(y_test, y_pred)
    print ("Confusion Matrix : \n", cm)

    from sklearn.metrics import accuracy_score

    print ("Accuracy : ", accuracy_score(y_test, y_pred))


    Confusion Matrix :
     [[16  1]
     [ 1 57]]
    Accuracy :  0.9733333333333334
```

Figure 10.3 Logistic regression result

10.5.2 Support vector machine

Another machine learning approach that is commonly used for classification problems is support vector machines (SVMs). The fundamental concept is finding a hyperplane that best differentiates the classes. The term "support vectors" refer to the points nearest to the hyperplane that, if removed, would modify the hyperplane's location. The margin is the distance between the support vector and the hyperplane.

We intuitively understand that the farther our classes are from the hyperplane, the more accurate our forecasts will be. As a result, as shown in Figure 10.4, while numerous hyperplanes can be found per problem, the purpose of the SVM method is to find the hyperplane with the highest margins [2].

10.5.3 Naïve Bayes

Naïve Bayes is a classifier machine learning algorithm based on Bayes' theorem, as shown in Figure 10.5. It can be used for binary and multiclass classification problems. The main point is based on the idea of handling each feature independently. The naïve Bayes method evaluates the probability of each feature independently, does not depend on any correlation, and makes predictions based on Bayes' theorem. That's why this approach is called "naïve"—in real-world problems, features often have some degree of correlation between them. To understand the naïve Bayes algorithm, it is necessary to first introduce class and conditional probability concepts:

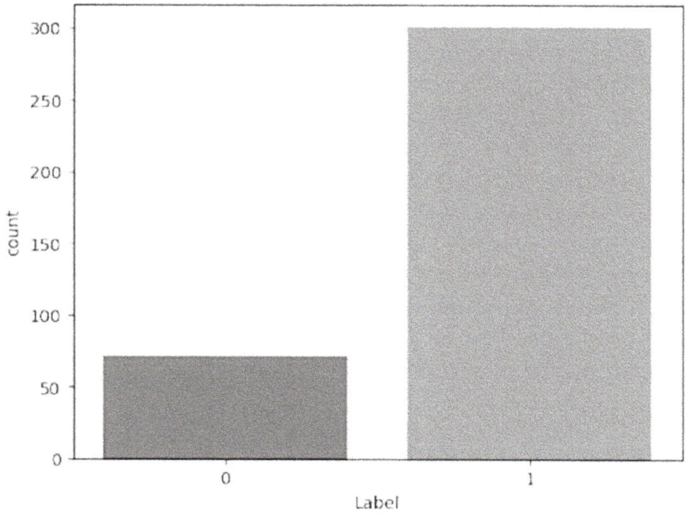

Figure 10.4 Result of SVM

```
[ ] data.isnull().sum()

    Label  0
    F_1    0
F_2    0
F_3    0
F_4    0
    ...
F_527  0
F_528  0
F_529  0
F_530  0
F_531  0
Length:532, dtype: int64

[ ] data['Label'] = data.Label.map({'non-malicicous':0, 'malicious':1})
```

Figure 10.5 Naïve Bayes

1. Class probability is the probability of a class in a dataset. In other words, if we choose a random element from the dataset, the probability that it belongs to a certain class is the probability.
2. The conditional probability is the probability of the characteristic value given to the class [2].

10.5.4 AdaBoost

AdaBoost (Adaptive Boosting) is an ensemble learning technique that combines multiple weak classifiers to create a strong classifier. It iteratively trains a weak model with differently weighted versions of the training data, and each successive model adds more weight to the misclassified samples from the previous model. The final prediction is made by aggregating the predictions of all weak models using weighted voting. AdaBoost is effective at processing complex datasets and can improve performance compared to a single weak model. However, it can be sensitive to noisy or outlier data and can overfit if the weak model is too complex. Learners are determined by accuracy or error rate. A more accurate and weaker learner usually receives a higher weight in the final prediction.

10.5.5 Linear regression

Linear regression is a simple but powerful supervised learning algorithm used for regression tasks. It models the relationship between the input and continuous target variables by fitting a linear equation to the data. Linear regression estimates coefficients that minimize the sum of squares of the difference between the predicted and actual values. It can handle both single and multiple input variables. Linear regression is interpretable, computationally efficient, and provides insight into the direction and magnitude of relationships between variables. However, it assumes a linear relationship between the input and the target variable, which may limit its ability to capture complex nonlinear patterns.

10.5.6 Confusion matrix

A confusion matrix is a table used to evaluate the performance of a classification model. It summarizes the predictions made by the model from a set of test data and compares them with the actual labels or ground truth. This matrix provides a detailed breakdown of the model's performance in terms of true positive, true negative, false positive, and false negative predictions.

Each cell in the matrix represents a specific combination of predicted and actual classes. Here is a breakdown of the terms used:

- True positive (TP): The model correctly predicted the positive class.
- True negative (TN): The model correctly predicted the negative class.
- False positive (FP): The model incorrectly predicted the positive class when it should have been the negative class (type I error).
- False negative (FN): The model incorrectly predicted the negative class when it should have been positive (type II error).

Confusion matrices provide insights into model performance such as accuracy, precision, recall, and F1 score. These metrics can be calculated using the confusion matrix values.

$$Accuracy = \frac{TP + TN}{TP + TN + FP + FN} \tag{10.1}$$

$$Precision = \frac{TP}{TP + FP} \tag{10.2}$$

$$Recall\,(also\,called\,sensitivity\,or\,true\,positive\,rate = \frac{TP}{TP + TN} \tag{10.3}$$

$$Final\,Score = 2 \times \frac{Precision \times Recall}{Precision + Recall} \tag{10.4}$$

Confusion matrix analysis helps identify the strengths and weaknesses of classification models, especially when distinguishing between different classes. This can help determine if the model is targeting a specific class or if there are specific areas where it struggles to make accurate predictions.

10.5.7 Prediction accuracy

Table 10.1 is a prediction accuracy table for malware detection through machine learning models. Each row represents a different machine learning model used to detect malware. The "Model" column lists the model's name, and the "Prediction Accuracy" column shows the corresponding accuracy values obtained by each model. It is important to note that these accuracy values are provided as examples and may not represent actual performance or actual scenarios. Exact values will usually be obtained by evaluating a particular dataset using suitable validation techniques. The results showed

Table 10.1 Prediction accuracy

Model	*Prediction accuracy*
Logistic regression	0.97333
Support vector machine	0.97333
Naïve Bayes	0.98666
AdaBoost	0.97333
Linear regression	1.0

```
[ ] X = data.drop(["Label"],axis=1)
    y = data['Label'].values
    X_train, X_test, y_train, y_test = train_test_split(X,y,test_size=0.2, random_state=42)
```

Figure 10.6 Graphical representation of accuracy of various models

that logistic regression, SVM, and AdaBoost achieved a detection accuracy of 97%, while naïve Bayes achieved a detection accuracy of 98%, and linear regression achieved a detection accuracy is 100%.

Plotting graphs in machine learning is an essential part of data analysis, model evaluation, and results visualization. We will import Matplotlib.pypl ot for plotting different kinds of graphs to analyze and evaluate the results. Here in this research work, we used different kinds of graphs, including bar graphs, pie charts, heat maps, and box plots, to understand the models in a better way. Figure 10.6 represents the accuracy graph of the model.

The research work focuses on detecting malware using machine-learning techniques. The main goal is to develop a robust and effective system for identifying and mitigating malware threats. The research work explored various machine learning algorithms, including logistic regression, linear regression, support vector machine, AdaBoost, and naïve Bayes, to generate software-detectable models. The performance of each model was assessed using measures with accuracy. The final report shows that based on information collected from the models and based on the accuracies, it seems that linear regression gives the best results in terms of detection accuracy. It achieved 100% perfect accuracy, which means it correctly classified all the cases in the dataset. However, it is important to note that linear regression is typically used for regression tasks, not classification tasks. Therefore, it is unusual to see linear regression applied in this context. When comparing classification models, naïve Bayes achieved the highest accuracy of 98%, outperforming logistic regression, SVM, and AdaBoost, which achieved 97% accuracy. Naïve Bayes is a probabilistic classifier based on Bayes' theorem, which assumes that all features are independent of each other. It is known for its simplicity and computational efficiency, and generally works well in situations with relatively small amounts of data.

The research work concluded that machine learning could significantly improve malware detection capabilities and contribute to a more robust cybersecurity framework. By combining machine learning algorithms with traditional security measures, organizations can achieve higher accuracy, early detection of threats, and efficient resource allocation. However, keeping up with evolving threats and maintaining the effectiveness of malware detection systems requires constant updating, monitoring, and collaboration with cybersecurity experts. Overall, this research work demonstrated the potential and promise of machine learning in the malware detection field, highlighting its advantages, limitations, and areas for further research and development.

Machine learning models can learn malware patterns and characteristics from large datasets, allowing them to detect and classify malware with greater accuracy than methods based on traditional rules. Machine learning models can adapt and learn from new malware patterns, allowing them to detect never-before-seen or zero-day malware. This adaptability is critical to combating attacks from rapidly evolving and sophisticated malware. Machine learning algorithms can automatically extract relevant features from malware samples, reducing the need for manual feature engineering. This helps manage the growing complexity and variety of malware. Machine learning techniques can efficiently process large-scale datasets, allowing large amounts of malware samples to be analyzed. This scalability is essential due to the exponential growth of malware versions. Machine learning models can be trained and deployed to perform real-time malware detection, enabling rapid identification and mitigation of malware threats as they arise. Machine learning in malware detection automates the process, reducing the manual effort required to analyze and identify malware. This allows security professionals to focus on more complex tasks and investigations. Machine learning algorithms can detect malware at an early stage, helping to prevent widespread damage. Early detection improves incident response and mitigation efforts.

10.5.8 LIMITATIONS OF THE WORK

The following restrictions must be observed for this research work to run properly: Google Collab is completely online and saves the projects to disk, so you need an active internet connection to work with Google Collab. Uploaded CSV files may be deleted if a stable internet connection cannot be established. Use Google Collab or Jupyter as the files are large and installing on many libraries in your system can consume a lot of memory on your hardware. Make sure your CSV file is fully uploaded before starting your research work. The output will change every 10% of the CSV file and the analysis will never be accurate.

This research work was carried out using historical time series data and does not claim real market validity. The comparative analysis was performed specifically for this dataset, and we do not claim that the accepted model is the most appropriate model for all other datasets. The process of creating two models is very cumbersome and time-consuming, which is helpful for long-term investors but can be frustrating for those looking for short-term investors.

10.6 CONCLUSION AND FUTURE SCOPE

The research work focused on malware detection using machine learning techniques and aimed to develop a robust and efficient system for identifying and stopping malicious software threats. Evaluation of various machine learning algorithms such as logistic regression, linear regression, support vector machine, AdaBoost, and naïve Bayes showed their effectiveness in detecting malware with varying accuracy. The results showed the following benefits of machine learning in malware detection: improved accuracy, adaptability to evolving threats, real-time detection, and automated analysis. The research work also identified limitations such as the need for high-quality labeled data, vulnerability to adversarial attacks, and challenges in model interpretability and explainability. Overall, this research work provided valuable insight into the benefits, limitations, and prospects of machine learning in malware detection. It stressed the importance of continued research and development to meet challenges, improve the effectiveness of malware detection systems, and ultimately contribute to a safer digital environment.

In this research work we used algorithms including logistic regression, support vector machine, naïve Bayes, AdaBoost, and linear regression. These are great models that suit the dataset. Further studies could focus on developing and implementing enhanced defense techniques, and developing methods and techniques to improve the interpretability of the models. This research work completely focused on traditional learning algorithms, so future work could explore the use of deep learning techniques such as convolutional neural networks (CNNs) or recurrent neural networks (RNNs) and neural networks for malware detection because they have promising results in various domains and can be adapted for more effective malware detection. They should emphasize the development of privacy-preserving techniques that ensure sensitive information is protected while still enabling effective malware detection.

REFERENCES

1. Chumachenko, K. (2017). Machine learning methods for malware detection and classification. *Lataukset*, 1, 16581.
2. Rathore, H., Agarwal, S., Sahay, S. K., & Sewak, M. [cs.CR] (4 April 2019). *Malware Detection Using Machine Learning and Deep Learning*, BITS, Pilani, Department of CS & IS, Goa Campus, Goa, India.
3. Senanayake, J., Kalutarage, H., & Omar Al- Kadri, M. (5 July 2021). A systematic review. *Electronics*, 10(13), 1606, doi.org/10.3390/electronics10131 606.

4. Shoaib Akhtar, M., & Feng, T. (2022). *School of Computer and Communication*, Lanzhou University of Technology, Lanzhou 730050, China, Symmetry.
5. Oliphant, T. E. (2006). *A Guide to NumPy* (Vol. 1, p.85). USA: Trelgol Publishing.
6. Lemenkova, P. (2019). Processing oceanographic data by Python libraries NumPy, SciPy, and Pandas. *Aquatic Research*, 2(2), 73–91.
7. Moruzzi, G., & Moruzzi, G. (2020). Plotting with matplotlib. *Essential Python for the Physicist*, 1, 53–69.
8. Bisong, E., & Bisong, E. (2019). Matplotlib and seaborn. *Building Machine Learning and Deep Learning Models on Google Cloud Platform: A Comprehensive Guide for Beginners*, 1, 151–165.
9. Vasani, V., Bairwa, A. K., Joshi, S., Pljonkin, A., Kaur, M., & Amoon, M. (2023). Comprehensive Analysis of Advanced Techniques and Vital Tools for Detecting Malware Intrusion. *Electronics*, 12(20), 4299. https://doi.org/10.3390/electronics12204299.
10. Vaibhav, S., & Haritha, D. (May 2019). Volatile memory acquisition and extracting of data using volatility framework and web GUI application. *International Journal of Innovative Technology and Exploring Engineering (IJITEE)*, 8(7), 1–16.
11. Amari, K. (2009). *Techniques and Tools for Recovering and Analyzing Data from Volatile Memory"-33049.* SANS Institute.
12. Cai, L., Sha, J., & Qian, W. (2013). Study on forensic analysis of physical memory. *International Symposium on Computer, Communication, Control and Automation,* 3, 221–224.
13. Rehman, Z. U., Ahmad, A., & Saleem, S. (2017). A brief survey of memory analysis tools. *NUST Journal of Engineering Sciences*, 10(2), 57–64.
14. Schuster, A. (2006). Searching for processes and threads in Microsoft Windows memory dumps. *Digital Investigation,* 3, 10–6.
15. Moser, A., Kruegel, C., & Kirda, E. (2007). Exploring multiple execution paths for malware analysis. *Proceedings of IEEE Symposium on Security and Privacy*, 17. https:// doi.org/10.1109/SP.2007.
16. Singh, A. K., Taterh, S., & Mitra, U. (2023). An efficient tactic for analysis and evaluation of malware dump file using the volatility tool. *SN Computer Science*, 4(5), 457.
17. Ravula, A. K., Ahmad, S. S., Singh, A. K., Sweeti, S., Kaur, A., & Kumar, S. (2023, June). Multi-level collaborative framework decryption-based computing systems. In *AIP Conference Proceedings* (Vol. 2782, No. 1). AIP Publishing.
18. Aslan, Ö. A., & Samet, R. (2020). A comprehensive review on malware detection approaches. *IEEE Access*, 8, 6249–6271.
19. Gaurav, A., Gupta, B. B., & Panigrahi, P. K. (2023). A comprehensive survey on machine learning approaches for malware detection in IoT-based enterprise information system. *Enterprise Information Systems*, 17(3), 2023764.

20. Shaukat, K., Luo, S., & Varadharajan, V. (2023). A novel deep learning-based approach for malware detection. *Engineering Applications of Artificial Intelligence*, 122, 106030.
21. Bhat, P., Behal, S., & Dutta, K. (2023). A system call-based android malware detection approach with homogeneous & heterogeneous ensemble machine learning. *Computers & Security*, 130, 103277.

Chapter 11

Eco-friendly operations in Reality

Analysis of key factors of sustainability performance in manufacturing companies

Vishakha Kuwar, Vandana Sonwaney, Neeta Bhide, Kamal Upreti, and Ramesh Chandra Poonia

11.1 INTRODUCTION

In recent years, the sustainability has become increasingly evident as the world grabs with challenges like resource depletion, climate change, and environmental degradation. These issues are more urgent as global temperatures rise and natural resources become scarcer. The issues are addressed through various approaches across different sectors, including product design, energy management, and waste handling. This literature review focuses on three critical areas: sustainable product design (SPD) [1], energy efficiency (EE) [2], and waste management (WM) [3]. Insights from recent academic research and reviews provide an inclusive understanding of these aspects and highlight the key findings, methodologies, and future directions.

SPD involves product design for minimal environmental impact throughout the lifecycle, from conception to disposal. Sustainable materials are used in production processes to reduce waste and encourage product reuse or recycling. A recent analysis identified 297 factors affecting sustainable product design, and out of those, 43 factors are frequently addressed [1]. These factors span economic, environmental, and social dimensions, giving SPD a multifaceted nature. Research suggests SPD is necessary because it addresses the environmental footprint of products from their conception to end of life. It also encourages a shift toward more eco-friendly practices in manufacturing and consumption. For instance, recycled materials can lower the carbon footprint, reduce the use of virgin resources, and encourage product design for disassembly to facilitate recycling and reduce waste [4].

EE refers to the optimization of energy consumption to achieve the same or best results with minimum energy. This is crucial for greenhouse gas emission reduction and lower energy costs across various sectors, including buildings, transportation, and industry. A critical assessment of energy efficiency research over the past 40 years noted that while programs have reduced energy use per economic output, there are still challenges related

DOI: 10.1201/9781003569251-11

to rebound effects at the macro level [2]. These underscore nuanced policy approaches for the benefits of EE. For example, EE for buildings can be achieved through efficient heating, better insulation, cooling systems, and the use of renewable energy sources [5]. The evidence leans toward EE being essential for climate change mitigation, but debates around rebound effects highlight the complexity of implementation.

WM encompasses strategies to reduce, reuse, recycle, and properly dispose of waste. Effective WM is essential for minimizing the impact of waste and promoting a circular economy. The World Bank said that the global waste generation is expected to increase by 70% by 2050 [6], underscoring the urgency of effective WM strategies. A literature review on solid waste management found that most developing countries emphasized the importance of community involvement and education in effective WM [3]. The study highlights that knowledge and participation are key drivers for successful WM practices. Implementation gaps persist due to various socio-economic factors [8]. WM needs strategies like recycling and source reduction, with developing countries facing unique challenges, and community involvement is crucial for success [7].

This review was conducted systematically searching academic databases such as Google Scholar, ScienceDirect, and MDPI for peer-reviewed papers within the last five years. Keywords used were "sustainable product design," "energy efficiency," and "waste management," along with related terms specific to each topic such as "lifecycle assessment in product design" for SPD, "energy-efficient buildings" for EE, and "recycling efficiency" for WM. The selection process involved the evaluation of the relevance and quality of each paper. It prioritized comprehensive analyses or synthesis of existing research. Inclusion criteria prioritized peer-reviewed journals, recent publications, and studies covering multiple sustainability dimensions. Exclusion criteria excluded non-academic sources or papers prior to 2018, ensuring a focus on current work. Approximately 46 papers were selected for this review and distributed across SPD (13), EE (12), and WM [21], providing a balanced review. For SPD, the papers were selected that discussed lifecycle assessments, material selection, and consumer behavior related to sustainable products [9]. For EE, the topics were policy impacts, technological innovations, and sector-specific case studies [9]. The WM papers covered waste reduction strategies, recycle efficiency, and community engagement [10].

Each selected paper was critically evaluated for its relevance, methodology, and contribution to the field. The findings from the papers were then synthesized to identify common themes, trends, and areas of contention. The SPD studies agree on the importance of lifecycle assessments but disagree on their practical utility at early design stages [11, 12]. For EE, consensus exists on cost savings, but controversy surrounds the magnitude of rebound effects, requiring nuanced policy. For WM, community involvement is supported, but implementation gaps in developing countries pose challenges [13].

11.1.1 Structure and purpose

The review is structured into three main sections: SPD, EE, and WM.

11.1.1.1 Sustainable product design (SPD)

This includes factors that influence SPD, tools and methods used in its implementation, and the challenges faced in sustainability integration into product design. Ljungberg [14] discusses the material choices and lifecycle assessments.

11.1.1.2 Energy efficiency (EE)

The current state of research in EE includes policy impacts, technological advancements, and sector-specific efficiency measures on industrial applications [15] and on buildings [16].

11.1.1.3 Waste management (WM)

WM includes strategies for effectiveness, community involvement, and policies for promoting sustainable waste practices. It covers topics like plastics recycling [17] and global strategies [18].

The latest research in SPD, EE, and WM literature gives valuable insights for researchers, policymakers, and practitioners interested in advancing sustainability across these domains. The literature review identified gaps in current knowledge and presents directions for future research, such as integrating AI in SPD, addressing EE rebound effects, and enhancing WM implementation in developing countries.

11.2 LITERATURE REVIEW ON SUSTAINABILITY

SPD, EE, and WM are critical components of environmental sustainability. Each address distinct yet interconnected challenges. SPD focuses on products designs with less environmental impact throughout their life cycle, from conception extraction to disposal [19]. EE involves energy consumption optimization to reduce costs and emissions, mostly through technological and policy interventions [20]. WM encompasses strategies for waste generation reduction and enhance recycling, crucial for urban and industrial waste management [21]. The VOS viewer network visualization represents key research topics related to sustainability, sustainable development, and associated themes.

11.2.1 Methodology for literature review

To conduct a thorough review, we began with systematic searches using academic databases like Google Scholar, ScienceDirect, and MDPI, focusing on peer-reviewed, recent publications (post-2018 for relevance).

- For SPD, search terms like "sustainable product design factors" yielded Mengistu et al. [21], identifying 297 factors across economic, environmental, and social dimensions.
- For EE, "energy efficiency academic papers" led to Zakari et al. [22] offering a 15-year review.
- For WM, "waste management literature review" revealed Jain et al. [23] focusing on MSMEs.

Inclusion criteria prioritized peer-reviewed journals, recent publications, and studies covering multiple sustainability dimensions. Exclusion criteria excluded non-academic sources or outdated pre-2018 papers, ensuring a focus on current research. The study selected 46 papers, distributed across SPD (10), EE (12), and WM (13), to provide a balanced review.

11.2.2 Organizing the literature

The chapter organizes findings thematically to enhance clarity.

- SPD: Themes included material selection, lifecycle assessment, design for disassembly, and consumer behavior [24].
- EE: Themes covered policy impacts, technological advancements, and sector-specific efficiency (buildings, transportation, industry), with William et al. [25] noting green building rating systems.
- WM: Themes considered collection methods, treatment technologies, and behavioral strategies, with Raghu and Rodrigues [26] emphasizing education's role.

The three critical areas of sustainability and their themes are represented in Table 11.1. SPD measures environmentally friendly product development, EE evaluates energy consumption optimization, and WM assesses waste reduction and recycling strategies. By synthesizing insights from 46 academic papers, the study aimed to highlight key findings, methodologies, and future directions, providing a comprehensive overview for researchers and practitioners.

11.2.2.1 Sustainable product design (SPD)

SPD focuses on designing products with minimal environmental impact, considering materials, production, and disposal. Research suggests life cycle assessments (LCAs) are crucial, with Mengistu et al. [27] identifying

Table 11.1 Key components of environmental sustainability

Component	*Focus area*	*Key themes*	*Challenges*	*Sustainable solutions*
Sustainable product design (SPD)	Designing products with minimal environmental impact	Material selection	Implementation challenges in business	Use of recycled/renewable materials
		Life cycle assessment (LCA)	High cost of sustainable materials	Ecofriendly manufacturing processes
		Design for disassembly	Early-stage design limitations	Circular economy principles
		Consumer behavior		
Energy efficiency (EE)	Optimizing energy use to reduce costs and emissions	Smart energy management	High initial investment for green technologies	AI-driven energy monitoring
		Renewable energy integration	Fossil fuel dependency	Smart grids and predictive analytics
		Energy efficient buildings and industry	Rebound effect in energy savings	Green building policies
Waste management (WM)	Reducing waste generation and enhancing recycling	Source sorting and composting	Inefficient waste collection and treatment	Reverse logistics and sustainable packaging
		Recycling strategies	Low consumer awareness	AI-based waste sorting and management
		Circular supply chain and logistics	High carbon footprint in logistics	Green transportation and optimized routing

297 factors, 43 frequently addressed, affecting economic, environmental, and social sustainability. Material selection is vital, emphasizing recycled or renewable materials [28]. Design for disassembly facilitates recycling [29], while consumer behavior influences market success [30]. Challenges include practical utility at early design stages and implementation gaps in business practices.

11.2.2.1.1 Sustainability in manufacturing

Sustainability in manufacturing has become a difficult focus for industries, which feel the increased pressure to reduce their environmental footprint. Traditional manufacturing processes are characterized by excessive waste generation, high energy consumption, and harmful emissions. Sustainable manufacturing emphasizes adoption of eco-friendly practices, resource efficiency, and renewable energy sources [1]. Minimizing waste, optimally utilizing resources, and reducing carbon emissions are helpful for manufacturers to achieve economic and long-term environmental benefits.

One of the key drivers of sustainable manufacturing is the shift toward green manufacturing. The common approach incorporates environmentally friendly materials, energy-efficient production methods, and the circular economy. This involves product designs with recyclability in mind using biodegradable materials, and implementing closed-loop supply chains. Advanced technologies such as cloud computing, Industry 4.0, and the Internet of Things (IoT) enhance sustainability with real-time monitoring of energy use, predictive maintenance, and data-driven decision-making to optimize operations [30].

The role of eco-friendly operations in manufacturing extends beyond environmental benefits and economic and social advantages. Sustainable manufacturing leads the cost savings, reduced energy and raw material consumption, production efficiency, and less waste disposal costs. Additionally, companies that adopt sustainable practices can enhance their regulatory requirements, brand reputation, and environmentally conscious consumers. Governments and industry leaders are encouraging sustainability through stricter regulations, carbon taxation, and green technology adoption incentives [31].

Despite the obvious benefits, implementing sustainable manufacturing practices poses several challenges. High initial investment costs, technology limitations, and change resistance are common barriers that manufacturers must overcome. Collaborative efforts among governments, industries, and research institutions are essential to accelerate changes.

11.2.2.1.2 Eco-friendly operations in Industry 4.0

The emergence of Industry 4.0 has revolutionized manufacturing by integrating smart technologies such as artificial intelligence (AI), the Internet of Things (IoT), big data analytics, cyber-physical systems (CPS) and cloud computing. This transformation not only enhances productivity and efficiency but also plays an important role in promoting eco-friendly operations. By leveraging digital technologies, Industry 4.0 enables sustainable manufacturing practices that reduce waste, optimize energy consumption, and environmental impact minimization [32].

One of the key aspects of eco-friendly operations in Industry 4.0 is smart energy management. Manufacturers allow monitoring of the real-time energy usage, inefficiencies identification, and automated control systems implementation and power consumption optimization. This results in carbon emissions reduction, energy costs minimization, and operational efficiency improvement [33]. Additionally, renewable energy integration such as wind and solar power is facilitated through smart grids and digital twin technologies and ensures a seamless transition to cleaner energy sources [34].

Another critical component is waste reduction and resource optimization. Industry 4.0 technologies include additive manufacturing (3D printing) and advanced robotics. Predictive analytics and machine learning algorithms assist in preventive maintenance, extend the life cycle of machinery, and reduce material wastage reduction. Cloud-based supply chain management ensures better resources tracking, overproduction reduction, and excess inventory, which often contribute to environmental degradation.

Smart manufacturing processes also play a role in environmental impact. The adoption of digital twins' virtual replicas of physical assets allows manufacturers to simulate and optimize production processes before implementation, reducing errors, material wastage, and energy consumption. Similarly, blockchain technology ensures transparent and sustainable supply chains by tracking the sourcing of raw materials and ensuring ethical and eco-friendly procurement practices. Moreover, Industry 4.0 fosters sustainable product design LCA incorporating tools for the environmental impact of a product from raw material extraction to end-of-life disposal [35]. This promotes the use of biodegradable materials, recyclable components, and energy-efficient product designs, contributing to a greener manufacturing ecosystem.

Despite these advancements, the transition to eco-friendly operations in Industry 4.0 faces challenges such as high initial investment costs, cybersecurity risks, and the need for workforce upskilling. However, with increasing global environmental regulations, consumer demand for sustainable products, and advancements in green technologies, industries are progressively moving toward more sustainable production models. Industry 4.0 is a key enabler of eco-friendly operations driving sustainable manufacturing practices that align with global climate goals [31]. To harnesses of the power of digital transformation, industries can reduce their ecological footprint while enhancing their operational efficiency and competitiveness. As businesses continue to embrace these technologies, the future of manufacturing is smarter, greener, and more resilient.

11.2.2.1.3 Digital transformation in achieving sustainability

Digital transformation is the driver for pursuing sustainability, which enables industries to optimize resources, minimize waste, and reduce their environmental footprint. The integration of advanced technologies such as AI, big data analytics, blockchain, IoT, and cloud computing can help organizations enhance operational efficiency and eco-friendly practices. Digital solutions provide predictive analytics, real-time monitoring, and automation, and allow businesses to adopt sustainable production models and make data-driven decisions supporting environmental conservation [36].

One of the most significant contributions of digital transformation to sustainability is smart energy management. IoT-enabled sensors and AI-driven analytics help industries track energy consumption, detect inefficiencies, and automate processes to optimize energy use.

11.2.2.2 Energy efficiency (EE)

EE optimizes energy consumption across sectors, reducing CO_2 emissions and costs. Studies suggest innovation lowers technology costs, with programs reducing energy use per economic output, though macro-level rebound effects warrant policy attention. For buildings, Huang et al. [37] highlight insulation and lighting upgrades, while Milewski and Milewski [38] address transportation efficiency. Industries benefit from process optimization, with policies like financial incentives playing a key role.

11.2.2.2.1 High carbon emissions and energy consumption

The manufacturing sector is one of the largest consumers of energy, with a significant dependency on fossil fuels such as natural gas, coal, and oil. Industrial operations require vast amounts of energy for machinery, heating, cooling, and production processes, making energy consumption a critical sustainability challenge. Fossil fuel reliance not only depletes finite natural resources but also contributes to air pollution, environmental degradation, and climate change [39].

11.2.2.2.2 Fossil fuel dependency in industrial operations

Traditional manufacturing heavily depends on non-renewable energy sources, with many industries still operating on outdated and energy-intensive processes. Fossil fuel use for power generation, transportation, and material processing results in high energy costs and environmental consequences. For example, steel and cement manufacturing rely on coal and petroleum-based fuels to power high-temperature furnaces. The fossil fuel burning releases large amounts of sulfur dioxide (SO_2), carbon dioxide (CO_2), and nitrogen oxides (NO_x), contributing to air pollution and climate change. Additionally, many factories and industrial plants operate with low energy efficiency [40].

11.2.2.2.3 Contribution of manufacturing to global greenhouse gas emissions

The manufacturing sector is a major contributor to global greenhouse gas (GHG) emissions, accounting for nearly 20% of total CO_2 emissions worldwide. Fossil fuel combustion in industrial processes releases methane (CH_4), carbon dioxide (CO_2), and other pollutants into the atmosphere and accelerates global warming. Some manufacturing processes, such as chemical production, cement manufacturing, and metal refining, generate emissions beyond energy consumption, with industrial process emissions affecting climate change [41].

The carbon footprint of manufacturing is particularly high in industries such as:

- Steel and cement production: Both industries are among the largest CO_2 emitters due to the use of coal in blast furnaces and kilns.
- Chemical and petrochemical industry: Generates GHGs from energy-intensive processes and the release of industrial by-products.
- Textile and apparel manufacturing: Contributes to emissions through the energy required for dyeing, fabric processing, and transportation.

To mitigate the impact of carbon emissions, industries must transition toward low-carbon energy sources such as wind, solar, and bioenergy. Energy efficiency measures, smart grid integration, carbon capture technologies, and sustainable supply chain management can further help reduce emissions. Governments and environmental agencies worldwide are enforcing emissions trading systems, carbon pricing, and stricter environmental regulations to encourage industries to adopt cleaner energy solutions [42].

The shift toward renewable energy, industrial electrification, and energy-efficient technologies is crucial for minimizing manufacturing's environmental footprint. As industries embrace sustainable energy strategies and digital transformation, the goal of achieving net-zero emissions and carbon neutrality in manufacturing becomes increasingly attainable.

11.2.2.3 Waste management (WM)

WM assesses waste reduction and recycling, crucial for sustainability. Research leans toward strategies like source sorting and composting, with Hopewell, Dvorak, and Kosior [17] noting recycling's eco-efficiency, though challenges persist in developing countries. Education and awareness are vital, with community involvement enhancing effectiveness [43]. Global strategies, like reducing plastic waste mismanagement, are proposed.

11.2.2.3.1 Supply chain issues

The global supply chain plays a hard role in modern manufacturing, ensuring the seamless movement of raw materials, components, and finished goods. However, traditional supply chain and logistics systems are often plagued by inefficiencies that lead to high carbon footprints, excessive fuel consumption, and increased emissions. The reliance on fossil fuel-powered transportation, inefficient distribution networks, and outdated logistics strategies contributes significantly to environmental degradation and climate change. Addressing these problems is essential for achieving sustainable and eco-friendly manufacturing.

11.2.2.3.2 Carbon footprint of global supply chains

Manufacturing industries operate in complex, globalized supply chains that require raw materials, parts, and products to be transported across multiple regions. This extensive transportation network, which relies heavily on diesel-powered trucks, cargo ships, and air freight, is a major contributor to GHG emissions [44].

- Freight transportation accounts for nearly 8% of global CO_2 emissions, with road transport being the largest emitter.
- Maritime shipping—responsible for moving nearly 90% of global trade—emits approximately 1 billion tons of CO_2 annually due to heavy fuel oil consumption.
- Air cargo transport, while faster, produces significantly higher emissions per ton-kilometer compared to other modes, making it one of the least sustainable logistics options.

The high carbon footprint of supply chains is further exacerbated by energy-intensive warehousing, inefficient inventory management, and excessive packaging waste. Many manufacturers continue to rely on just-in-time (JIT) models, which, while cost-effective, often lead to increased shipment frequency, raising emissions levels.

11.2.2.3.3 Inefficiencies in transportation and distribution contributing to emissions

Logistical inefficiencies, such as poor route planning, underutilized cargo capacity, and outdated transportation technologies, further exacerbate the environmental impact of supply chain operations.

- Empty or partially loaded trucks: A significant percentage of transport vehicles operate below full capacity, leading to unnecessary fuel consumption and emissions.
- Last-mile delivery challenges: The rise of e-commerce and demand for fast delivery has increased carbon emissions and traffic congestion

in urban areas, making last-mile logistics a significant environmental concern.

- Outdated fleet and fuel dependence: Many logistics providers continue to use older, fuel-inefficient vehicles instead of transitioning to electric, hybrid, or hydrogen-powered transport solutions.

Additionally, inefficient warehouse operations, such as excessive energy use for heating, cooling, and lighting, further contribute to the overall carbon footprint of supply chains [45].

11.2.2.3.4 Sustainable solutions for greener supply chains

Logistics and supply chain operations have an environmental impact. Industries must adopt digital technologies and sustainable practices for efficiency and lower emissions. Key strategies include:

- Green transportation technologies: Implementing electric trucks, hydrogen-powered cargo ships, and biofuel alternatives to reduce fuel dependence.
- Optimized route planning and AI-driven logistics: Using big data, AI, and IoT to optimize routes, improve load capacity, and reduce unnecessary trips.
- Smart warehousing and automation: Adopting energy-efficient warehouses, robotic automation, and AI-driven inventory management to lower energy consumption.
- Sustainable packaging and reverse logistics: Reducing packaging waste and implementing circular supply chain models to promote recycling and product reuse.

The integration of Industry 4.0 technologies like blockchain for transparent supply chain tracking, IoT-enabled fleet monitoring, and AI-powered demand forecasts so manufacturers can build resilient, eco-friendly supply chains contribute to long-term sustainability goals [46].

11.3 RESEARCH METHODOLOGY

This section provides an in-depth analysis of the research methodology for a study that investigates SPD, EE, and WM on overall sustainability performance (OSP). The methodology used descriptive statistics, multiple linear regression analysis, and correlation analysis to test three hypotheses:

H1: SPD has a positive impact on OSP.
H2: EE significantly contributes to OSP.
H3: WM enhances OSP.

11.3.1 Background and importance

Sustainability is a critical area of study for climate change and resource depletion. SPD focuses on eco-friendly products, EE optimizes energy use, and WM reduces waste, and all contribute to the OSP. The study quantifies these relationships, providing insights for companies and policymakers. The hypotheses suggest positive impacts, but the complexity of real-world data requires robust statistical methods for valid conclusions.

11.3.2 Research design

The study adopts a cross-sectional quantitative research design and utilizes a dataset from a survey of 124 companies across various industries. The design is appropriate for relationship analysis at a single point and aligns with the summarized trend's goal and impacts testing. The cross-sectional approach allows for a snapshot of sustainability practices and performance, though it limits causal inferences to the longitudinal studies.

11.3.3 Data collection and variables

The survey targets companies to assess their sustainability practices. The sample size of 124 ensures sufficient statistical power, for the analysis of representativeness depends on the industries included. The variables are defined as follows:

- Sustainable product design (SPD): Measured on a 5-point Likert scale (1 = very poor, 5 = excellent) to assess the extent to which companies incorporate environmentally friendly design principles such as sustainable materials and designing for disassembly.
- Energy efficiency (EE): EE on a 5-point Likert scale evaluates the company's efforts to optimizing energy consumption through efficient lighting or process optimization.
- Waste management (WM): Measured on a 5-point Likert scale to assess strategies for waste reduction, reuse, and recycling such as source sort and compost.
- Overall sustainability performance (OSP): A composite index derived from various sustainability metrics measured on a 5-point scale and reflects overall environmental and economic performance.

These variables are treated as continuous for analysis and a common practice in social sciences for Likert scales with five or more points, though technically ordinal. This assumption facilitates regression analysis, acknowledges the debate around ordinal data as continuous, but aligns with standard practice for simplicity.

11.3.4 Descriptive statistics

Descriptive statistics summarize the key trends in the dataset. They provide an overview of the central tendencies and variability. They include calculation of means, medians, standard deviations, and ranges for each variable (SPD, EE, WM, OSP). These statistics help to distribute responses, detect outliers, and the general landscape of sustainability practices. A high mean for EE might indicate widespread adoption, while a large standard deviation could suggest variability across industries. Additionally, skewness and kurtosis are checked against the data, which is approximately normally distributed, with linear regression assumption, though minor deviations are often acceptable. The descriptive statistics might reveal industry-specific trends such as higher SPD scores in manufacturing versus services, which could inform targeted interventions. This insight is not immediately obvious and enhances the analysis of potential sector differences.

11.3.5 Correlation analysis

Correlation analysis assesses the relationships among the variables using Pearson's correlation coefficient, calculated for all pairs (SPD with EE, SPD with WM, EE with WM, and each with OSP). This step is crucial for understanding how variables move together, whether positively or negatively, and for identifying potential multicollinearity among independent variables. For instance, a high correlation between SPD and WM might suggest they are closely related practices, potentially affecting regression results. The correlation matrix was examined, and if any pair shows a correlation above 0.7, it is flagged for potential multicollinearity, which could be addressed by excluding one variable or using techniques like ridge regression, though for now, we assume moderate correlations.

The dataset has 90 observations without any missing values out of 124. Here are some key insights from the descriptive statistics:

- The mean values for sustainable product design (2.89), energy efficiency (2.89), and waste management (3.15) indicate that the data is fairly balanced.
- The overall sustainability performance has a mean of 2.98 and a standard deviation of 0.69.
- The range for all variables is from approximately 1 to 5, suggesting a rating or scoring system.

This analysis is vital for ensuring the regression model's validity, as high multicollinearity can inflate standard errors and make coefficient estimates unstable.

11.3.6 Multiple linear regression analysis

Multiple linear regression analysis quantifies the impact of SPD, EE, and WM on OSP, testing the hypotheses. The model is specified as:

$$OSP = \beta_0 + \beta_1 SPD + \beta_2 EE + \beta_3 WM + \epsilon$$

where

- β_0 is the intercept.
- $(\beta_1, \beta_2, \beta_3)$ are the coefficients representing the impact of SPD, EE, and WM on OSP, respectively.
- ϵ is the error term, capturing unexplained variation.

The coefficients $(\beta_1, \beta_2, \beta_3)$ indicate the change in OSP for a one-unit increase in each independent variable, holding others constant. This analysis determines the magnitude and significance of each variable's contribution, aligning with the hypotheses that SPD, EE, and WM positively affect OSP.

11.3.7 Hypothesis testing

The study tests the following hypotheses, interpreted as directional (one-tailed tests) given the positive impact implied:

- H1: Sustainable product design has a positive impact on overall sustainability performance (i.e., $\beta_1 > 0$).
- H2: Energy efficiency significantly contributes to overall sustainability performance (i.e., $\beta_2 > 0$).
- H3: Waste management enhances overall sustainability performance (i.e., $\beta_3 > 0$).

To test these, the t-statistics for each coefficient are examined to determine if they are significant at a 5% level ($p < 0.05$), and the sign checked to ensure positivity. The overall model fit was assessed using the F-test to ensure it explained significant variance in OSP. This approach acknowledges the complexity, as real-world data may show varying effects, and implementation challenges could affect results.

11.4 DISCUSSION

The findings highlight the role of integrated sustainability strategies in manufacturing. The strong positive correlation between sustainable product design and sustainability performance suggests that companies invest in eco-friendly design approaches so as to achieve better overall sustainability outcomes. This aligns with existing literature, which emphasizes that early-stage design interventions can significantly reduce environmental impacts throughout a product's life cycle.

For other variables, the positive correlation between waste management and sustainability performance underscores the importance of efficient waste reduction techniques such as resource optimization and recycling. Manufacturers that implement robust waste management practices are likely to enhance their economic and environmental sustainability.

Energy efficiency, while showing a positive relationship with sustainability performance, exhibited a weaker correlation compared to other factors. This finding suggests that energy efficiency alone may not be sufficient to drive comprehensive sustainability improvements. Instead, a holistic approach incorporating the multiple sustainability dimensions is required.

The weak negative correlation between energy efficiency and waste management suggests that energy-saving measures might lead to trade-offs in waste generation. Energy consumption optimized in certain processes might increase waste by-products. This highlights that balanced strategies need to integrate both energy and waste management to ensure long-term sustainability.

11.5 IMPLICATIONS

The study's findings provide valuable information for manufacturing companies to enhance their sustainability performance. Key recommendations are as follows:

- Prioritize sustainable product design: Companies should integrate eco-friendly design principles early in the product development process to maximize sustainability benefits.
- Enhance waste management practices: Implementing effective waste reduction strategies, including circular economy models, significantly contributes to sustainability goals.
- Adopt a holistic approach: Rather than focusing on isolated sustainability measures, companies should integrate energy efficiency, waste management, and product design to achieve optimal outcomes.

11.6 CONCLUSION

The results demonstrate that sustainable product design, waste management, and energy efficiency each play a role in sustainability in manufacturing. The study underscores the significance of a comprehensive approach toward sustainability to balance strategies and address multiple sustainability factors simultaneously. Future research could further explore potential trade-offs between different sustainability measures and identify the best practices for strategies in diverse manufacturing contexts.

REFERENCES

1. Delaney, E., Liu, W., Zhu, Z., Xu, Y., & Dai, J. S. (2022). The investigation of environmental sustainability within product design: a critical review. *Design Science*, 8, e15.
2. Patterson, M. G. (1996). What is energy efficiency? Concepts, indicators and methodological issues. *Energy Policy,* 24(5), 377–390.
3. Amasuomo, E., & Baird, J. (2016). The concept of waste and waste management. *Journal of Management and Sustainability*, 6, 88.
4. Camilleri, M. A., Cricelli, L., Mauriello, R., & Strazzullo, S. (2023). Consumer perceptions of sustainable products: a systematic literature review. *Sustainability*, 15(11), 8923.
5. Hafez, F. S., Sa'di, B., Safa-Gamal, M., Taufiq-Yap, Y. H., Alrifaey, M., Seyedmahmoudian, M., ... & Mekhilef, S. (2023). Energy efficiency in sustainable buildings: a systematic review with taxonomy, challenges, motivations, methodological aspects, recommendations, and pathways for future research. *Energy Strategy Reviews*, 45, 101013.
6. World Bank. (2018). What a waste 2.0: A global snapshot of solid waste management to 2050. https://openknowledge.worldbank.org/handle/10986/30317
7. Zhang, Z., Chen, Z., Zhang, J., Liu, Y., Chen, L., Yang, M., ... & Yap, P. S. (2024). Municipal solid waste management challenges in developing regions: a comprehensive review and future perspectives for Asia and Africa. *Science of the Total Environment,* 930, 172794.
8. Bhamra, T., Lilley, D., & Tang, T. (2011). Design for sustainable behaviour: using products to change consumer behaviour. *The Design Journal*, 14(4), 427–445.
9. Milewski, D., & Milewska, B. (2023). Efficiency of the consumption of energy in the road transport of goods in the context of the energy crisis. *Energies*, 16(3), 1257.
10. Debrah, J. K., Vidal, D. G., & Dinis, M. A. P. (2021). Raising awareness on solid waste management through formal education for sustainability: a developing countries evidence review. *Recycling*, 6(1), 6.
11. Ahmad, S., Wong, K. Y., Tseng, M. L., & Wong, W. P. (2018). Sustainable product design and development: a review of tools, applications and research prospects. *Resources, Conservation and Recycling*, 132, 49–61.
12. Gillingham, K., Newell, R. G., & Palmer, K. (2009). Energy efficiency economics and policy. *Annual Review of Resource Economics*, 1(1), 597–620.
13. Derhab, N., & Elkhwesky, Z. (2023). A systematic and critical review of waste management in micro, small and medium-sized enterprises: future directions for theory and practice. *Environmental Science and Pollution Research*, 30(6), 13920–13944.
14. Ljungberg, L. Y. (2007). Materials selection and design for development of sustainable products. *Materials & Design*, 28(2), 466–479.

15. Renna, P., & Materi, S. (2021). A literature review of energy efficiency and sustainability in manufacturing systems. *Applied Sciences*, 11(16), 7366.
16. Ranđelović, D., Jovanović, V., Ignjatović, M., Marchwiński, J., Kopyłow, O., & Milošević, V. (2024). Improving energy efficiency of school buildings: a case study of thermal insulation and window replacement using cost-benefit analysis and energy simulations. *Energies*, 17(23), 19961073.
17. Hopewell, J., Dvorak, R., & Kosior, E. (2009). Plastics recycling: challenges and opportunities. *Philosophical Transactions of the Royal Society B: Biological Sciences*, 364(1526), 2115–2126.
18. Pottinger, A. S., Geyer, R., Biyani, N., Martinez, C. C., Nathan, N., Morse, M. R., ... & McCauley, D. J. (2024). Pathways to reduce global plastic waste mismanagement and greenhouse gas emissions by 2050. *Science*, 386(6726), 1168–1173.
19. Awan, U., Arnold, M. G., & Gölgeci, I. (2021). Enhancing green product and process innovation: Towards an integrative framework of knowledge acquisition and environmental investment. *Business Strategy and the Environment*, 30(2), 1283–1295.
20. Saunders, H. D., Roy, J., Azevedo, I. M., Chakravarty, D., Dasgupta, S., de la Rue du Can, S., ... & Wei, T. (2021). Energy efficiency: what has research delivered in the last 40 years?. *Annual Review of Environment and Resources*, 46(1), 135–165.
21. Mengistu, A. T., Dieste, M., Panizzolo, R., & Biazzo, S. (2024). Sustainable product design factors: a comprehensive analysis. *Journal of Cleaner Production*, 463, 142260.
22. Zakari, A., Khan, I., Tan, D., Alvarado, R., & Dagar, V. (2022). Energy efficiency and sustainable development goals (SDGs). *Energy*, 239, 122365.
23. Jain, M., Kumar, D., Chaudhary, J., Kumar, S., Sharma, S., & Verma, A. S. (2023). Review on E-waste management and its impact on the environment and society. *Waste Management Bulletin*, 1(3), 34–44.
24. Majer, J. M., Henscher, H. A., Reuber, P., Fischer-Kreer, D., & Fischer, D. (2022). The effects of visual sustainability labels on consumer perception and behavior: a systematic review of the empirical literature. *Sustainable Production and Consumption*, 33, 1–14.
25. William, M. A., Suárez-López, M. J., Soutullo, S., & Hanafy, A. A. (2021). Building envelopes toward energy-efficient buildings: a balanced multi-approach decision making. *International Journal of Energy Research*, 45(15), 21096–21113.
26. Raghu, S. J., & Rodrigues, L. L. (2020). Behavioral aspects of solid waste management: a systematic review. *Journal of the Air & Waste Management Association*, 70(12), 1268–1302.
27. Mengistu, A. T., & Panizzolo, R. (2022). Tailoring sustainability indicators to small and medium enterprises for measuring industrial sustainability performance. *Measuring Business Excellence*, 27(1), 54–70. https://doi.org/10.1108/mbe-10-2021-0126
28. Ljungberg, L. Y. (2007). Materials selection and design for development of sustainable products. *Materials & Design*, 28(2), 466–479.
29. Rees, W., Tremma, O., & Manning, L. (2019). Sustainability cues on packaging: the influence of recognition on purchasing behavior. *Journal of Cleaner Production*, 235, 841–853.

30. Islam, Q., & Ali Khan, S. M. F. (2024). Assessing consumer behavior in sustainable product markets: a structural equation modeling approach with partial least squares analysis. *Sustainability*, 16(8), 3400.
31. Machado, C. G., Winroth, M. P., & Ribeiro da Silva, E. H. D. (2020). Sustainable manufacturing in Industry 4.0: an emerging research agenda. *International Journal of Production Research*, 58(5), 1462–1484.
32. Sharma, R., Jabbour, C. J. C., & Lopes de Sousa Jabbour, A. B. (2021). Sustainable manufacturing and industry 4.0: what we know and what we don't. *Journal of Enterprise Information Management*, 34(1), 230–266.
33. Yang, Z., Wang, Q., & Jia, M. (2023). Integrating Industry 4.0 and the Internet of Things (IoT) for eco-friendly manufacturing. *The International Journal of Advanced Manufacturing Technology*, 128, 1–10.
34. Matin, A., Islam, M. R., Wang, X., Huo, H., & Xu, G. (2023). AIoT for sustainable manufacturing: overview, challenges, and opportunities. *Internet of Things*, 24, 100901.
35. Zhong, R. Y., & Ge, W. (2018). Internet of things enabled manufacturing: a review. *International Journal of Agile Systems and Management*, 11(2), 126–154.
36. Dyatlov, S. A., Didenko, N. I., Lobanov, O. S., & Kulik, S. V. (2019, July). Digital transformation and convergence effect as factors of achieving sustainable development. In *IOP Conference Series: Earth and Environmental Science* (Vol. 302, No. 1, p. 012102). IOP Publishing.
37. Huang, H., Wang, H., Hu, Y. J., Li, C., & Wang, X. (2022). The development trends of existing building energy conservation and emission reduction—A comprehensive review. *Energy Reports*, 8, 13170–13188.
38. Milewski, D., & Milewska, B. (2023). Efficiency of the consumption of energy in the road transport of goods in the context of the energy crisis. *Energies*, 16(3), 1257.
39. Cheikh, N. B., Zaied, Y. B., & Chevallier, J. (2021). On the nonlinear relationship between energy use and CO2 emissions within an EKC framework: evidence from panel smooth transition regression in the MENA region. *Research in International Business and Finance*, 55, 101331.
40. Zhukovskiy, Y. L., Batueva, D. E., Buldysko, A. D., Gil, B., & Starshaia, V. V. (2021). Fossil energy in the framework of sustainable development: analysis of prospects and development of forecast scenarios. *Energies*, 14(17), 5268.
41. Zhong, X., Hu, M., Deetman, S., Steubing, B., Lin, H. X., Hernandez, G. A., ... & Behrens, P. (2021). Global greenhouse gas emissions from residential and commercial building materials and mitigation strategies to 2060. *Nature Communications*, 12(1), 6126.
42. Oreggioni, G. D., Ferraio, F. M., Crippa, M., Muntean, M., Schaaf, E., Guizzardi, D., ... & Vignati, E. (2021). Climate change in a changing world: socio-economic and technological transitions, regulatory frameworks and trends on global greenhouse gas emissions from EDGAR v. 5.0. *Global Environmental Change*, 70, 102350.
43. Zhang, Z., Chen, Z., Zhang, J., Liu, Y., Chen, L., Yang, M., ... & Yap, P. S. (2024). Municipal solid waste management challenges in developing regions: a comprehensive review and future perspectives for Asia and Africa. science of the total environment, 930, 172794.

44. Desa, A., Ba'yah Abd Kadir, N., & Yusooff, F. (2012). Waste education and awareness strategy: towards solid waste management (SWM) program at UKM. *Procedia-Social and Behavioral Sciences*, 59, 47–50.
45. Bleviss, D. L. (2021). Transportation is critical to reducing greenhouse gas emissions in the United States. *Wiley Interdisciplinary Reviews: Energy and Environment*, 10(2), e390.
46. Meng, Q., Li, M., Liu, W., Li, Z., & Zhang, J. (2021). Pricing policies of dual-channel green supply chain: Considering government subsidies and consumers' dual preferences. *Sustainable Production and Consumption*, 26, 1021–1030.

Chapter 12

AI-driven defense mechanisms for Sybil and DDoS attacks in cloud networks

Arnav Sharma, Kamal Upreti, Safia Farooqui, Rituraj Jain, and Pratham Nijhawan

12.1 INTRODUCTION

12.1.1 Escalating cyber threats in India

Over the last few years, India has witnessed a considerable growth in cyber threats, suggesting the urgency of advanced research to turn into reliable defense strategies. In 2024 (as given in Table 12.1) India ranked as the second most targeted nation in the world with 95 organizations becoming victims of data theft. The banking and financial sectors were the most affected sectors with 20 cases, followed by the Government of India with 13 cases, telecommunications sector with 12 cases, and healthcare sector with 10 cases. Rapid digitalization across the country has expanded the attack surface, leaving critical infrastructure more exposed to advanced cyberattacks. In particular, the Data Security Council of India's Cyber Threat Report 2025 indicated that over 369 million unique malware were detected, highlighting the ever-changing threat landscape. This growing cyber threat landscape underlines the increasing intricacy and advancements of attacks,

Table 12.1 Cyber threat statistics and financial losses in India (2024)

Cyber threat	*Number of victims*	*Financial losses*
Ransomware	64% of Indian organizations in 2023	Average ransom demand was INR 398.4 crore, and the median ransom paid was INR 166 crore
Phishing	2.4 million incidents were recorded in just 10 months of 2024	On an average INR 424.5 crore was the financial loss
DDoS attack	114,000 DDoS attacks recorded throughout 2024	4.73 lakh crore
Overall cybercrimes in India	More than 2.7 million complaints were registered in 2024	INR 11,333 crore in the first nine months of 2024

 DOI: 10.1201/9781003569251-12

emphasizing the importance of thoroughly analyzing specific threats such as Sybil attacks and DDoS attacks.

12.1.2 Growing threats of Sybil and DDoS attacks

Out of all the cyberattacks, threats like Sybil attacks and DDoS (Distributed Denial of Service) attacks, which were rare anomalies at a time, now pose a severe risk to global stability, threatening the privacy of personal data to critical public data. DDoS attacks are malicious acts that are launched from a group of compromised devices, that is, a botnet, to flood the network with false traffic so that the legitimate users cannot access the target website.

12.1.3 Understanding a DDoS attack

DDoS attacks are divided into three types: (1) volumetric attacks (VA), which include User Datagram Protocol (UDP) floods and spoofing packets measured in bits per second (bps); (2) protocol attacks (PAs), which include synchronization (SYN) floods and the Ping of death; and (3) application layer attacks (ALAs), which include GET and POST flood attacks that attack the OpenBSD vulnerabilities.

Following are some examples of DDoS attacks. In October 2023, Google experienced the largest DDoS attack at the time. An HTTPS/2 Rapid Reset attack peaked at 398 million requests per second. An HTTP/2 Rapid Reset is an attack that exploits flaws in the HTTP/2 protocol. The HTTP/2 protocol is used in browsers to request text, images, and other content from other sites. Cloudflare, one of the biggest network operators, experienced the largest DDoS attack of the time in October 2024, of 3.8 terabits per second (Tbps). In both attacks, the attacker launched a volumetric DDoS attack. Volumetric DDoS attacks are used to consume all the available resources of the target, draining the capacity of the target.

12.1.4 Understanding a Sybil attack

A Sybil attack uses a targeted node that operates multiple fake entities at the same time within a peer-to-peer network. It allows a single node to create and manipulate multiple fake identities. The attack is named after the book *Sybil*, which was about a woman who was diagnosed with dissociative identity disorder. It term was introduced in 2002 by Brian Zill, and subsequently discussed in a paper by John R. Douceur, both of Microsoft Research. The goal of a Sybil attack in a blockchain network is to influence the decisions made in the network.

An instance of a Sybil attack is, in 2016, Ethereum Classic, a cryptocurrency, experienced 51% attack, which is a sort of Sybil attack where the malicious user gets access to more than 50% of the network. Here, the hacker simulated multiple fake nodes and gained majority access over the network. This attack led to almost $1 million worth of ETC (Ethereum

Classic) spent during the double-spending, which caused millions of dollars of financial loss. This impacted ETCs reputation, and users and the exchanges stopped transactions made using ETC coins. Tor Network (The Onion Router), known for anonymous communication, experienced a Sybil attack in 2020 and 2021. It was the most expensive attack on the Tor Network, in which the threat actor created thousands of malicious nodes in the network, which compromised almost 23% of the total network at one point. This attack led to resource drainage, severe vulnerabilities in the network. and privacy threats.

This research aims to use advanced machine learning (ML) algorithms for mitigation of cyberattacks, such as Sybil and DDoS attacks. Using support vector machine and random forest, this chapter will help in discovering a systematic approach for detecting the patterns and differences between the fake traffic initialized to overload the network and legitimate users of the network. The detection of these threats is done using large datasets on specific attacks. A code is designed in a way to process all the large datasets at the same time, while classifying and identifying the data. The key parameters used for DDoS attacks are the number of requests and IP address of the sender, whereas the key parameters used for Sybil attacks are ICMP, TCP, and UDP attacks. The accuracy level of the defense system will be identified according to the key parameters only.

12.2 LITERATURE REVIEW

The current research underscores the ongoing and dynamic characteristics of DDoS and Sybil attacks, particularly in relation to the Internet of Things (IoT) and decentralized networks. A variety of studies have explored the mechanism of these attacks, the vulnerabilities they exploit, and the strategies for mitigation, thus providing valuable insights into the threats to network security and their corresponding defense. While some researchers emphasize blockchain-based solutions to improve security, others suggest detection methods that utilize machine learning. This review of the literature consolidates these different perspectives, revealing gaps in current defensive systems and highlighting the necessity for robust and adaptive security frameworks. Through a comprehensive analysis of these studies, this research seeks to create a cohesive approach that utilizes advanced threat detection techniques to enhance network resilience.

Table 12.2 highlights the critical findings, attack focus, and research gaps from existing literature on DDoS and Sybil attacks. It reveals that while numerous studies have explored mitigation strategies—particularly in IoT, SDN, and blockchain environments—many still lack real-time validation, scalability solutions, or comprehensive ML integration. This underscores the pressing need for more robust, adaptable, and empirically tested security frameworks.

Table 12.2 Summary of literature review on Sybil and DDoS attacks

Reference	*Key findings*	*Type of attack*	*Research gap*
[1]	IoT systems face significant DDoS threats, where attackers exploit vulnerabilities like weak authentication to turn devices into botnets, disrupting network availability. This paper examines DDoS attack types, IoT security flaws, and defense mechanisms. Addressing these threats is crucial as IoT adoption grows, making DDoS mitigation a key focus in IoT security research.	DDoS attack	This research paper does not cover real-time prevention methods and it didn't do an in-depth investigation on defense systems based on AI. Also, this paper did not cover the recent advancements in the attackers' methods.
[2]	Because of fewer resources, IoT devices are sensitive to DDoS attacks and thus serve as targets for botnets. Blockchain technology could help mitigate these attacks. This survey classifies solutions based on blockchain technology into distributed architecture, access management (AM), traffic control (TC), and Ethereum platform (EP), assessing their mechanisms, strengths, weaknesses, and potential future research directions.	DDoS attack	This paper has multiple research gaps, such as it only covered blockchain-based solutions, the study was limited to network layer attacks and excluded the application and sensor layers. Moreover, it did not address privacy issues, which is important for security.
[3]	The increasing use of the internet has led to a rise in cyber threats, such as DDoS attack, which is a very significant threat. In a DDoS attack, the bot herder controls the compromised computers to execute the attacks frequently exploiting network vulnerabilities. DDoS attacks such as HTTP floods, which is an application layer DDoS attack, have surged nearly 700% from 2010 to 2011, mostly targeting government websites.	DDoS attack	This research paper only focuses on application layer threats, leaving network layer threats aside. There are no in-depth details about developing botnet methods and prevention measures.
[4]	In this era, DDoS attacks are exploiting IoT security flaws so that the attackers can flood the target network or system, which causes a long outage due to which users cannot access the network. There are various detection techniques such as machine learning techniques. This review paper categorizes the techniques and then lists out the strengths and weaknesses and also explores the emerging threats in IoT, and routing protocols, giving a view for future research.	DDoS attack	This research paper lacks in some major areas as it has a limited dataset, which may affect the result. ML-based identification systems have faced resource limitations.

(*Continued*)

Table 12.2 (Continued)

Reference	*Key findings*	*Type of attack*	*Research gap*
[5]	The rise of DDoS attacks in the terms of scale and impact has increased the need for advanced mitigation systems. In this survey, the main focus is on the software defined networking (SDN)-based solutions, which would offer the detection and mitigation of cyber threats. It classifies the detection techniques and has proposed "ProDefense," which is an SDN framework to secure smart city applications by expressing the key challenges and insights for future research.	DDoS attack	This paper's limitations include relying on software defined networks that introduce single point failure. Large-scale environments remain unverified.
[6]	This study shows a security solution which is entropy-based, having a stateful SDN architecture for the identification and prevention of DoS and DDoS attacks in real IoT traffic. This study has used datasets like Dot-IoT and BigFlows. The experiment confirms that the approach was effective, showing SDN's limitations and helping in enhancing network security with improvement in detection, mitigation, and scalability.	DDoS attack	This paper is mainly dependent on OpenState, which may not be feasible as it does not provide wide support. It is not incorporated with ML parameter tuning.
[7]	DDoS attacks in the time span of 10 to 12 years have targeted large and major platforms like GitHub in 2018 with the incoming traffic of 1.3 Tbps and AWS in 2020 with the incoming traffic of 2.3 Tbps. Attacks grew nearly 154% in 2020 due to IoT vulnerabilities, weak passwords, and unsecure networks. This review paper focuses on strengthening the measures and mitigating large-scale disruptions that can happen in the near future.	DDoS attack	This study shows not being able to completely prevent rapidly growing DDoS attacks because of socioeconomic hindrance, distributed attack type, and less support for universal collaboration.
[8]	This study shows that DDoS attacks have disrupted the networks by using botnets and DDoS-as-a-Service, the major incidents which happened during 2019 and 2020 were AWS with incoming traffic of 2.3 Tbps and Google with incoming traffic of nearly 2.5 Tbps. The attacks have risen nearly 341% from 2019 to 2020, showing the need for advanced mitigation systems across smart city infrastructure, blockchain, IoT, etc.	DDoS attack	This study shows DDoS defense as a challenge, which includes delayed attack identification (DAI). High-quality data and resources were not available, as well as deep learning had increased costs.

(Continued)

Table 12.2 (Continued)

Reference	*Key findings*	*Type of attack*	*Research gap*
[9]	In this study, the focus is on DDoS attack solutions based on blockchain by highlighting IoT, software-defined networking (SDN), and machine learning integrations. As DDoS attacks have surged rapidly after the COVID period, with the incoming traffic exceeding 2 Tbps. Blockchain is offering secured threat intelligence sharing and mitigation.	DDoS attack	This paper has its complete reliance on a customized dataset for SDN. The 100% accuracy shows that there could be overfitting.
[10]	SDNs integrated with machine learning algorithms are facing challenges in mitigating DDoS attacks. This study focuses on ML based on dynamic DDoS attack detection, which has a 99.2% detection rate. The model has outperformed the benchmarks and commercial solutions, with enhancement in network security.	DDoS attack	In this paper, scalability is a major limitation in spite of proposed solutions. Some security concerns just like data leakage are unsolved.
[11]	This review paper focuses on the recent ML techniques used for detection and mitigation of DDoS attacks in the fields such as cloud computing, software-defined networking, and Internet of Things. This paper shows the shift from traditional to smart defense systems, providing insights for future research in improving DDoS attack preventions.	DDoS attack	This paper doesn't validate real-time networks, which restricts its practical usability. Cloud-based methods require future investigations .
[12]	This research paper focuses on increasing network traffic and congestion, which is helping SDN in improving control and automation. SDN divides the controlling and forwarding planes, which promotes network programmability.	DDoS attack	This research paper has multiple limitations such as reliance on an outdated dataset, it couldn't differentiate flash crowd from the attacks, no early identification, and vulnerabilty to botnet-driven attacks.
[13]	In this research paper the CIC-DDoS2019 dataset is used to identify DDoS attack. It evaluates the performance of different machine learning algorithms such as Gaussian naïve Bayes, random forest, K-nearest neighbor, and AdaBoost for detection. This paper shows that the binary representation of DDoS attacks improves detection accuracy.	DDoS attack	This paper mainly relied on Mininet simulations instead of real-world network environments, which then limited the attack scope evaluation.

(Continued)

Table 12.2 (Continued)

Reference	*Key findings*	*Type of attack*	*Research gap*
[14]	This research paper focuses on proposing ML-based methods for detecting Sybil attacks in vehicular networks. In this research paper support vector machine (SVM) is used to evaluate the similarity of vehicular driving patterns to differentiate malicious nodes. For dynamic environment detection, the approach uses driving pattern matrices (DPM).	Sybil attack	This research paper mainly works on eigenvalues of driving pattern matrices (DPM), which may not capture driving behaviors. The SVM algorithm is used in this paper, so the performance depends on it.
[15]	In this review paper, the main focus is on how a malicious node during a Sybil attack illicitly manipulates the system and acts like multiple entities within a network, which violates the one-to-one entity-identity correspondence. Due to this P2P reputation, systems get compromised and help the attacker to manipulate the numbers. In extreme situations, a single device can act and generate an infinite number of fake entities, which breaches the network security.	Sybil attack	This paper gives an overview about Sybil attacks and its prevention measures, but it does not provide an evaluation for the real-world effectiveness of Sybil attacks. Also, this paper did not address the types of attacks and the preventive measures.
[16]	In this research paper the main focus is on designing strategies for Sybil-proofing, and this paper covers identity stability and challenges faced when differentiated. While P2P networks are mainly based on logical addresses, Sybil attacks can manipulate the identifier assignment, which will weaken the replication and control over the system. This paper has proposed a self-registration system that analyzes the resistance.	Sybil attack	This paper does not address the detection and mitigating methods on large-scale network scalability.
[17]	In this research paper, the focus is on wireless sensor networks facing security threats such as data modification and data interference. Sybil attacks have been a major threat to wireless sensor networks, where a single entity initiates multiple fake identities to take over the network. During this attack, unauthorized access is given, which threatens network security and integrity.	Sybil attack	This study depends on a limited dataset, which may impact real-world scenarios. No hybrid approach has been explored to enhance the detection of the attacks.
[18]	This research paper has focused on developing SybilSocNet, which is a machine learning-based algorithm such as support vector machine (SVM), random forest (RF), and K-nearest neighbor (KNN), to detect Sybil nodes over the social networks. This research claims that it achieved 99.9% in the SVM algorithm, 99.6% in the random forest algorithm, and 97% in the KNN algorithm after optimizing the computational demands by data cleaning and data partitioning.	Sybil attack	This research paper's performance is impacted due to low specs hardware because of which large-scale testing cannot happen. The SVM algorithm could not be trained on all four matrices, which also reduced its effectiveness.

(Continued)

Table 12.2 (Continued)

Reference	*Key findings*	*Type of attack*	*Research gap*
[19]	This research paper mainly focuses on cyber threats, including black hole, Sybil attacks, and DDoS attacks for security threats in VANETs. These types of cyberattacks are the reason for packet drops and energy inefficiencies. This paper highlights the necessity of better security systems to ensure safety in vehicular networks.	DDoS attack and Sybil attack	This study did not address the prevention techniques for Sybil attack, black hole, and DDoS attacks. High mobility of UAVs and VANET topology can produce network disruptions.
[20]	The research presents a CAM (comparing and matching) strategy designed to thwart Sybil attacks in IoT environments by validating the locations and IDs of nodes during data exchanges. This technique strengthens the security of WSNs against unicast and multicast assaults, and it assesses network performance indicators, including end-to-end delay, packet delivery, and throughput, utilizing Cooja within the Contiki OS framework for effective intrusion detection.	Sybil attack	The CAM depends on complete node information for verification, which may not be always workable in WSN environments. Scalability issues are also not addressed.
[21]	This paper focuses on the Sybil attacks targeting public blockchains, where an attacker initiates multiple fake entities to gain access through the network. This research paper demonstrates the impact of the attack on throughput and gives a solution to the problem by monitoring the nodes and blacklisting suspicious behavior, which will help in restricting Sybil attack.	Sybil attack	This study depends on the NSL-KDD dataset, which may not be enough to portray MANET scenarios. This may struggle with feature selection biases.
[22]	In this research paper, the main focus is on the Sybil attack detection systems in self-organizing, social networking, and peer-to-peer, after which the paper analyzes mitigation systems. The evaluation parameters are detection rates, false positives, false negatives, and non-trustworthy rate, which helps in focusing on improving the detection accuracy and minimizing errors in detection and mitigation of Sybil attacks.	Sybil attack	In this study, computational overhead as well as scalability challenges have not been addressed for large-scale networks.
[23]	This research paper focuses on Sybil attacks happening in the field of IoT by classifying them into SA-1, SA-2, and SA-3, depending on the attackers' capabilities. This paper reviews defense schemes like behavior classification detection, mobile Sybil detection, and social graph detection by comparing them according to their effectiveness.	Sybil attack	This paper depends upon theoretical models without any thorough assessment.

(Continued)

Table 12.2 (Continued)

Reference	*Key findings*	*Type of attack*	*Research gap*
[24]	This research paper focuses on three approaches: (1) trusted certifications (centralized and decentralized), (2) resource testing (IP and puzzles), and (3) social network-based mitigation using random walk theory. This helps in finding the limitations in present schemes and identifies the research gaps.	Sybil attack	This research paper limits the historical context and mainly focuses on post-classification literature. It doesn't have empirical validations, which may affect its applicability.
[25]	This study focuses on Sybil attacks in P2P networks, where attackers fabricate multiple identities to compromise redundancy. It posits that without a central governing body, Sybil attacks cannot be avoided unless unrealistic conditions of resource equivalence and collaboration among entities are achieved, highlighting the importance of trusted identity verification.	Sybil attack	In this research paper, the development of network protocols such as IPv6 privacy extensions could compromise implicit certification methods.
[26]	Intelligent embedded systems incorporate computational and physical spheres, securing instantaneous coordination. This research contrasts DDoS and Sybil attack identification in cyber-physical systems (CPS). Machine learning models improve safeguarding by reducing threats, enhancing strength. A combined method of anomaly identification and verification fortifies CPS protecting against developing cyber threats and liabilities.	DDoS attack and Sybil attack	The vulnerabilities in CPS security exploration incorporate deprivation of systematized techniques resulting in disjointed security executions. Privileged outcomes often depend on unreliable methods, and mobile CPS devices enhance weakness. Instantaneous risks feedback is demanding due to the high dataset, while restricted coordination minimizes the effectiveness of efforts for reducing threats.

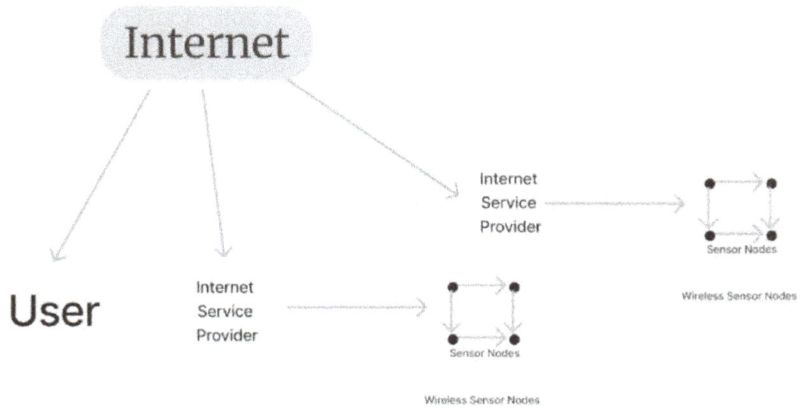

Figure 12.1 Illustration of a Sybil attack in a network environment

12.3 SYBIL ATTACK

A Sybil attack, as shown in Figure 12.1, is a security threat that creates multiple fake entities to get access to the system. Sybil attacks are mostly seen in P2P, blockchains, and wireless sensor networks (WSNs), and exploit and control the network operations. In a Sybil attack, the attacker starts with Sybil nodes, which are used to initiate multiple fake identities. The attacker generates these identities by taking advantage of vulnerabilities in the identity creation process of the system. Once the fake identities are made, the attacker gains control within the network. This can happen by the interaction of fake identities with the legitimate nodes and then manipulating the network [27]. For example, if we talk about peer-to-peer networks, fake identities can disturb the message routing and create problems in the network operations. The ultimate goal of the attacker is to manipulate the network and control the decision-making processes. For example, in the Tor Network attack, the attackers compromised users' privacy. The attacker had control over the entry and exit nodes, which helped the attacker to monitor and manipulate the traffic accordingly.

12.4 DDOS ATTACK

A DDoS attack is an attack type where an attacker overwhelms the server with heavy traffic to cause legitimate users to not be allowed access to the online services. A DDoS attack is a wider category of DoS attack, as the main difference between a DoS and DDoS attack is that a DDoS attacks use multiple compromised machines to hit the target instead of just one.

The number of remote employees and devices is continuously increasing in IoT. In a DDoS attack, the internet network of the target is so full of malicious traffic that legitimate users cannot use it. The types of DDoS attacks include volumetric attacks, protocol attacks, and application-layer attacks. In the case of volumetric attacks, the attacker controls all the bandwidth that can be used between the victim to the network. One such example of a volumetric attack is UDP floods, in which the attacker sends a large volume of UDP packets to random ports on the target system in order to exhaust the resources of the target. An example of a volumetric attack is DNS amplification, which is when the victim's address is spoofed by the attacker, and a DNS lookup request to open DNS server with the spoofed address is made.

In protocol attacks, the attacker tries to exhaust the available capacity of web servers like firewalls. In this attack, the third and fourth layer of the OSI model, that is, the network layer and transport layer, is exposed by the attacker. An example of a protocol attack is a SYN flood, wherein the attacker floods TCP handshake requests of the same Transmission Control Protocol (TCP) to the target with pseudo IP addresses. Every request is responded to by the target, but they never get to complete the last handshake.

An attack can occur in the seventh layer of the OSI model, which is the application layer, where the web pages are generated in response to the Hypertext Transfer Protocol (HTTP) requests. For instance, an example of an application layer attack is the HTTP flood, which causes the server to be flooded with HTTP requests such that the resources get exhausted [28].

Representation of various types of DDoS attacks is given in Figure 12.2.

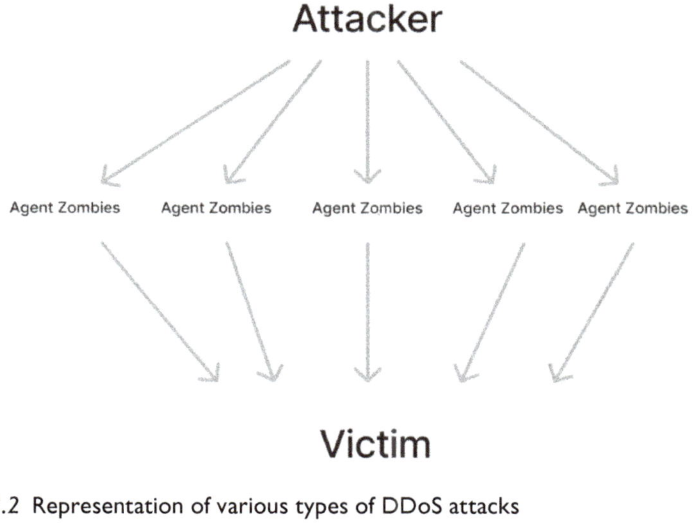

Figure 12.2 Representation of various types of DDoS attacks

12.5 METHODOLOGY

12.5.1 Data collection

We used a public dataset—the DDoS SDN dataset—for the DDoS attack analysis, and we used the KDD dataset for the Sybil attack analysis. The features for DDoS attack detection were packet size, traffic volume, source IPs, destination IPs, and connection durations. For Sybil attack detection, the features were node identifiers, communication patterns, and connection frequency. During data preprocessing, missing values were addressed, uniform scaling applied for consistency, and irrelevant features eradicated using correlation analysis.

12.5.2 Attack simulation

The KDD dataset was used for Sybil attack detection to analyze and detect patterns focusing on communication patterns indicating malicious behavior. This analysis is done to identify and mitigate threats that can happen due to the Sybil attack, which compromises the integrity of the network environment [29]. On the other hand, for the DDoS attack analysis, we replicated the high-volume traffic using advanced network simulation tools such as the Cisco Packet Tracer. This step gave a better approach and understanding of attack behaviors under realistic conditions.

12.5.3 Machine learning (ML) models

We used the random forest (RF) classifier algorithm, support vector machine (SVM) algorithm, decision tree (DT) algorithm, and logistic regression (LR) algorithm for both the DDoS attack and Sybil attack. We divided the data 80:20, that is, 80% of the dataset for training and 20% of the dataset for testing. We used cross-validation for robust model evaluation, and to optimize the model performance we used hyperparameter tuning. The assessment of the efficiency of the model was made on various metrics, such as accuracy, precision, recall, F1 score, and AUC-ROC.

12.5.4 Implementation

During the execution, identification was made using the RF algorithm, SVM algorithm, DT algorithm, and LR algorithm, which evaluated the characteristics of the corresponding dataset to detect and signify them. The main characteristics of DDoS attacks were packet size, traffic volume, and connection duration, while the main attributes of Sybil attacks were node references and communication patterns. We used minimization strategies such as traffic filtering for DDoS attacks and node validation techniques for Sybil attacks to maintain network safety and minimize disruptions.

12.5.5 Evaluating and testing

We have optimized and evaluated the model on the DDoS SDN dataset for DDoS attacks by using the random forest algorithm, support vector machine algorithm, decision trees algorithm, and logistic regression algorithm, where the random forest classifier attained the highest precision. In the same way, we used the KDD dataset for training and evaluation for Sybil attack detection, where random forest surpassed all the other ML algorithms. When executing a comparative analysis for the models where the highs and lows of both models were emphasized, random forest provided a higher level of precision for both cyberattacks.

12.6 RESULTS

In this section, we outline the analysis of the random forest classifier algorithm, support vector machine algorithm, decision trees algorithm, and logistic regression algorithm used for identification of DDoS attacks and Sybil attacks. We examined the outcomes on metrics like accuracy, precision, F1 score, recall, and AUC-ROC.

12.6.1 DDoS attack

The DDoS attack segment describes the implementation of DDoS SDN datasets for training (80%) and testing (20%). Accuracy, precision, recall, F1 score, and AUC-ROC were the indicators used to analyze. The corresponding Table 12.3 and Figure 12.3 demonstrate the performance of four ML algorithms: random forest, support vector machine, decision tree, and logistic regression. The random forest algorithm obtained 99.42% in terms of accuracy, 98.7% in precision, 97.3% in recall, 99.5% in F1 score, and 0.99 in AUC-ROC. Support vector machine attained 93.6% in accuracy, 96.9% in precision, 97.3% in recall, 98.6% in F1 score, and 0.94 in terms of AUC-ROC.

Decision tree achieved 98.22% in accuracy, 98% in precision, 98.30% in recall, 98% in F1 score, and 0.98 in AUC-ROC. Logistic regression achieved 76.64% in terms of accuracy, 75% in precision, 76% in recall,

Table 12.3 Performance comparison of ML models for DDoS attack detection

Model	*Accuracy (%)*	*Precision (%)*	*Recall (%)*	*F1 score (%)*	*AUC-ROC*
Random forest (RF)	99.42	98.7	97.3	99.5	0.99
Support vector machine (SVM)	97.7	96.9	97.3	98.6	0.94
Decision tree (DT)	98.22	98	98.30	98	0.98
Logistic regression (LR)	76.64	75	76	77	0.76

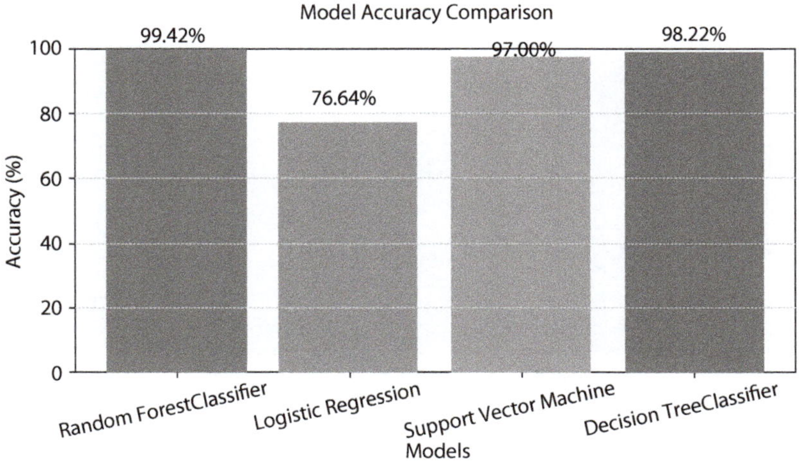

Figure 12.3 Accuracy comparison of ML models for DDoS attack detection

77% in F1 score, and 0.76 in AUC-ROC. These findings indicate that random forest and decision tree outperformed SVM in all evaluation metrics while utilizing a specified dataset.

12.6.2 Sybil attack

The Sybil attack identification section emphasizes the use of the KDD dataset, which is available on Kaggle, for the purpose of training and testing. Accuracy, precision, recall, F1 score, and AUC-ROC are the metrics considered. The comparative analysis, as shown in Table 12.4 and Figure 12.4, reveals that the random forest and decision tree algorithms consistently outperformed support vector machine and logistic regression across all metrics. Random forest achieved 98.04% in terms of accuracy, 98.5% in precision, 97.8% in recall, 98.6% in F1 score, and 0.97 in AUC-ROC. Support vector machine achieved 76.16% in terms of accuracy, 84.7% in precision, 73.63% in recall, 73.92% in F1 score, and 0.72 in AUC-ROC. Decision tree achieved 98.01% in terms of accuracy, 98.7% in precision,

Table 12.4 Performance comparison of ML models for Sybil attack detection

Model	*Accuracy (%)*	*Precision (%)*	*Recall (%)*	*F1 score (%)*	*AUC-ROC*
Random forest (RF)	98.04	98.5	97.8	98.6	0.97
Support vector machine (SVM)	76.16	84.7	73.63	73.92	0.72
Decision tree (DT)	98.01	98.7	98.3	97.5	0.98
Logistic regression (LR)	88.18	88.2	87.9	89	0.88

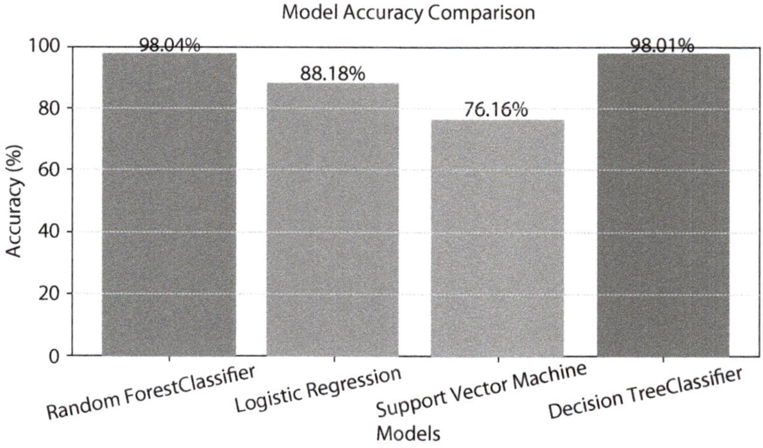

Figure 12.4 Accuracy comparison of ML models for Sybil attack detection

98.3% in recall, 97.5% in F1 score, and 0.98 in AUC-ROC. Logistic regression achieved 88.18% in terms of accuracy, 88.2% in precision, 87.2% in recall, 89% in F1 score, and 0.88 in AUC-ROC. These findings indicate that random forest is more effective in the detection of a Sybil attack within a specified dataset.

12.6.3 Key observations

The key finding shows that the random forest and decision tree algorithms excelled in the detection of DDoS and Sybil attacks, with random forest achieving the highest accuracy of 99.42% and an F1 score of 99.5%. This model demonstrated exceptional reliability in detecting even subtle anomalies, classifying it as a prime candidate for a Sybil attack. This analysis underscored the significance of node identifiers and communication frequencies in the detection of malicious behavior. Support vector machine also performed commendably, attaining an accuracy of 97.7%, although it required extensive fine-tuning and had a lower AUC-ROC of 0.94, suggesting a likelihood for enhancement in kernel optimization. In the context of DDoS attacks, both the random forest and decision tree algorithms effectively managed a large volume of network traffic. On the other hand, logistic regression delivered suboptimal results, achieving only 76.6% precision, which made it less efficient for detecting large volume attack patterns. Generally, the random forest algorithm surfaced as a competent algorithm, while simultaneously decision tree provided strong competition, although with a lower recall score and F1 score. Support vector machine showed future scope, but was computationally complicated, and logistic regression did not meet the performance forecasts.

12.6.4 Visualization of results

The outcomes derived from the confusion matrix evaluation show that the random forest model reliably surpassed all the ML algorithms in the identification of both Sybil and DDoS attacks. It attained higher true positive rates (TPR) and lower false negative rates (FNR), highlighting that the random forest algorithm is more accurate in identifying actual attacks while decreasing the risk of unnoticed detections. In contrast, SVM was associated with higher false positive rates in DDoS detection, which could lead to abundant alerts and limitations in real-world applications. Furthermore, the testing of feature importance in addition supports the dominance of random forest, detecting the packet size, traffic volume, and association duration as essential factors for DDoS detection, thus assisting in the clear classification between normal and malicious traffic. For Sybil attacks, the model emphasized node identifiers and communication trends as the most important features, displaying its capability to retrieve valuable insights from network behavior. These results strengthen the relevance of random forest for cybersecurity applications, primarily in managing large datasets and accurately identifying complicated attack patterns.

12.7 CONCLUSION

The study centered around the competency of the ML models in the identification and minimization of cyber threats such as Sybil and DDoS attacks. The technique includes data preprocessing, modulation of real-world attacks, and implementation of the support vector machine, random forest, decision tree, and logistic regression algorithms. The results demonstrated that the random forest algorithm outperformed all the other algorithms in both types of attacks. Random forest performed more efficiently while detecting DDoS attacks, as it is beneficial while working on large datasets and can recognize complicated traffic patterns. Similarly, random forest had higher consistency while identifying Sybil attacks. These results highlight the importance of incorporating machine learning-based detection systems within cybersecurity approaches to enhance real-time threat detection. Nevertheless, complications related to computational performance remain, especially for SVM, which demands significant hyperparameter optimization. Future discoveries should evaluate aggregated models that combine deep learning strategies with conventional machine learning techniques to further enhance detection accuracy. Furthermore, improving SVM for large datasets and developing real-time outlier detection systems could improve the practical applications of these techniques. The continuous improvements in cyber risks demand the creation of versatile security solutions to ensure effective network security.

REFERENCES

1. Kumari, P., & Jain, A. (2023). A comprehensive study of DDoS attacks over IoT networks and their countermeasures. *Computers & Security, 127*, 103096. https://doi.org/10.1016/j.cose.2023.103096
2. Shah, Z., Ullah, I., Li, H., Levula, A., & Khurshid, K. (2022). Blockchain-based solutions to mitigate Distributed Denial of Service (DDoS) attacks in the Internet of Things (IoT): A survey. *Sensors, 22*(3), 1094. https://doi.org/10.3390/s22031094
3. Alomari, E. S., Manickam, S., Gupta, B. B., Karuppayah, S., & Alfaris, R. (2012). Botnet-based Distributed Denial of Service (DDoS) attacks on the web. *International Journal of Computer Applications, 49*(10), 10–18. https://doi.org/10.5120/7640-0724
4. Adedeji, K. B., Abu-Mahfouz, A. M., & Kurien, A. M. (2023). DDoS attack and detection methods in internet-enabled networks: Concept, research perspectives, and challenges. *Journal of Sensor and Actuator Networks, 12*(4), 51. https://doi.org/10.3390/jsan12040051
5. Bawany, N. Z., Shamsi, J. A., & Salah, K. (2017). DDoS attack detection and mitigation using SDN: Methods, practices, and solutions. *Arabian Journal for Science and Engineering, 42*(2), 425–441. https://doi.org/10.1007/s13369-017-2414-5
6. Galeano-Brajones, J., Carmona-Murillo, J., Valenzuela-Valdés, J. F., & Luna-Valero, F. (2020). Detection and mitigation of DoS and DDoS attacks in IoT-based stateful SDN: An experimental approach. *Sensors, 20*(3), 816. https://doi.org/10.3390/s20030816
7. Rahamathullah, U., & Karthikeyan, E. (2021). Distributed Denial of Service attacks prevention, detection, and mitigation – A review. *Proceedings of the International Conference on Smart Data Intelligence (ICSMDI 2021).* https://doi.org/10.2139/ssrn.3852902
8. Singh, A., & Gupta, B. (2022). Distributed Denial-of-Service (DDoS) attacks and defense mechanisms in various web-enabled computing platforms: Issues, challenges, and future research directions. *International Journal on Semantic Web and Information Systems, 18*(1), 1–43. https://doi.org/10.4018/IJSWIS.297143
9. Chaganti, R., Bhushan, B., & Ravi, V. (2023). A survey on blockchain solutions in DDoS attack mitigation: Techniques, open challenges, and future directions. *Computer Communications, 197*, 96–112. https://doi.org/10.1016/j.comcom.2022.10.026
10. Alashhab, A. A., et al. (2024). Enhancing DDoS attack detection and mitigation in SDN using an ensemble online machine learning model. *IEEE Access, 12*, 51630–51649. https://doi.org/10.1109/ACCESS.2024.3384398
11. Aljuhani, A. (2021). Machine learning approaches for combating Distributed Denial of Service attacks in modern networking environments. *IEEE Access, 9*, 42236–42264. https://doi.org/10.1109/ACCESS.2021.3062909
12. Kannan, C., Muthusamy, R., Srinivasan, V., Chidambaram, V., & Karunakaran, K. (2023). Machine learning-based detection of DDoS attacks in software-defined networks. *Indonesian Journal of Electrical Engineering and Computer Science, 32*(3), 1503–1511. https://doi.org/10.11591/ijeecs.v32.i3.pp1503-1511

13. Ferdous, F., Biswas, T., & Jony, A. (2024). Enhancing cybersecurity: Machine learning approaches for predicting DDoS attacks. *Malaysian Journal of Science and Advanced Technology, 4*, 249–255. https://doi.org/10.56532/mjsat.v4i3.306
14. Gu, P., Khatoun, R., Begriche, Y., & Serhrouchni, A. (2017). Support Vector Machine (SVM)-based Sybil attack detection in vehicular networks. *2017 IEEE Wireless Communications and Networking Conference (WCNC)*, 1–6. https://doi.org/10.1109/WCNC.2017.7925783
15. Balachandran, N., & Sanyal, S. (2012). A review of techniques to mitigate Sybil attacks. *International Journal of Advanced Networking and Applications*. https://arxiv.org/pdf/1207.2617
16. Dinger, J., & Hartenstein, H. (2006). Defending the Sybil attack in P2P networks: Taxonomy, challenges, and a proposal for self-registration. *First International Conference on Availability, Reliability and Security (ARES'06), 8*, 763. https://doi.org/10.1109/ARES.2006.45
17. Mandala, M., Vijayasaraswathi, R., & Raparthi, V. (2021). Detecting Sybil attack in wireless sensor networks using machine learning algorithms. *IOP Conference Series: Materials Science and Engineering, 1042*, 012029. https://doi.org/10.1088/1757-899X/1042/1/012029
18. Cárdenas-Haro, J. A., Salem, M., Aldaco-Gastélum, A. N., López-Avitia, R., & Dawson, M. (2024). Enhancing security in social networks through machine learning: Detecting and mitigating Sybil attacks with SybilSocNet. *Algorithms, 17*(10), 442. https://doi.org/10.3390/a17100442
19. Thakur, A., Sharma, S. K., Bhogey, R. K., & Kumar, B. (2024). Evaluating ad-hoc network performance against black hole, Sybil, and DDoS attacks. (2024). *Machine Intelligence Research, 18*(2), 161–173. Retrieved from https://machineintelligenceresearchs.com/index.php/mir/article/view/145
20. Singhal, P., Sharma, P., & Rizvi, S. N. (2018). Thwarting Sybil attack by CAM method in WSN using Cooja simulator framework. *International Journal of Engineering & Technology, 7*, 589–594. https://doi.org/10.14419/ijet.v7i4.19.27965
21. Swathi, P., Modi, C., & Patel, D. (2019). Preventing Sybil attack in blockchain using distributed behavior monitoring of miners. *2019 10th International Conference on Computing, Communication and Networking Technologies (ICCCNT)*, 1–6. https://doi.org/10.1109/ICCCNT45670.2019.8944507
22. Bhise, A., & Kamble, S. (2016). Review on detection and mitigation of Sybil attack in the network. *Procedia Computer Science, 78*, 395–401. https://doi.org/10.1016/j.procs.2016.02.080
23. Zhang, K., Liang, X., Lu, R., & Shen, X. (2014). Sybil attacks and their defenses in the Internet of Things. *IEEE Internet of Things Journal, 1*(5), 372–383. https://doi.org/10.1109/JIOT.2014.2344013
24. Mohaisen, D., & Kim, J. (2013). The Sybil attacks and defenses: A survey. *The Smart Computing Review, 3*, 9–21. https://doi.org/10.48550/arXiv.1312.6349
25. Douceur, J. R. (2002). The Sybil attack. In P. Druschel, F. Kaashoek, & A. Rowstron (Eds.), *Peer-to-Peer Systems* (Vol. 2429, pp. 251–260). Springer, Berlin, Heidelberg. https://doi.org/10.1007/3-540-45748-8_24
26. Baliyan, Anupam, Kaswan, Kuldeep, Kumar, Naresh, Upreti, Kamal, & Kannan, Ramani. (2022). Cyber physical systems: concepts and applications. https://doi.org/10.1201/9781003220664

27. Kamal, U., Prashant, V., Aravindan, S., Daya Sagar, K. V., Mahaveer Kannan, R., & Charles Babu, G. (Aug, 2023). Detection of banking financial frauds using hyper-parameter tuning of DL in cloud computing environment. *International Journal of Cooperative Information Systems*, World Scientific Publishing Co. https://doi.org/10.1142/S0218843023500247.
28. Kamal, U., Sheng-Lung, P., Pravin Ramdas, K., Prasun, C., Al-Alshaikh Halah, A., Sharma, A. K., & Ramesh Chandra, P. (2023). A multi-model unified disease diagnosis framework for cyber healthcare using IoMT-cloud computing networks. *Journal of Discrete Mathematical Sciences and Cryptography*, 26(6), 1819–1834. https://doi.org/10.47974/JDMSC-1831
29. Bhatnagar, S., Dayal, M., Singh, D., Upreti, S., Upreti, K., & Kumar, J. (May 2023). Block-Hash Signature (BHS) for transaction validation in smart contracts for security and privacy using blockchain. *JMM*, 19(4), 935–962.

Chapter 13

A comparative study of machine learning

Models for web tracker detection

Calvin Mathew and Ramesh Chandra Poonia

13.1 INTRODUCTION

Web trackers are used to collect, store, and analyze user activity on a website. Web trackers were created to provide personalized experiences online. The rapid increase of web trackers on websites has sparked growing concerns over online privacy and the unauthorized collection of user data. Many web trackers collect data without the user's explicit consent or awareness. This includes browsing patterns, search terms, location information, and sometimes even personal information such as email addresses or IP addresses. In most cases, users have little to no idea just how much their activity is being tracked from one website to another. It is through web trackers that advertisers and companies are able to engage in behavioral targeting, or ads and content created based on a user's browsing history. This may make it more applicable but leaves behind a nagging feeling of being surveilled, which decreases the degree to which the user owns their browser. While some tracking methods claim to collect anonymous data, the reality is that it's often possible to deanonymize users by combining various data points collected over time. As trackers gather more information, they can reidentify individuals, even if they never provided personal details directly.

Traditional tracker-blocking tools, such as browser-based extensions like Adblock Plus and Ghostery, rely on predefined blacklists to block known trackers. These blacklists are community driven, whereby users submit newly found trackers. While these tools are effective to some extent, they are limited by the constant evolution of tracking techniques. Newer methods, such as CNAME cloaking, fingerprinting, and script-based obfuscation, allow trackers to bypass these blacklists, making it increasingly difficult to maintain effective blocking strategies. As a result, there is a pressing need for more dynamic and adaptable solutions.

As a possible solution for these problems, machine learning (ML) is emerging. Trained on large-scale web traffic datasets and understanding tracker behavior patterns, machine learning enables the recognition of both known and new tracking techniques. Several ML techniques have been employed in recent studies to automate the detection of web trackers.

DOI: 10.1201/9781003569251-13

Coupled with feature engineering techniques, including HTTP request analysis, JavaScript behavior extraction, and profiling of domains, these algorithms suggest a method for scalable and adaptive tracker detection.

Despite the promise machine learning holds in this domain, no consensus has been reached in the determination of the most effective algorithm or technique in web tracker detection. Various studies have reported different levels of success across different performance metrics. Further, some complications, like the difficulty of feature selection, the imbalance of classes, and continually evolving tracking techniques, make it challenging to build robust detection models. Most datasets depend on community-maintained blacklists for labeling trackers because these lists cannot be exhaustive, and accuracy may vary.

Here we present a general study of several machine learning techniques for web tracker detection, evaluating their performance from the key metrics and computational efficiency. By leveraging a comprehensive dataset that includes web traffic and tracking behavior, we are trying to find the strong points and weak points of different algorithms and gain insight into the most effective approaches for real-world deployment. Furthermore, this study highlights the importance of feature engineering, dataset prepossessing, and the challenges posed by new and evolving tracking methods. These results will aid in developing better web tracker detection systems, ultimately enhancing online privacy for users worldwide.

13.2 LITERATURE REVIEW

The growing prevalence of web trackers has sparked anxiety over the security of user's personal and sensitive data. Here we review findings from recent studies that used various machine learning techniques for web tracker detection.

The first step is to collect a dataset to create the ML models. There are many ways to collect a dataset. including OpenWPM [1, 2], Selenium WebDriver [3], Alexa top sites datasets, and popular datasets from Adblock Plus, Ghostery, and Privacy Badger. Datasets are used to train and evaluate different ML models. They have collected data on HTTP requests and responses, JavaScript files, server logs, and browser network logs, which are then labeled as tracking or non-tracking resources based on established blacklists for supervised learning or unlabeled for unsupervised learning.

One study proposed a hybrid approach that integrates traditional blacklisting techniques with machine learning classifiers to enhance detection capabilities [3]. This approach not only leverages existing knowledge from community-driven filter lists but also enhances detection capabilities through the application of predictors such as random forest, multilayer perceptron (MLP), support vector machine (SVM), and k-nearest neighbors (k-NN). The comparative analysis of these classifiers allows researchers to identify the most effective techniques for detecting web trackers.

Another study takes an approach that uses wildcard matching with lists for blocking trackers, from major tracker mitigation lists, to identify CNAME cloaking-based tracking. The method generates patterns derived from recognized tracking domains that are used for matching CNAMEs associated with various subdomains. In contrast, this research also introduces a supervised machine learning approach to identify CNAME cloaking-based tracking without depending on real-time DNS lookups. It focuses on extracting features from requests linked to CNAME cloaking trackers and applies a classification algorithm to identify tracker requests [1].

One paper uses machine learning classifiers such as ADTree and AdaBoost M1 to detect cross-site scripting [4].

The tools and techniques utilized in these studies are diverse and sophisticated. Commonly employed tools include web measurement platforms and data collection frameworks that facilitate the extraction of relevant features from web resources. For instance, the use of abstract syntax tree (AST) parsing has been highlighted as an effective technique for extracting JavaScript API calls, which are critical for identifying tracking behaviors.

Machine learning classifiers play a central role in these studies, with researchers testing various algorithms to determine their effectiveness in detecting web trackers. We used accuracy, precision, recall, and F-measure to evaluate the classifiers' performance, providing an in-depth analysis of their strengths and weaknesses.

Despite the development of ML approaches for web tracker detection, several limitations and gaps have been identified. One major challenge is the reliance on existing labeling systems, which may not always be comprehensive or accurate. This may lead to potential biases in the training datasets and affect the overall classifier efficacy.

Furthermore, many studies focus primarily on specific types of tracking, such as JavaScript-based tracking, potentially overlooking other sophisticated tracking methods, including fingerprinting techniques. This narrow focus may limit the generalization of the discoveries and the relevance of the proposed solutions.

Another significant challenge is the need for continuous research to keep pace with rapidly evolving web tracking technologies. As new tracking methods emerge, it becomes essential to regularly update and retrain machine learning models to ensure they remain effective.

One specific study investigated the application of machine learning in labeling third-party resources as functional or tracking to improve user privacy through disconnection from static blacklists [5]. The authors harvested a dataset of 1000 web resources with an equal number of functional and tracking resources based on browsing history from top-ranked websites. The extracted features were different HTTP request parameters like cookie duration, URL length, and payload attributes. Some classifiers like naïve Bayes, random forest, multilayer perceptron, and support vector machines were tested, and random forest produced the maximum classification accuracy of 91% [5]. The method proved the feasibility of ML-based detection systems in dynamically excluding tracking resources.

The research has also presented major strengths and weaknesses of ML-based web tracking detection. On the one hand, the model was highly accurate and resilient in separating tracking and functional resources and showed a huge improvement compared to conventional blacklist-based approaches [5]. In addition, the diversity of extracted features enriched detection by comprehensively capturing tracking behaviors. Nonetheless, the research recognized issues like the small size of the dataset, dependence on expert-annotated data, and the possible computational burden in real-time scenarios [5]. These issues imply the necessity for scalable datasets and adaptive models that can adapt to evolving tracking methods.

Advancements in the area have also been toward building hybrid models that integrate supervised and unsupervised learning methods, enhancing classification accuracy in changing environments. With tracking detection, privacy-preserving protocols like federated learning have been suggested for boosting tracking detection without infringing on user information [5]. While web tracking techniques keep changing, future research should concentrate on real-time detection, explainability, and adaptive learning to maximize the effectiveness of ML-based countermeasures against third-party tracking.

One of the earliest works in this area is Mayer and Mitchell's study, which presents a comprehensive examination of third-party web tracking from both policy and technological standpoints. The authors discuss tracking methods, regulatory policies, and the possible risks involved in data gathering. Utilizing the FourthParty measurement platform, the study presents a comprehensive empirical analysis of tracking practices, including stateful and stateless tracking mechanisms [6].

The research indicates that third-party tracking services often obtain user browsing histories from HTTP referrers and embedded scripts, which can pose privacy threats, including the disclosure of sensitive personal data. The authors point out user concerns, referencing survey findings that show broad resistance to tracking-based advertising. They also address the economic implications of tracking, raising the question of whether such activities are actually necessary to maintain free web services [6].

One of the key strengths of this research is its holistic approach, combining technical analysis with policy debate to present a complete picture of third-party tracking. One of the limitations is the temporal nature of the results, since the web tracking environment changes very quickly, which may make some of the insights obsolete. Additionally, although the research concentrates mainly on web tracking, it recognizes the necessity for additional research on mobile tracking mechanisms [6].

This study fills the gap left by machine learning-based detection methods by shedding light on the underlying mechanisms and policy implications of third-party tracking. Whereas ML-based studies aim to automate tracker discovery, research such as this provides significant background information on the wider implications of tracking technologies. Future studies ought to combine both technical innovation and policy concerns in order to create comprehensive solutions for online privacy protection.

A significant contribution in this domain is Wang's study on using deep learning for blocking web tracking. This research explores how MLP-based models can distinguish between tracker and non-tracker domains using domain name characteristics. By utilizing Peter Lowe's Ad and Tracking Server List for tracker domains and Alexa's top 100,000 websites for non-tracker domains, the study constructs a robust training dataset to improve classification accuracy [7].

The process of working includes converting the domain names to numerical forms utilizing the word2vec method prior to training the neural network [7]. To utilize the model, a proxy server based on Python intercepts HTTP/HTTPS requests and blocks any captured trackers in real time. The outcome is evidenced by a high accuracy of classification at 98%.

One of the significant advantages of this research is its novel use of AI, which learns more effectively to emerging tracking methods than fixed rule-based approaches. One of its significant drawbacks is that it relies on manually predefined datasets, which can be difficult to generalize to new or changing tracker domains. Second, the model might face the issue of overfitting, mislabeling some domains based on biases in the training data [7].

By combining deep learning for tracker detection, this study extends existing machine learning methods, for example, those developed by Guarino et al., who employed random forest classifiers to distinguish tracking and non-tracking resources with high precision [5]. In contrast to conventional methods using manually labeled features, deep learning methods such as those outlined by Wang provide more flexibility and automation in tracker identification.

In general, web tracking detection utilizing AI has an evolving impact to provide good potential directions for developing future privacy-related technologies. A blend of real-time deployment strategy with deep learning models, that is, proxy-based filtering, is an essential milestone to provide more powerful and adaptive third-party tracking countermeasures.

Before we move on, let's briefly explore different ways people block web trackers. One method is through privacy-focused browsers such as Brave, Firefox, and TOR. People are now moving from generic browsers like Google Chrome, Safari, and Microsoft to these privacy-based browsers. Brave blocks web tracking by default. It also includes features like blocking fingerprinting and HTTPS for secure connections. Firefox offers enhanced tracking protection where they block known trackers, third-party cookies, and fingerprinting. They balance privacy and user experience efficiently.

Tor provides the highest level of privacy by routing through the Tor network, preventing surveillance and tracking by default, but it sacrifices speed and usability for privacy. Another method is through browser extensions like uBlock Origin, Ghostery, and Privacy Badger. Also, most VPN providers give an option to block known web trackers. DNS-based prevention blocks web trackers at the DNS level, and by using firewalls we can add rules to block web trackers as shown in the summary of existing literature provided in Table 13.1.

13.3 METHODOLOGY

Here we are describing the mode and approach used for detecting web trackers, focusing on data collection, prepossessing, feature engineering, training the model, and assessing its performance. The objective is to compare multiple machine learning techniques and identify the most constructive procedure for web tracker detection. An overview of the methodology pipeline for web tracker detection is provided in Figure 13.1.

13.3.1 Data collection

For this experiment, we used the TrackSign-labeled web tracking dataset developed by Ismael Castell-Uroz and Pere Barlet-Ros in their paper "TrackSign-labeled web tracking dataset" [12]. This dataset offers a full suite of web traffic gathered from URL-profiling of both tracking and non-tracking resources, offering a rich source of data for tracking mechanisms on the web.

- HTTP requests: HTTP headers, cookies, and metadata of the request.
- Domain and URL information: These identify domains and enable the separation of tracking based on domain reputation.
- Tracking labels: Each resource in the dataset is labeled as "istracking" based on established tracking detection methodologies and blacklists, such as Adblock Plus and Ghostery.

This dataset contains approximately 29,00,000 entries of URLs and tracking conditions hash values, country code requests, responses, etc. We took a sample with an equal amount of tracker data and non-tracker data.

13.3.2 Data preprocessing

Before feeding the data into the machine learning models, several prepossessing steps had to be made to verify the dataset prepared for training.

Table 13.1 Summary of research studies

Author	*Year*	*Title*	*Tool and technique used*	*Limitations*
Georg Merzdovnik, Markus Huber [8]	2017	Block Me If You Can: A Large-Scale Study of Tracker-Blocking Tools	AdBlock Plus, Ghostery, analysis of websites	Incomplete blocking of stateless fingerprinting, mobile device challenges, risk of breaking websites
Ha Dao [1]	2021	CNAME Cloaking-Based Tracking on the Web: Characterization, Detection, and Protection	Measurement platforms, OpenWPM, network-based blocking, UglifyJS	Incompleteness of tracking filter lists, potential missed cases, unstable crawling results
Syed Rizvi [9]	2022	Defending Application of Artificial Intelligence to Network Forensics: Survey Challenges and Future Directions	Ensemble learning, hybrid learning approach, behavioral analysis	Data quality and volume, standardization issues
Umar Iqbal [2]	2021	Fingerprinting the Fingerprinters: Learning to Detect Browser Fingerprinting Behaviors	Static analysis, feature selection, dynamic analysis, combining models	Script dispersion, partial fingerprinting, API usage overlap
Rui Wang [4]	2017	Machine Learning-Based Cross-Site Scripting Detection in Online Social Network	Feature extraction, ADTree, AdaBoost M1 algorithms	May generate false positives for benign but complex JavaScript code
Mohammad Saidur Rahman [10]	2021	Mockingbird: Defending Against Deep-Learning Based Website Fingerprinting Attacks with Adversarial Traces	Burst molding, WTF-PAD, application-layer defenses	Bandwidth and latency overhead, vulnerability to adversarial training
Alfonso Guarino [5]	2020	On Analyzing Third-Party Tracking via Machine Learning	Random forest, J48, naïve Bayes, multilayer perceptron	Manual labeling of samples was required
Jonathan R. Mayer, John C. Mitchell [6]	2012	Third-Party Web Tracking: Policy and Technology	FourthParty web measurement platform, network traffic monitoring	Technical complexity, dependence on browser capabilities
Qianru Wu, Qixu Liu [11]	2015	TrackerDetector: A System to Detect Third-Party Trackers Through Machine Learning	Dynamic JavaScript analysis, feature extraction	False positives/negatives, evolving tracking techniques
Jianyi Wang [7]	2020	Using Deep Learning to Block Web Tracking	AI deep learning, multilayer perceptron	Only mitigates threats higher than business impact, complex implementation

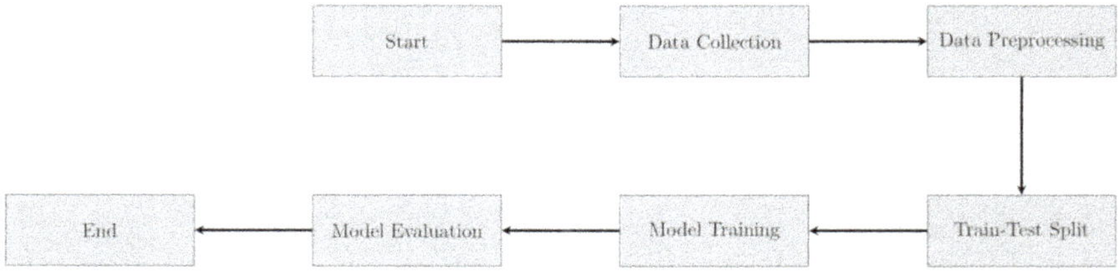

Figure 13.1 Overview of the methodology pipeline for web tracker detection

- Handling missing data: Any data gaps, such as incomplete HTTP headers or missing JavaScript execution logs, were removed in cases where the missing information was not critical. Unwanted columns, like server ID, country code, from cache, and timestamps, were removed as these were not needed.
- Categorical encoding: These were transformed into the appropriate pattern for ML by using one-hot encoding or label encoding
- Scaling: Values such as hostid, mimetypeid, and resourceid are scaled after using logarithms on them.
- df[columnstoscale] = df[columnstoscale].apply(lambda x: np.log1p(x))
- Feature selection: To reduce noise and improve model accuracy, variable importance was assessed using random forest and mutual information methods. Features that did not contribute significantly to the detection task were removed.

Because of the limited computational resources, we used the sampling method to randomly select rows from the dataset. As this dataset had class imbalance, this approach fixed both problems of limited computational resources and class imbalance, as it only took a specific number of tracker and known tracker data. The steps of data preprocessing are illustrated in Figure 13.2.

13.3.3 Models

Here we are trying to employ different combinations of ML models to detect web trackers with the same large dataset. Given the varied nature of ML models, we are trying to find out which are effective. Each model has its own advantages, and we are trying to find which of these advantages can be used by us to find the web trackers. We are testing both supervised learning and unsupervised learning.

As mentioned in Figure 13.3, in unsupervised learning we are trying out isolation forest, which is an anomaly detection model, and DBSCAN, which is a clustering model that tries to cluster anomalies and normal data. In supervised learning, we are testing random forest, which is based on

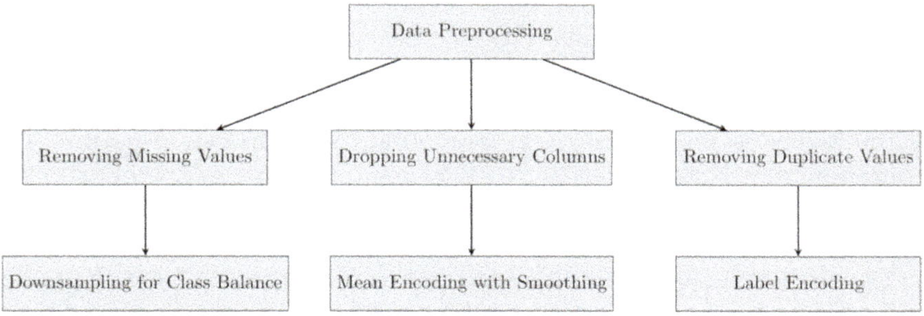

Figure 13.2 Steps involved in the data preprocessing phase

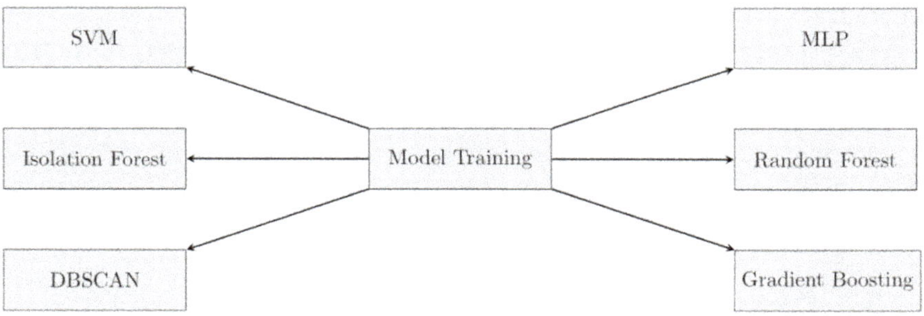

Figure 13.3 Model training techniques used in the project

decision trees and combines the output of multiple trees to increase predictive accuracy and robustness. SVM is effective in binary tasks. Gradient boosting is an ML model that learns from the errors of previous models, creating a powerful ensemble model. Artificial neural networks (ANNs) help in capturing nonlinear relationships, making them suitable for detecting sophisticated tracking behaviors. We are not using convolutional neural networks (CNNs) or recurrent (RNNs) because CNNs are best used for images and RNNs are best used for time series data.

13.3.4 Isolation forest

Isolation forest is an unsupervised model used for outlier identification. It functions by recognizing data points that substantially deviate from normal patterns, classifying them as anomalies. The term "isolation" refers to the process of separating an individual data instance from the rest [13]. This model works by recursively partitioning the data, creating shorter paths for anomalies (such as trackers) compared to regular data points.

In simpler terms, isolation forest constructs a pool of isolation trees for the specified dataset. Instances having minimized mean path lengths across these trees are highlighted as abnormalities. The method relies on only two key criteria: the quantity of trees to build and the extent of the subsamples [13].

In this implementation, isolation forest is used to detect web trackers without relying on labeled data, leveraging its unsupervised nature. The contamination parameter is set to 0.5, assuming an equal proportion of trackers and non-trackers in the dataset. This configuration allows for greater flexibility in identifying unusual tracking behavior.

The dataset includes features (denoted as X) separated from the target column (tracksign), which contains binary labels: 1 for trackers and 0 for non-trackers. The isolation forest model is trained solely on the features (X), as it does not require labels (y) for training. This makes it extremely acceptable for tasks for identifying irregularities where labeled data may be limited.

Once trained, the model predicts anomalies in the dataset, outputting values of –1 for outliers and 1 for expected data. These predictions are then mapped to a binary format consistent with the dataset's labeling, where 1 represents a tracker and 0 represents a non-tracker. This transformation simplifies the comparison between model predictions and the original labels, enabling a more intuitive assessment of the model's execution.

13.3.5 DBSCAN

The DBSCAN algorithm identifies clusters by estimating the minimum density required for clustering. It does this by setting the minimum required total count of neighboring entities (minPts) inside a defined range, ϵ, based on a chosen distance metric. If a data point has more than minPts adjacent points within the ϵ range, it is categorized as a core point. The key idea of DBSCAN is to locate regions with sufficient density that are divided by regions of sparse areas. As a substitute for estimating density between points, DBSCAN groups every entry within the ϵ radius of a central point into the same cluster, a concept known as "direct density reachability." When adjacent points of a central point are also core points, their neighborhoods belong to the cluster through "density reachability" [14]. Points within the cluster that are not core points are referred to as boundary points, and all entries in the cluster are considered "density connected." Entries that are not connected to any cluster are treated as noise and excluded [14]. DBSCAN does not rely on any specific data distribution, making it a non-parametric algorithm.

In this implementation, GPU-accelerated libraries are used for dimensionality reduction and clustering to detect anomalies. The workflow begins by loading a dataset from a CSV file into a cuDF DataFrame, where the features (X) are separated from the target variable (y). PCA is applied to X to diminish its dimensionality, thereby improving the clustering process. The number of PCA components is set to 14, determined through prior optimization. The reduced dataset, represented as Xpca, is clustered using the DBSCAN algorithm.

DBSCAN is configured with an epsilon (ϵ) value of 0.1, defining the maximum distance for neighborhood inclusion, and a min_samples value of 2, specifying the minimum number of points essential to establish a compact grouping. Cluster labels are then merged with the original dataset, with points labeled –1 flagged as noise (potential anomalies) and other labels indicating the identified clusters.

13.3.6 Random forest

Random forest is a potent group of training procedures built on decision trees. It is carried out by combining the outputs of multiple tree-based classifiers, with each tree casting a vote on the prevalent class. The final classification is determined by aggregating these votes. Known for its high veracity, solidity, and ability to handle outliers and noise, random forest is resistant to overfitting and remains one of the most popular techniques in data mining [15]. In our research, random forest acts in the capacity of a reliable benchmark model for detecting web trackers by classifying instances as tracker or non-tracker with high precision.

To optimize performance, we utilize a GPU-accelerated random forest classifier via cuML, a GPU-optimized machine learning library. The dataset is loaded into a cuDF DataFrame, cuML's equivalent of a Pandas DataFrame specifically designed for GPU-based processing. The target variable, tracksign, is separated from the feature set, X, for training purposes. Since cuML does not provide its own train_test_split function, we use the train_test_split function from scikit-learn, converting the split datasets back to cuDF format for GPU-based training.

We set up a random forest model using cuML's RandomForestClassiffier, configuring it with 100 decision trees (nestimators = 100) and (maxdepth = 10). These settings are chosen to find the optimal balance between model accuracy and computational cost. The model is developed using the training dataset (Xtrain and ytrain), and estimates are derived for the test dataset (Xtest).

To analyze the model's effectiveness, the predictions are transferred back to CPU memory to be compatible with scikit-learn's evaluation functions.

13.3.7 Support vector machine

SVM is a supervised learning model that is highly effective for binary classification tasks. As one of the classical machine learning techniques, SVM remains a strong option for solving large-scale classification problems [16]. The model is based on the simple equation $y = \omega x + \gamma$, but it can be adjusted to enable linear separation of data into different categories.

The goal of SVM is to establish a decision boundary that best distinguishes between tracker and non-tracker data points, offering a reliable baseline for accurate classification. SVM performs exceptionally well with high-dimensional data when the right features are provided. However, it can be computationally expensive, and its performance may suffer if the data is noisy, especially when the classes cannot be separated by a straight line.

This implementation leverages GPU-accelerated SVM for binary classification of web tracker data. First, the dataset is loaded as a cuDF DataFrame (a GPU-optimized format) from a CSV file. The target variable, tracksign, is separated from the feature set X, which contains the input data used for training and predictions. Since the cuML library (the GPU-accelerated machine learning library used here) does not yet support train_test_split, the code uses scikit-learn's equivalent function to split the data, then converts the resulting training and testing sets back to cuDF format to take advantage of GPU acceleration.

The model with a radial basis function (RBF) kernel is initialized using the cuML SVC class, with hyperparameters like C (regularization parameter) and gamma (kernel coefficient) set for optimization. After training the model with the training data (Xtrain and ytrain), estimations are created on the basis of the test set (Xtest). These predictions are transferred to CPU memory for evaluation, using scikit-learn's accuracy score, classification report, and confusion matrix functions.

13.3.8 Gradient boosting

Boosting algorithms integrate multiple weak learners—models that achieve performance marginally above random chance—into a more robust model through an iterative process [17]. Gradient boosting is a specific type of boosting algorithm designed for tasks like regression [17]. Simply put, gradient boosting is a group approach that builds several weak learners, typically decision trees, in a sequential manner. Each successive learner focuses on correcting the errors of the previous ones.

For a training dataset D = xi, yiNi = 1. The intent of gradient boosting is to estimate a function F(x) that associates input features x to corresponding outputs y by reducing the anticipated loss [17]. The algorithm incrementally constructs F(x) as a value-adjusted total of simpler functions [17]. Gradient boosting excels at handling complex datasets by capturing nonlinear relationships, which are often present in tracking patterns. Its ability to address both bias and variance makes it particularly useful for tasks like tracking detection. However, it requires careful hyperparameter tuning to avoid overfitting and can be computationally intensive.

In this study, a gradient boosting classifier is applied to detect web trackers using labeled data. The dataset is loaded into a pandas DataFrame, with the features (X) and target variable (y) separated. The target variable, tracksign, indicates whether an instance represents a tracker or a non-tracker. The dataset is divided into training and testing sets in a 75:25 ratio using the train_test_split function. The GradientBoostingClassifier from scikit-learn is used to build the model. For this task, the classifier is configured with 100 estimators (nestimators = 100), a learning rate of 0.1, and a maximum depth of 6 for each tree, balancing model complexity with interpretability.

The model is fitted with the dataset (Xtrain, ytrain) to learn distinguishing features between trackers and non-trackers. Once training is complete, predictions are made on the test set (Xtest). This implementation showcases gradient boosting as a highly effective method for identifying subtle patterns in data, making it ideal for complex classification tasks.

13.3.9 Neural networks

We are utilizing multilayer perceptron (MLP) in the ANN family to test the model's ability to recognize sophisticated structures in web tracking data. The concept behind ANNs is inspired by the way the human brain processes information, where it establishes a simple model and connects various "neurons" in different configurations [18]. An ANN serves as a computational framework made up of numerous linked elements, or neurons, where each node performs a specific function, known as the activation function [18]. The reason we chose MLP is because the features in the dataset are tabular in form. It is well-suited for tabular or structured data. While neural networks like CNNs depend on spatial data and web tracking features, MLPs do not require these inputs to function. CNNs are mainly used for image processing and 2D feature maps, and RNNs are designed for sequential data or text data where context from previous steps is important. MLPs are computationally less expensive than other neural networks like CNNs or RNNs. MLPs require fewer parameters and are easier to train on moderate-sized datasets, making them a very good choice for web tracker detection.

In this model, we use an ANN to explore deep learning's potential to detect intricate patterns that simpler models may miss. While ANNs are powerful tools, they often demand extensive data preprocessing and substantial computational resources, especially when dealing with large datasets. To balance performance and computational feasibility, we've implemented a relatively simple ANN architecture.

This implementation uses a fully connected neural network built in PyTorch to classify web tracking data, distinguishing trackers from non-trackers. The dataset is first loaded and split into features (X) and labels (y), using train_test_split to create training and testing sets. These sets are then converted into PyTorch tensors, and a DataLoader is created to handle batch processing, making model training more efficient.

Once training is complete, it is evaluated on the test set without calculating gradients, and the predicted class labels are obtained using torch.max. This setup demonstrates how a neural network can be applied to binary classification tasks, shedding light on its ability to differentiate tracker patterns within the data.

13.3.10 Inferences

See Table 13.2.

13.3.11 Individual model performance analysis

To assess the performance of various machine learning models for the purpose of web tracker detection, several models were tested and validated using standard classification metrics, including precision, recall, and F1 score. The results provide significant insights into how supervised and unsupervised models behave when used in the context of detecting web trackers. The evaluation metrics for all models are summarized in Table 13.2.

Random forest proved to be one of the highest-performing models, whose precision, recall, and F1 score measurements were 0.975. This high value is due to its ensemble structure, which makes strong and stable predictions by voting over the multiple decision trees and their outputs. The model tolerates feature variation well and overfits less when compared to one-tree approaches. Its balance between bias and variance positions makes it favorable for use in tasks where interactions between features are nonlinear and complex, as they typically are in tracker detection situations.

Gradient boosting performed slightly better than random forest with all three measures of evaluation at 0.98, which represents a very accurate classification performance. Gradient boosting is recognized for minimizing error by iteratively refining model accuracy by concentrating on hard-to-classify instances, so it is especially effective in enhancing model accuracy. Its capacity to learn from residuals of previous

Table 13.2 Performance comparison of different models

Model	*Precision*	*Recall*	*F1 score*
Isolation forest	0.68	0.68	0.68
DBSCAN	0.43	0.86	0.573
Random forest	0.975	0.975	0.975
SVM	0.97	0.97	0.97
Gradient boosting	0.98	0.98	0.98
MLP	0.97	0.97	0.97

models is a big strength in identifying nuanced patterns that simpler models may not catch. In addition to its overall excellence in different domains, its impressive performance here reinforces its aptitude for high-stakes classification problems.

SVM also performed very well with a precision, recall, and F1 score of 0.97. The strength of SVM is that it can identify an optimal hyperplane that classifies the data into separate classes with maximum margin. SVM is especially good in high-dimensional spaces and is robust against overfitting, particularly when well-regularized. This performance indicates that the tracker/non-tracker boundary in the decision space is fairly well defined in the used feature space, and so the SVM can perform well.

MLP attained a set of metrics similar to SVM, with 0.97 on all three classes. MLP, as a deep neural network type, can learn complex, nonlinear relationships among features. The robust performance confirms that the domain data feature set offers a rich enough signal for deep learning algorithms to leverage, although it might take a higher computational cost and training process than for basic models.

Isolation forest, a type of unsupervised learning algorithm for anomaly detection, performed fairly well with 0.68 precision, recall, and F1 score. This technique tends to be better applied to situations involving outlier detection than binary classification. Its lower performance here indicates that the model might not perform well in separating intricate patterns in legitimate versus tracking domains, since it is mostly based on the fact that anomalies (such as trackers in this instance) are rare and distinct in form from the majority of data.

DBSCAN (Density-Based Spatial Clustering of Applications with Noise) demonstrated an unusual performance graph, attaining a very good recall value of 0.86, yet a much lower precision value of 0.43, and a cumulative F1 score of 0.573. This indicates that DBSCAN correctly identified many actual tracker domains (as seen through its good recall), yet in doing so it incorrectly marked several valid (non-tracker) domains as trackers (seen through its poor precision). Such false positives might be a concern in real-world usage, where blocking non-tracker domains would cause reduced functionality or user unhappiness. Sensitivity of the model to parameter settings (e.g., the radius and minimum number of samples for cluster definition) might further affect its consistency and generalizability.

Overall, supervised learning models like random forest, gradient boosting, SVM, and MLP greatly outperformed unsupervised models like isolation forest and DBSCAN on this classification problem. This implies that with the availability of labeled data, supervised methods deliver much higher accuracy and reliability in web tracker detection. These findings also underscore the necessity of selecting suitable algorithms depending on the type of problem, the availability of labeled data, and the deployed environment's practical constraints like proxies or browsers.

13.3.12 Statistical significance analysis

To confirm whether the observed performance difference between all the models was statistically different, we conducted a one-way ANOVA test and Kruskal–Wallis test. One-way ANOVA is a statistical test used to determine whether there is any significant difference between the independent variables. In this context ANOVA helps determine whether the results of the models highly differ. The test operates under the assumption that data is normally distributed and variance is equal across the groups. The Kruskal–Wallis test is a nonparametric alternative to ANOVA, which does not assume they are normally distributed and have equal variance. This test is particularly useful for skewed data. The ANOVA results showed an F-statistic of 10.77 and a p-value of 0.0004, and the Kruskal–Wallis test results showed a p-value of 0.026. Because the p-values of both tests are less than 0.05, we can reject the null hypothesis, that is, there is no significant difference between the results of the model. This statistical validation strengthens the reliability of our model selection and suggests that unsupervised methods may not be optimal for web tracker detection in high-dimensional settings.

13.4 CONCLUSION

As seen in the bar chart in Figure 13.4, the comparative analysis of ML techniques for web tracker detection reveals a big advantage of supervised learning approaches over unsupervised techniques. Models like random forest, SVM, gradient boosting, and neural networks give high results, indicating their effectiveness in identifying web trackers accurately. Their robust performance is due to their capability to learn complex patterns within labeled data, giving us a highly accurate classification of tracker and non-tracker instances.

In contrast, unsupervised methods like isolation forest and DBSCAN show limited efficacy, with lower precision and recall. While DBSCAN demonstrates high recall, its lower precision highlights challenges in distinguishing tracker patterns without labeled guidance, often leading to false positives. Isolation forest also exhibits balanced but low performance across metrics, suggesting that purely anomaly-based detection is insufficient for capturing the nuanced patterns of modern web trackers.

These findings support the conclusion that supervised models, which leverage labeled data, offer a more reliable and scalable solution for web tracker detection. This study underscores the need for more complex, adaptive supervised approaches in tackling evolving tracking techniques,

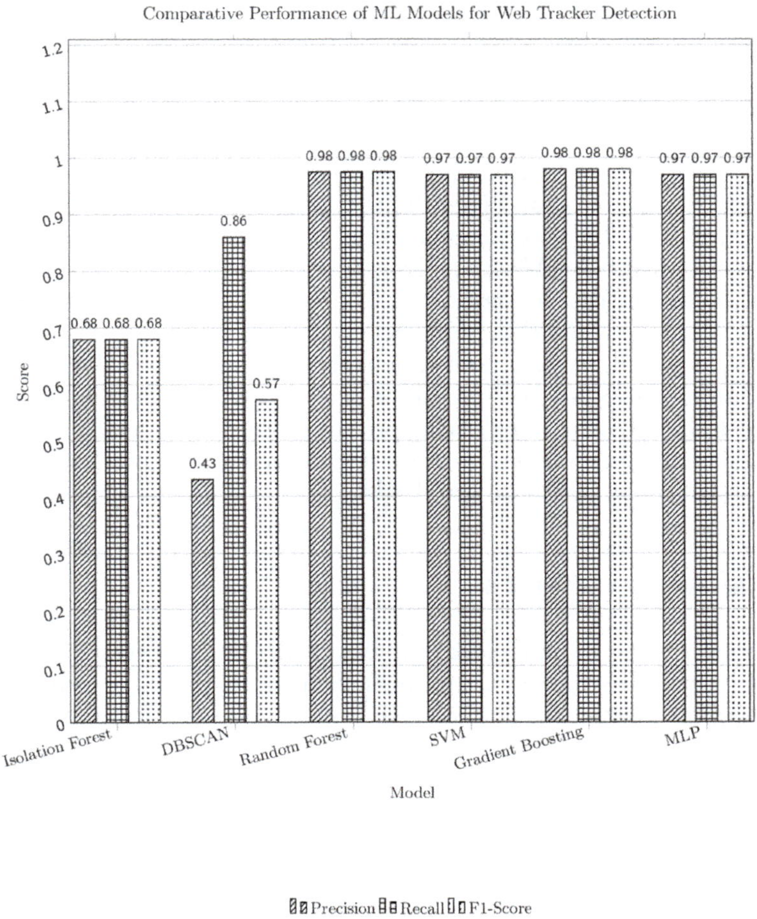

Figure 13.4 Comparison of model performance

positioning these models as viable solutions for enhancing privacy protection in digital environments.

13.5 FUTURE WORK

Expanding upon the discovery and addressing the shortcomings of this study, several potential exploration opportunities in web tracker detection are identified. Future research in web tracker detection will focus on improving model adaptability, accuracy, and resilience against evolving tracking methods. Enhancing detection models through hybrid learning, ensemble techniques, and self-supervised learning can improve precision while minimizing false positives. Additionally, real-time detection capabilities across platforms, including mobile and IoT, remain a key challenge. Efficient, scalable models using GPU acceleration and distributed computing will allow real-time, large-scale analysis. Privacy-preserving techniques, such as federated learning and differential privacy, will ensure tracker detection without compromising user data. NLP-based models like BERT, GPT, and LlaMA could enhance detection by analyzing privacy policies, classifying tracker-related URLs, and identifying tracking patterns in JavaScript and network traffic. However, computational constraints remain a challenge. Furthermore, explainable AI will enhance transparency in detection decisions, fostering trust among users. Continuous model updating through online learning will help counteract newly emerging tracking methods. Finally, legal and ethical considerations surrounding tracker detection must be addressed to guide policymaking and regulatory compliance. Future research should focus on balancing privacy protection with effective tracker identification, ensuring a safer browsing experience for users while maintaining adherence to ethical standards.

ACKNOWLEDGMENTS

I (Calvin Mathew) would like to express my deepest gratitude to my guide, Dr. Ramesh Chandra Poonia, for his invaluable guidance, encouragement, and support throughout this research. I would also like to sincerely thank Ismael Castell Uroz and Pere Barlet-Ros for their invaluable contribution in creating the TrackSign-labeled web tracking dataset [12], which has been a critical resource for this work. Additionally, I would like to thank all those who supported me throughout this journey, whose encouragement and assistance have been invaluable in completing this research.

REFERENCES

1. Ha Dao, Johan Mazel, and Kensuke Fukuda. Cname cloaking-based tracking on the web: Characterization, detection, and protection. *IEEE Transactions on Network and Service Management,* 18(3):3873–3888, 2021.
2. Umar Iqbal, Steven Englehardt, and Zubair Shafiq. Fingerprinting the fingerprinters: Learning to detect browser fingerprinting behaviors. In *2021 IEEE Symposium on Security and Privacy (SP)*, pages 1143–1161, 2021.
3. Federico Cozza, Alfonso Guarino, Francesco Isernia, Delfina Malandrino, Antonio Rapuano, Raffaele Schiavone, and Rocco Zaccagnino. Hybrid and lightweight detection of third party tracking: Design, implementation, and evaluation. *Computer Networks*, 167:106993, 2020.
4. Rui Wang, Xiaoqi Jia, Qinlei Li, and Shengzhi Zhang. Machine learning based cross-site scripting detection in online social network. In *2014 IEEE Intl Conf on High Performance Computing and Communications, 2014 IEEE 6th Intl Symp on Cyberspace Safety and Security, 2014 IEEE Conf on Embedded Software and Syst (HPCC,CSS, ICESS)*, pages 823–826, 2014.

5. Guarino, Alessandro, Domenico Malandrino, Roberto Zaccagnino, Francesco Cozza, and Antonio Rapuano. On analyzing third-party tracking via machine learning. In *Proceedings of the 6th International Conference on Information Systems Security and Privacy - ICISSP*, 2020. SciTePress, pages 532–539. DOI: 10.5220/0008972005320539
6. Jonathan R. Mayer and John C. Mitchell. Third-party web tracking: Policy and technology. In *2012 IEEE Symposium on Security and Privacy*, pages 413–427, 2012.
7. Jianyi Wang. Using deep learning to block web tracking. *Computer and Information Science*, 13:27, 2020.
8. Georg Merzdovnik, Markus Huber, Damjan Buhov, Nick Nikiforakis, Sebastian Neuner, Martin Schmiedecker, and Edgar Weippl. Block me if you can: A large-scale study of tracker-blocking tools. In *2017 IEEE European Symposium on Security and Privacy (EuroS&P)*, pages 319–333, 2017.
9. Syed Rizvi, Mark Scanlon, Jimmy McGibney, and John Sheppard. Application of artificial intelligence to network forensics: Survey, challenges and future directions. *IEEE Access*, 10:110362–110384, 2022.
10. Mohammad Saidur Rahman, Mohsen Imani, Nate Mathews, and Matthew Wright. Mockingbird: Defending against deep-learning-based website fingerprinting attacks with adversarial traces. *IEEE Transactions on Information Forensics and Security*, 16:1594–1609, 2021.
11. Qianru Wu, Qixu Liu, Yuqing Zhang, and Guanxing Wen. Trackerdetector: A system to detect third-party trackers through machine learning. *Computer Networks*, 91:164–173, 2015.
12. Ismael Castell-Uroz and Pere Barlet-Ros. TrackSign-labeled web tracking dataset. *Computer Networks*, 226:109687, 2023.
13. Fei Tony Liu, Kai Ming Ting, and Zhi-Hua Zhou. Isolation forest. In *2008 Eighth IEEE International Conference on Data Mining*, pages 413–422, 2008.
14. Erich Schubert, Jörg Sander, Martin Ester, Hans Peter Kriegel, and Xiaowei Xu. DBSCAN revisited, revisited: Why and how you should (still) use DBSCAN. *ACM Transactions on Database Systems*, 42(3): 1–14, July 2017.
15. Yanli Liu, Yourong Wang, and Jian Zhang. New machine learning algorithm: Random forest. In Baoxiang Liu, Maode Ma, and Jincai Chang, editors, *Information Computing and Applications*, pages 246–252, Berlin, Heidelberg, 2012. Springer Berlin Heidelberg.
16. Shan Suthaharan. *Support Vector Machine*, pages 207–235. Springer US, Boston, MA, 2016.
17. Candice Bentéjac, Anna Csörgo, and Gonzalo Martínez-Muñoz. A comparative analysis of gradient boosting algorithms. *Artificial Intelligence Review*, 54(3):1937–1967, Mar 2021.
18. Yu-chen Wu and Jun-wen Feng. Development and application of artificial neural network. *Wireless Personal Communications*, 102(2):1645–1656, Sep 2018.

Chapter 14

Bridging the present and future

Emerging trends in smart cyber-physical systems

Urvashi Sugandh, Arvind Panwar, Sunita Yadav, and Anupam Sharma

14.1 INTRODUCTION

14.1.1 Overview of SCPS

Cyber-physical systems (CPS) are an emerging, transformative class of systems that integrate computation and physical processes [1]. These systems have the capability to perceive, sense, and interact with the physical world in real-time using advanced sensors, actuators, and complex algorithms. Smart CPS (SCPS) is essentially where the digital and physical meet, where intangible processes and software-driven computation have a symbiotic relationship at their heart [2].

The importance of SCPS in today's technology cannot be overstated. They are at the heart of many mission-critical applications across a wide variety of industries, including manufacturing, healthcare, transportation, and energy management [3, 4]. SCPS unlocks the potential to deliver efficient, safe, and innovative physical processes in a way that has not been possible before with its ability to monitor, control, and optimize the precision of physical processes. Examples can be seen in smart manufacturing, where SCPS enables adaptive production lines that could reconfigure themselves on the fly based on market demand and resource availability, contributing considerably to productivity improvement and waste reduction [5].

SCPS introduces a convergence of computational processes with physical processes, constituting one of our most radical transformations in our approach to interfacing and managing an immersive environment. Such integration enables new levels of automation, decision-making, and response to changing conditions [6, 7]. Processing of data obtained from physical sensors is carried out by computational elements that use advanced mathematical computation methods, including machine learning and artificial intelligence in SCPS to make decisions or for predictions. These decisions are then physically acted out through actuators, which in turn continue to feed back cyber-physical world interaction [8–10].

 DOI: 10.1201/9781003569251-14

It can be seen in a number of ways, whether through self-driving cars with giant sensor arrays attempting to navigate the physical world or smart grids adjusting energy supply and demand on the fly [11, 12]. This dynamic feedback among computational intelligence and physical processes provides SCPS with the ability to adjust to variations in its environment, retain information on long-term trends as well as immediate events, and ultimately enhance performance through operational experience—an advancement of a different order than that manifested by device-scale systems [13, 14].

14.1.2 Purpose of the chapter

The goal in this chapter is principally to demystify the answer to how upcoming developments are amalgamating current SCPS technologies with prospective enhancement. With the exponential growth in SCPS, it is important to look at where this trend is taking us, how different standards and specifications can influence industry or the society we live in. This look will give some context to why we need better, more efficient, and next-level capable technology with the new evolution for modern computing.

In going through those new trends, we aim to give a full view of what the future of SCPS looks like. This entails evaluating progress being made in such things as artificial intelligence, edge computing, Internet of Things (IoT), and cutting-edge connectivity solutions like 5G and beyond. This section will look at how these technologies are being bolted onto the current SCPS frameworks, making up a whole litany of new capabilities.

In addition, this chapter illustrates the need for vigilance in tracking said trends to remain competitive and innovative. Given the rapid pace of technology, organizations and individuals who do not keep up with advancing SCPS will be left behind in a world that is moving forward. The importance of these trends goes far beyond simple technological curiosity and will instead be crucial to your continued competitive advantage in the increasingly digital world we live in.

Identifying SCPS may help businesses become more productive, offer better product and service solutions, or deliver an improved customer experience. In academia, this trend-awareness-oriented approach with an eye on interdisciplinary collaborations helps in deciding the directions in research. It is important for policymakers to anticipate the evolution of SCPS in order to promote patentable innovations as well as preempt new types of risks.

This chapter is designed to provide the reader with an understanding of SCPS that bridges current realities and future potential, in order to help manage the complexity found within today's SCPS environment. It will lay the groundwork for truly grasping what is possible today—and tomorrow; in that spirit, making sure engagement with such revolutionary technologies happens ahead of schedule.

14.2 THE PRESENT LANDSCAPE OF SCPS

14.2.1 Current technologies and applications

The SCPS landscape is multifaceted, with a number of technologies and applications currently deployed throughout several sectors. Such systems are at the forefront of marrying computational intelligence with hard, industrial reality, promising unprecedented levels of automation as well as efficiency mixed with adaptability [15, 16].

SCPS has ushered in the era of Industry 4.0, wherein smart factories use machine-to-machine communication and data analytics for manufacturing automation to streamline production processes across various verticals in the manufacturing industry [17]. With a combination of advanced robotics and automated guided vehicles (AGVs) teaming up with human operators, you will have one powerful productivity machine that also protects workers. Learn how predictive maintenance systems, driven by machine learning algorithms and sensor data, can predict when the equipment will fail even before it does to avoid downtime and higher maintenance expenses [18].

The use of SCPS applications has brought substantial advancements to the transportation industry. Advanced driver assistance systems (ADAS) use sensors, cameras, and AI to make driving safe in the modern vehicle [19, 20]. These systems come with advanced driver assistance features, including adaptive cruise control, lane departure warnings, and automated parking. Next, an ultimate specification in SCPS integration is autonomous driving, where sensor fusion and real-time data processing are mixed with decision-making algorithms for movement through a domain of convoluted environments [21].

In healthcare there are SCPS applications for the emergence of smart medical devices and telemedicine solutions. Wristwatch health monitors monitor vital signs 24/7 and send data to healthcare providers, which helps real-time patient monitoring as well as early intervention [22]. These systems support highly precise sensors and intelligent control algorithms to help surgeons in complex surgeries with increased accuracy while minimizing human invasion.

SCPS has been adopted by the energy sector through smart grids that are used to efficiently distribute and protect their power. SCPS leverages smart meters and sensors to access real-time data, effectively balance supply with demand, integrate renewable resources on the grid, and rapidly adjust to shifts in consumption patterns. SCPS is integrated into smart buildings to optimally control lighting, HVAC maintenance, and ventilation systems, and reduce energy costs [23].

SCPS is used for various purposes such as monitoring crop health, soil conditions, and weather patterns in agriculture [24]. Precision irrigation systems based on soil moisture sensors and weather forecasts have started to save a lot of water. Multispectral cameras and AI algorithms work together on drones to determine crop health, identify pest infestations, and target interventions.

SCPS applications have also been radically shuffling the urban infrastructure. Smart city applications use connected sensors and data analytics to optimize traffic flows, monitor air quality, and help prevent crime. Smart traffic management systems change signal timings on the fly according to the current status of vehicles that result in uncongested roadways and easy passage across the city [25].

14.3 EXISTING CHALLENGES

Although many improvements and proliferation have been witnessed in SCPS, there remain several challenges in the current scenario that are stymying their complete potential as well as broad-scale deployment.

A key factor in the SCPS area is that interoperability and standardization are still part of a challenge. The biggest problem with SCPS is the assorted collection of devices, protocols, and platforms that they are running on because all these solutions do not cooperate well together causing many compatibility issues [26]. This lack of harmony gives rise to challenging issues for interoperability across different manufacturers and sectors, hampering the possibilities for encompassing large-scale SCPS realizations. Standardization is currently ongoing to reach common standards and agree on protocols; however, the pace at which technology grows has hindered standardization that typically takes a couple of years longer [27].

And once established SCPS security and privacy will then raise its head as the next major hurdle for us all. Given the increasing value of data transmitted and stored within complex systems connected to one another, such entities have become rather vulnerable to cyberattacks [28]. Concerns about security threats in SCPS can be significant, ranging from data theft to physical damage or potential harm to human lives. Balancing strong cybersecurity measures with system performance and ease of use is a difficult challenge, one that needs constant attention to keep up to date [29].

The more involved and expansive your SCPS, the higher the potential for scalability problems to arise. Millions of sensors and devices are generating data that must be managed and analyzed in real time, which presents some major technical hurdles. Edge computing has arisen as a possible answer, but the integration of edge capabilities into traditional cloud infrastructure and maintenance of unified performance across distributed systems is still complicated [30].

There is also the question of its critical application of reliability and resilience of SCPS. Their importance increases dramatically as they move into mission-critical organizations like healthcare, transportation, and energy management, so the continuous availability of these systems must be maintained along with a commitment to bug-free operation. One of the continuous challenges is to develop fault-tolerant systems that can perform in a safe manner where components fail or environmental conditions are unexpected [31].

The ethical and legal issues associated with SCPS are complex. Questions persist, such as who owns the data, how consent can be achieved for capturing and using that data, and where responsibility lies when systems fail. Autonomous decision-making by sophisticated SCPS also raises questions of accountability and the ethics surrounding AI-based choices, especially when it comes to human safety.

An additional significant challenge faced by many industries is the integration of legacy systems with modern SCPS technologies. For many things, the cost of upgrading or outright replacing existing infrastructure is far too costly and disruptive to be replaced immediately [32].

There is another challenge to the widespread adoption of SCPS, which includes workforce adaptation and skills. Developing and maintaining such features also requires certain skill sets that are often scarce. The interdisciplinary expertise in physical systems, computer science, and data analytics is increasingly beneficial [33]. Education and training programs to bridge this skill gap are critical for sustained SCPS innovation.

The upfront cost of many SCPS initiatives may be challenging for organizations to absorb, particularly those with limited budgets or profit margins. Although the major efficiency and cost-saving advantages of SCPS are evident in the long term, upfront investment to upgrade hardware, software, and infrastructure can be significant. Wider adoption will require more affordable solutions that clearly demonstrate return on investment.

The challenges have to be tackled by industry, academia, and policymakers working together. Meeting this challenge is key to unlocking the full benefits of these disruptive technologies for society as SCPS matures.

14.4 ARCHITECTURE AND WORKING OF SCPS

The SCPS architecture is a multilayered solution for managing complex industrial operations. The configure layer, at the top of the stack, overarches it and is responsible for supervisory control to decide what needs to be done to ensure no disruptions on the RCS (resilient control system) execution level. This layer is responsible to prepare the system: Controls and Actions of data transactions in order for some operational scenarios. The first layer which is responsible for decision support system (DSS), which allows for analyzing data for processing and making decisions determined by prioritization, leading to more informed, optimized choices.

In the cyber layer, the digital dimension is merged with the physical realm by means of CPS. It consists of adaptive analysis, peer-to-peer monitoring, and time-machine snapshots—the "learning blockchain" so that the system can compare against itself. It also analyzes in real time how effective its strategy setting is and adjusts strategies on a more frequent basis. The conversion layer specializes in the physical part, controlling machines and components such as implementing prognostics and health management (PHM) to forecast breakdowns, enabling system health maintenance before main/final failure. The connection layer places an emphasis on data acquisition and sensing with sensors because it enables us to pick out efficient types of sensors as well as establish a condition-based monitoring (CBM) that predicts when maintenance is required accurately to drive effective operation [34]. All of these layers work together to form a unified system that can adjust and respond to the changing nature of industrial environments.

14.5 EMERGING TRENDS BRIDGING PRESENT AND FUTURE

14.5.1 Integration of advanced artificial intelligence and machine learning

14.5.1.1 Adaptive and intelligent systems

The combination of next-generation artificial intelligence (AI) and machine learning (ML) has matured SCPS to become intelligent and adaptive. And today, these systems have the ability to learn from their surroundings and adapt independently while delivering better performance at an increasing rate. For example, in manufacturing, AI-driven supply chain management planning, SCPS can predict failures of machines and provide a schedule with the best output given the available production tools while being aware of any condition altering unpredictability [35, 36]. In smart cities with AI, algorithms process huge amounts of data from numerous sensors to reduce traffic congestion and energy consumption, and manage waste. Being able to process vast amounts of complex, multidimensional data and derive valuable intelligence from it is slowly revolutionizing some key aspects on how we are interfacing with our physical environment as a whole.

14.5.1.2 Reinforcement learning and deep learning

AI advancements in SCPS are being made primarily with the paradigms of reinforcement learning (RL) and deep learning (DL). This property makes RL algorithms well-suited for systems that must perform a sequence of actions in an uncertain and possibly dynamic environment, where hard rules or specific decision paths may be incomplete. An example of this is in autonomous vehicles, where RL algorithms can enable the vehicle to navigate through difficult traffic situations by learning from past data [37].

DL serves as a fuel enhancing the cognition sensing and decision-making potential of SCPS by analyzing vast data. DL models can also be trained in the healthcare field, such as analyzing medical images with human-level accuracy, which helps with early disease detection and diagnosis. This combination of RL and DL is currently expanding the limits of what can be done in SCPS, allowing for more advanced and autonomous systems.

14.5.2 Edge and fog computing

14.5.2.1 Shift toward decentralized computing

The adoption of edge computing now adds to the shift by bringing computation closer than ever before in SCPS. This new decentralized paradigm circumvents the constraints of centralized cloud-centric models, like bandwidth challenges and latency problems. Edge computing is when the data processing takes place on or near the device from which it originated, and fog computing extends cloud services to be closer to where they are needed at the edge of a network [38]. This is a very important step for applications like autonomous vehicles and industrial control systems (i.e., robotic handling), which need real-time processing and decision-making. Edge and fog computing improve response time and efficiency by lessening the need to transmit a massive amount of datacentric workload back to a centralized server.

14.5.2.2 Implications for SCPS

The adoption of edge and fog computing will bring about an SCPS revolution. Faster response time, greater reliability, improved privacy, and security edge computing can help monitor and control production processes in real time, which greatly benefits smart manufacturing because systems will experience less downtime, reducing quality issues. In healthcare, processing of sensitive patient data can be done locally—thus addressing issues about privacy while enabling rapid analysis and response. But this new paradigm comes with its own set of problems—distributed systems, consistency across many nodes, and building an algorithm that is resilient enough to run on a resource-constrained edge device [39, 40].

14.5.3 Expansion of the Internet of Things (IoT) and Industrial IoT (IIoT)

14.5.3.1 Massive device connectivity

The marketplace is broadening and getting more intelligent as IoT and IIoT devices rapidly proliferate the scope, scalability, and capabilities that SCPS can provide. This vast array of device connectivity is enabling an unprecedented level of monitoring and control over physical processes within a data-driven ecosystem. Sensors can measure soil conditions and crop health, and weather patterns; smart agriculture enables precision farming. Within industrial environments, IIoT devices monitor in real-time nearly every aspect of the production from logistics and supply chain to equipment performance, allowing for predictive maintenance and higher resource utilization [41, 42].

14.5.3.2 Interoperability and standardization efforts

Standard interconnected devices are a challenge because of proprietary protocols. Makers are struggling to stake out ground, working on common protocols and standards aimed at making sure the devices created by different companies can talk with one another. The Industrial Internet Consortium (IIC) with the Open Connectivity Foundation (OCF) will standardize IoT and IIoT. Standardization is needed to unlock the full benefits that interconnected SCPS can deliver across applications, and for development in general, more complete/solution-based efforts [43].

14.5.4 Advanced connectivity with 5G/6G networks

14.5.4.1 Enabling technologies

5G networks produce vastly higher peak data rates, and considerably reduce latency and the capacity to connect a large number of devices almost concurrently. This reported capability is allowing new real-time data processing and control applications in SCPS. For example, 5G networks would be used to help build large sensor grids in smart cities such as traffic management systems, environmental condition monitoring, and public safety centers. For healthcare, 5G allows remote surgery and real-time patient monitoring along with high-fidelity data transmission [44].

14.5.4.2 Future network technologies

In the future, 6G networks will offer many more groundbreaking features. 6G may offer speeds up to 100 times faster than those of 5G and have near-zero latency, unlocking, for example, truly immersive AR/VR experiences, holographic communications (human-to-human), and ultra-precise localization within centimeters. Those advancements could expand the realm of what's feasible today in SCPS, for example, with fleets of autonomous drones responding to disasters or fully immersive remote collaboration on industry sites [45].

14.5.5 Digital twins and virtual modeling

14.5.5.1 Creating virtual counterparts

Among the most exciting developments in SCPS is that of digital twins—a virtual copy or an avatar of any machine, environment, process, or system for better visualization and monitoring. The digital twins remain in sync with their real-world counterparts, so they can be used to perform complex simulations (for example), monitoring and predictive analysis based upon the latest data about what is going on. When it comes to manufacturing, digital twins for production lines are used for process optimization, predictive maintenance, and virtual commissioning of new assets. In urban planning, digital twin cities enable simulation of most aspects of life in a city and allows for optimal operation from traffic modeling up to building energy consumption.

14.5.5.2 Bridge between physical and digital

In SCPS, digital twins act as a vital bridge between the physical and digital worlds. They allow engineers and operators to see, simulate, and control a complex system without getting in the way of what is happening physically. This is especially useful for high-risk applications such as aerospace or nuclear power plants where testing this kind of software in the real world could be expensive, dangerous, and even impossible. Digital twins will evolve with advanced models including those powered by AI, which can handle autonomous decision-making and optimization [46].

14.5.6 Enhanced cybersecurity measures

14.5.6.1 Adaptive security solutions

With SCPS becoming more sophisticated and interconnected, it has never been more important to protect them with strong cybersecurity capabilities [47]. Some of the emerging trends include the adoption of adaptive security solutions, where AI and ML algorithms are being leveraged for real-time threat detection and response. In doing so, these systems can reflect upon previous attacks, identify abnormal network behavior, and respond at a level more granular than ever before [48]. Such adaptive security measures are essential for maintaining operational integrity and public safety in critical infrastructure SCPS such as power grids or water treatment facilities.

14.5.6.2 Emerging threats and countermeasures

SCPS includes a threat landscape that shifts rapidly and is continually evolving with more advanced cyber threats. Cyber-physical attacks arise from the current cyber-based penetration paradigm and target with increasing frequency individual SCPS physical components. With threats rising day by day, cybersecurity paradigms are evolving as well. These range from blockchain-based security protocols for data integrity, quantum cryptography to secure communication, and AI-powered threat intelligence systems. There is also a trend toward considering security-by-design principles, which means integrating cybersecurity considerations as part of the SCPS development from its inception [49].

14.5.7 Blockchain and distributed ledger technologies

14.5.7.1 Secure and transparent transactions

There is a broad movement to integrate blockchain and distributed ledger technologies into SCPS. They are powerful tools useful for security network management that bring transparency to ensure trust. These technologies create a distributed system of accounting to maintain an immutable record within the network. Blockchain offers end-to-end traceability of products in supply chain management, which ensures quality control and minimizes fraud. Within energy systems, it enables peer-to-peer energy trading in microgrids and helps create more efficient processes for a distributed power grid.

14.5.7.2 Smart contracts

SCPS and blockchain are all about smart contracts, which are automated self-executing contracts with the terms of agreement directly written in lines of code. These smart contracts can also be programmed to automatically enforce rules, validate parameters, and trigger state changes without the involvement of trusted intermediaries. In the context of SCPS, smart contracts will be able to automatically manage complex interactions between multiple parties, for example, interoperable workflows in manufacturing supply chains or ensuring access rights are properly granted over a shared resource [50]. This automation not only decreases errors but enables them to work more efficiently, and also imposes adherence to predefined rules and regulations.

14.5.8 Human–machine collaboration

14.5.8.1 Advanced interfaces

In SCPS, human–machine interfaces are necessary for intentionally designed interactions that aim to support an intuitive, efficient, and natural use of them. In addition, next-generation interfaces using augmented reality (AR) and virtual reality (VR) are providing novel ways for humans to operate complex cyber-physical systems. AR interfaces are used in commercial settings to display real-time data on factory floors or maintenance information onto physical equipment. This way, these interfaces can serve as VR technologies that provide users with full immersion of virtual representations of their surroundings and enable people to interact remotely with physical systems [51].

14.5.8.2 Collaborative robots (cobots)

Cobots allow advancement in human-manufacturing collaboration. Cobots differ from traditional industrial robots, which are meant to work separately and independently of humans. The robots are built with new sensors and AI algorithms that enable them to respond more thoughtfully around people, working side by side on tasks. For example, in manufacturing, cobots work alongside human beings on production lines undertaking repetitive or physically demanding tasks, leaving the more cerebral challenges to be completed by humans. This joint solution increases productivity, enhances safety for workers, and enables more versatile production processes [52].

These trends are continuing to evolve and intersect in ways that together represent the future of SCPS. The marrying of these technologies is insufficient to close the chasm between what we can do today and what tomorrow may bring, but opens new doors for innovation and efficiency in numerous verticals [53]. The challenge is mastering these trends while managing the technologies, ethics, and society they come with.

14.6 FUTURE PROSPECTS OF SCPS

14.6.1 Holistic integration of emerging trends

The future of SCPS in general seems to depend on the holistic integration of these emerging trends. This fusion results in smarter and more efficient systems that are also much more flexible and resilient. Integration and convergence of advanced AI, edge computing, IoT, 5G/6G networks, and digital twins enhanced by cybersecurity blockchain human–machine collaboration redefine the SCPS capabilities paradigm [54].

But, we would rather be looking at SCPS where decision-making by algorithms in an AI-driven framework happens on edge devices using real-time data from a network of thousands or millions (yes!) IoT sensors. The systems will be connected by ultrafast, low-latency 5G, or future 6G networks to ensure response times are near instantaneous when conditions change. Then we have the digital twins, which enable a “holistic, virtual replica of an entire system” using simulation and predictive analytics. Through the blockchain, all data and transactions will be transparent by having their integrity confirmed with adaptive cybersecurity measures to guard against evolving threats [55].

This sweeping integration across applications will provide SCPS levels of autonomy and intelligence that were heretofore unattainable. These systems will self-optimize, adapt to changing demands and opportunities in real time, and heal from failures autonomously. For instance, in the operation of a traffic management system within a smart city use case, it can do more than simply respond to current conditions by dynamically changing lists as programmatically determined; for example, practicing historical data sets and models for situations such as forecasted weather that could influence current or future conditions and seasonal events where action is recommended prior to scheduled encounters [56].

This will also lead to the integration of various SCPS domains and thus break down silos. Although we are likely to see more cross-sector applications in which manufacturing, energy, transportation, and healthcare systems communicate with one another—sharing data as well as resources to increase the performance of all components—there is little doubt that AI-based technologies will remain indispensable for ensuring a competitive edge through advancing operational efficiency [57].

14.6.2 Transformative impact on industries

From a manufacturing perspective, Industry 4.0 will (one hopes) mature to the use cases now being called Industry 5.0, facilitated through ultrapersonalization and blurred human–machine interactions in industry as we know it today. That future becomes possible when the SCPS is utilized, as it will allow factories to be completely autonomous and also reshape themselves in real time according to customer requirements, thereby allowing not only taking and customizing product orders on demand but manufacturing too. Full digital twins of supply chain networks will enable end-to-end optimization, minimizing waste and increasing sustainability.

We will move to a prescriptive and personalized version of healthcare. Wearable devices, AI-driven diagnostics, and real-time health monitoring in SCPS will result in early diagnosis of ailments through individual screening. AI combined with AR-motivated robotic surgeries will see a steep rise in the next few years, solving problems of hospital conversion rates as well as permitting remote surgery [58].

In the area of energy, SCPS will enable a 100% renewable power system. Smart grids will transform into local, self-regulating energy networks that balance supply and demand at the molecular level while integrating multiple forms of clean power. Accelerated development will lead to all the possibilities unlocked by blockchain for peer-to-peer energy trading, turning everyone into a producer and consumer of energy.

Autonomous cars and intelligent traffic management systems will revolutionize transportation. We foresee the rise of Mobility as a Service platforms, with different types of transport (ground, air, and maybe hyperloop-based systems) working together in an integrated way to deliver efficient door-to-door transit [59].

Farmers will gain from the further increased precision of farming techniques. Drones and robots that work with AI analytics will be self-directed to monitor specific plants in crops, utilizing resources efficiently, resulting into increased productivity while ensuring the preservation of nature [60].

14.6.3 Societal and economic implications

The implications of widespread SCPS adoption and development are vast, changing society as we know it.

Economically, SCPS will pump up significant productivity increases in countless sectors, and hence it has the potential to drive larger economic growth. But it would be naïve to believe that this will not result in job displacement for some sectors, mainly because those jobs are routine or repetitive [61]. Novel types of jobs will also be created that contemplate the design, operation, and upkeep of these intertwined systems, thus there is a growing need for skills in AI, data analytics, robotics, and systems integration.

Job orientation, meanwhile, will change fundamentally, with human–machine collaboration becoming more important. Workers will more and more be supervising intelligent systems, interfacing with them rather than directly performing tasks themselves. This transition would require extensive reskilling and upskilling programs for the workforce to be ready in an SCPS-driven economy.

SCPS have the potential to greatly enhance the quality of life societally. It seems that smart cities will be able to offer with integrated SCPS improved public services efficiency, decreased road traffic congestion, and lower levels of pollution, and increased security for all users. SCPS could usher in improved health, increased longevity, and personalized care on an individual level as far as healthcare is concerned.

But these advances also pose crucially pertinent ethical and societal considerations. The growing autonomy of AI-driven decision-making within SCPS may be a source of accountability and transparency concerns. As SCPS continue to collect and process significant personal data, substantive issues of data privacy and security will be even more pressing [62].

As digitalization progresses, existing inequalities will deepen further, widening the gap in socioeconomic development with those who have access to advanced SCPS utilities, gaining substantial health education and economic advantages over others. Fair access to these technologies will be a big issue for policymakers [63].

A positive influence on environmental sustainability was expected from this role of the optimization capabilities that SCPS provides, leading to a more efficient use of resources and less waste. Managing the energy demands of these systems and minimizing their environmental impact, as well as managing how hardware components are produced and waste are at odds to this issue [64].

For example, the omnipresent influence of SCPS may fundamentally change human behavior and social relations. We still are years away from the technology that actually makes control over reality responsive and personalized, but once it exists (and I anticipate this will happen in our lifetimes), how we view ourselves and interact with physical things could begin to change. This would have a broad potential impact for social structures, urban planning, and even cognitive development [65].

Overall, the future of SCPS has great promise in varied areas to enhance efficiency, sustainability, and quality of life. But to achieve such benefits while minimizing any ill effects will take thoughtful foresight, moral work, and advanced regulation. However, as we march toward such SCPS, future deployment and development must be done in an inclusive manner, which ensures sustainability and societal value creation [66].

14.7 CASE STUDIES: BRIDGING PRESENT AND FUTURE

14.7.1 Case study 1: Autonomous vehicles

The autonomous vehicle is an archetype example of SCPS, minding the gap between the present and future. They are at the intersection of several emerging technologies and showcase the game-changing capabilities of SCPS in transportation.

14.7.1.1 Present state

The first of them is driver-assistance systems supplemented with conditional automation. Modern vehicles commonly feature advanced driver assistance systems (ADAS), which encompass adaptive cruise control, lane-keeping assistance, and automated parking. These systems use an array of sensors (such as cameras, LIDAR, and radar) incorporating GPS and AI algorithms together to perceive the environment and driver assistance tasks [67].

14.7.1.2 Emerging trends and future prospects

Several major trends are pushing SCPS toward a fully autonomous vehicle (level 5 autonomy):

1. Advanced AI and machine learning: Deep learning algorithms are improving the vehicle's capacity to comprehend intricate driving situations, anticipate the behavior of other types of road users, and make

real-time decisions. They are using reinforcement learning to make better decisions for rare or unexpected scenarios [68].
2. Edge computing: With our data trained until October 2023, more and more autonomous vehicles are using edge computing to lower latency in critical decision-making processes. As such, it enables processing of the sensor data in real time, facilitating a quick response to changing road conditions.
3. 5G connectivity: 5G networks are rolling out, and with the help of these networks, the vehicle-to-everything (V2X) communication would be established, which enables autonomous vehicles to exchange or collect information in real-time between other vehicles, infrastructures, and pedestrians [69].
4. Digital twins: Virtual models of autonomous vehicles and traffic systems are being used for testing and optimization. These digital twins allow for extensive simulation of various scenarios, enhancing safety and performance.
5. Enhanced cybersecurity: In addition, there are developing strong cybersecurity measures to mitigate any risk of hacking or similar interference as connected vehicles and AVs become more commonplace [70].

14.7.2 Case study 2: Smart grids

SCPS that are changing critical infrastructure systems include the energy category represented by the smart grid. They signify a transition from traditional, one-way distribution of power to a two-way flow not just of electricity but also of information.

14.7.2.1 Present state

Key nowadays to smart grid implementations are advanced metering infrastructure (AMI; providing two carbon-free instances of communication, in addition to intermediate agents). Demand response systems exist in certain areas, where the cost per unit of electricity varies throughout the day based on current demand. The grid is being integrated with renewable energy, although it is not without intermittency management issues.

14.7.2.2 Emerging trends and future prospects

The evolution of smart grids is being shaped by several SCPS trends:

1. IoT and IIoT integration: The proliferation of smart meters, sensors, and IoT devices is enabling granular monitoring and control of the grid. This extends to home appliances and industrial equipment, allowing for more precise demand management [71].

2. AI and predictive analytics: Machine learning algorithms are being employed to predict energy demand, optimize distribution, and anticipate potential failures or maintenance needs [72].
3. Blockchain technology: Blockchain is being explored for peer-to-peer energy trading, enabling prosumers (consumers who also produce energy) to sell excess energy directly to neighbors [73–75].
4. Edge computing: Distributed intelligence at the grid edge is enabling faster response to local changes in supply and demand, improving grid stability and efficiency.
5. Advanced energy storage: Integration of large-scale batteries and other storage technologies is being optimized through AI-driven management systems.

14.8 CHALLENGES AND CONSIDERATIONS

SCPS, as they grow and develop, enter into each part and facet of our culture, and pose plenty of challenges and considerations that have to be resolved for their successful and accountable implementation.

14.8.1 Technical challenges

In order for SCPS to reach its potential, it must overcome several significant technical challenges concerning development and deployment:

1. Complexity management: SCPS is inherently complex, comprising multiple components in both cyber and physical domains. Management through air, sea, and land means that understanding structures and their environments becomes extremely difficult. Complexity management is one of the most challenging aspects in the supply chain as cyber and physical systems interact with each other. The complexity of managing this is a pretty big task, especially as systems are scaled more widely. Researchers and engineers are constantly trying to solve these issues: how to have multiple components working without a hitch, keep the SCPS stable, and how to predict emergent behaviors in large-scale SCPS.
2. Real-time performance: Many SCPS applications (e.g., autonomous vehicles, industrial control systems) are real-time (or at least near-real-time) performance-oriented. Ensuring that the messages are processed with little delay, which becomes complicated when they're sent over random network conditions (distributed systems), is still a considerable technical challenge.
3. Reliability and fault tolerance: SCPS security and ease of maintenance are critical hallmarks of the architecture, as SCPS is, at some point in time, linked to critical applications that cannot go down. There are a

lot of technical challenges involved in designing and building systems that can function safely and effectively in the presence of components that may fail, loss of service including network outages, and unpredictability of the environment that the machines are working in.
4. Interoperability and standardization: SCPS is inherently heterogeneous, with multivendor components across multiple domains of SCPS. In this case, strong interoperability standards become a necessity. A key challenge for the industry will be to create and implement these standards while enabling innovation and flexibility.
5. Energy efficiency: SCPS technologies, especially those that are comprised of multiple distributed sensors or actuators, often consume a large amount of energy. This is particularly important in resource-limited environments, as energy-efficient algorithms, hardware, and system architectures will be essential for the sustainable deployment of SCPS.
6. Scalability: As SCPS rapidly accelerates in scale and complexity, one of the foremost technical challenges is ensuring that systems can be scaled efficiently without impacting performance or reliability. This includes problems concerning data governance, network bandwidth, and the computations needed.

14.8.2 Security and privacy concerns

The model for SCPS is complex given its interconnected nature and fundamental integration with physical processes, which raises significant security and privacy concerns:

1. Cybersecurity threats: SCPS is susceptible to a variety of cyberattacks, including data breaches, denial-of-service attacks, and malware injections. Such attacks could be devastating in nature, causing physical destruction, economic impairment, or even threatening lives.
2. Data privacy: SCPS handles lots of data, including potentially sensitive personal data. Ensuring the privacy of this data, complying with data protection regulations, and managing user consent are challenge.
3. Secure communication: Because SCPS needs to manage and operate across multiple entities and frameworks, including cooperative operation on heterogeneous devices, developing secure and trusted communication methods between the structural layers of SCPS is highly anticipated. This also means protecting your data in transit and at rest.
4. Physical security: A strategy for safeguarding cyber assets is more easily defined than in a purely physical system environment. The need for physical security of these components adds another level of complexity to the security environment.

5. Trust and authentication: SCPS services and users need to authenticate each other and establish a chain of trust. Building strong authentication mechanisms that are effective across cyber and physical domains becomes imperative.

14.8.3 Ethical and legal implications

SCPS has raised myriad ethical and legal issues that need to be tackled by society:

1. Accountability and liability: The more autonomous SCPS, the more difficult it is to answer the questions of who will be accountable and legally liable in the event of system failure or accident. Assigning blame in cases of AI-enabled decision-making poses a complex legal and ethical dilemma.
2. Algorithmic bias: AI and machine learning algorithms are central to many SCPS. They could also feed this data into algorithms that determine who receives a loan, who gets a job, etc.
3. Human autonomy and control: As self-controlled property systems become more prevalent and autonomous in nature, there are concerns around a potential failure to maintain human autonomy and control over such systems. Finding the right mix of automation and human control is an important ethical challenge.
4. Transparency and explainability: Many SCPS, particularly those utilizing advanced AI, function as "black boxes," rendering their reasoning and decision-making processes unclear. Transparent and explainable systems are critical for a level of trust and oversight within these systems.
5. Regulatory frameworks: The swift development of SCPS often outruns current regulatory structures. One of the biggest challenges for policymakers is how to bring appropriate safety and ethical use to this technology while not stifling innovation.

14.9 CONCLUSION

14.9.1 Summarization of key points

This chapter discussed future trends of SCPS that can connect our present technology to the upcoming SCPS. Having covered a whirlwind course through the history, present, and future of SCPS, there are a number of key topics worthy of discussion. Advanced artificial intelligence and machine learning have established themselves as a fundamental constituent for developing SCPS adaptive behavior and intelligent systems capable of learning and optimizing operations. At the same time, edge and fog computing paradigms are moving computational processes near to where data is generated with the aim of providing time-sensitive decisions in SCPS applications.

As both IoT and IIoT continue to proliferate, the scale and scope of SCPS are being exponentially expanded into much more connected and data-driven ecosystems. Enabling this growth are advanced forms of connectivity, particularly 5G and later generation 6G networks, giving the high-speed, low-latency communication backbone necessary for complex SCPS operations. Simultaneously, the application of digital twins and virtual modeling techniques provides unparalleled abilities for simulation, optimization, and predictive maintenance within SCPS.

With increasing complexity of SCPS ecosystems, cybersecurity measures are evolving with these systems, gaining more sophisticated communication and interconnectivity. Over the long term, blockchain and distributed ledger technologies (building blocks and lenses of SCPS) would be used to create new paradigms for mutually trusted transactions across SCPS. In addition, human–machine collaboration is being transformed by new interfaces and collaborative robots, which transform how humans work and benefit from SCPS.

14.9.2 Vision for the future

During this period, technological development in the field of SCPS holds critical importance because with the help of the technology of SCPS, a very big turnaround for society and the economy can be brought about. What does the future hold? We dream of SCPS being omnipresent and integrated into everyday life in ways that will improve effectiveness, safety, and quality in many areas. For instance, in manufacturing we expect, fully autonomous and intelligent factories capable of reconfiguring the production line on the fly to meet current manufacturing demand requirements. These intelligent factories won't just scale up manufacturing; they will also reduce the amount of waste created along the way and the energy used to streamline those processes, leading to greener industry.

In cities, we envision SCPS continually connected throughout the functional urban system. Traffic movements are optimized on the fly, power demand and supply balance are inherent and maximized, and public services become increasingly dynamic with respect to citizens' needs. SCPS holds great potential and promises to revolutionize the healthcare sector, paving the way for personalized medicine, real-time remote monitoring, and early disease detection like never before. Operating in conjunction with other factors, autonomous vehicles within an intelligent transportation system could greatly minimize accidents and congestion—and increase mobility for everyone involved.

Systems will be not just smart and functional but also resilient and self-healing. We expect SCPS that can sense the unexpected, learn from their environments, and evolve over time. Such adaptability will be key to combating global challenges like climate change, resource scarcity, and health crises.

REFERENCES

1. O. Ozyurt, H. Özköse, and A. Ayaz, "Evaluating the latest trends of Industry 4.0 based on LDA topic model," *J. Supercomput.*, vol. 80, no. 13, pp. 19003–19030, May 2024, doi: 10.1007/s11227-024-06247-x.
2. M. Khosravy, N. Gupta, A. Pasquali, N. Dey, R. G. Crespo, and O. Witkowski, "Human-collaborative artificial intelligence along with social values in Industry 5.0: a survey of the state-of-the-art," *IEEE Trans. Cogn. Dev. Syst.*, vol. 16, no. 1, pp. 165–176, 2024, doi: 10.1109/TCDS.2023.3326192.
3. A. Panchadi, B. Bastakoti, P. Raghava, and P. K. Jha, "Digital twins: a next-gen solution for agricultural sustainability," *CSA News*, vol. 69, no. 9, pp. 32–36, Aug. 2024, doi: 10.1002/csan.21374.
4. T. H. Szymanski, "A quantum-safe software-defined deterministic Internet of Things (IoT) with hardware-enforced cyber-security for critical infrastructures," *Inf.*, vol. 15, no. 4, p. 173, Mar. 2024, doi: 10.3390/info15040173.
5. P. Kamat and A. S. Gautam, "Recent trends in the era of cybercrime and the measures to control them," in *Handbook of e-Business Security*, Auerbach Publications, 2018, pp. 243–258. doi: 10.1201/9780429468254-10.
6. R. Saravanakumar et al., "Advancing smart cyber physical system with self-adaptive software," *Int. Res. J. Multidiscip. Scope*, vol. 5, no. 3, pp. 571–582, 2024, doi: 10.47857/irjms.2024.v05i03.01013.
7. M. A. Torres-Hernández, M. H. Escobedo-Barajas, H. A. Guerrero-Osuna, T. Ibarra-Pérez, L. O. Solís-Sánchez, and M. del R. Martínez-Blanco, "Performance analysis of embedded multilayer perceptron artificial neural networks on smart cyber-physical systems for IoT environments," *Sensors*, vol. 23, no. 15, p. 6935, Aug. 2023, doi: 10.3390/s23156935.
8. M. Ryalat, H. ElMoaqet, and M. AlFaouri, "Design of a smart factory based on cyber-physical systems and internet of things towards Industry 4.0," *Appl. Sci.*, vol. 13, no. 4, p. 2156, Feb. 2023, doi: 10.3390/app13042156.
9. A. Marchisio, G. Nanfa, F. Khalid, M. A. Hanif, M. Martina, and M. Shafique, "SeVuc: a study on the security vulnerabilities of capsule networks against adversarial attacks," *Microprocess. Microsyst.*, vol. 96, p. 104738, 2023, doi: 10.1016/j.micpro.2022.104738.
10. Z. Noor, S. Hina, F. Hayat, and G. A. Shah, "An intelligent context-aware threat detection and response model for smart cyber-physical systems," *Internet of Things (Netherlands)*, vol. 23, p. 100843, 2023, doi: 10.1016/j.iot.2023.100843.
11. K. Ahmad, A. Kamal, K. A. Bin Ahmad, M. Khari, and R. G. Crespo, "Fast hybrid-MixNet for security and privacy using NTRU algorithm," *J. Inf. Secur. Appl.*, vol. 60, p. 102872, 2021, doi: 10.1016/j.jisa.2021.102872.
12. M. A. Hanif and M. Shafique, "A cross-layer approach towards developing efficient embedded deep learning systems," *Microprocess. Microsyst.*, vol. 88, p. 103609, 2022, doi: 10.1016/j.micpro.2020.103609.

13. D. Weyns, J. Andersson, M. Caporuscio, F. Flammini, A. Kerren, and W. Löwe, "A research agenda for smarter cyber-physical systems," *J. Integr. Des. Process Sci.*, vol. 25, no. 2, pp. 27–47, May 2021, doi: 10.3233/JID210010.
14. L. F. Rivera, M. Jiménez, G. Tamura, N. M. Villegas, and H. A. Müller, "Designing run-time evolution for dependable and resilient cyber-physical systems using digital twins," *J. Integr. Des. Process Sci.*, vol. 25, no. 2, pp. 48–79, May 2021, doi: 10.3233/JID210014.
15. T. P. Nguyen, H. Nguyen, and H. Q. Thinh Ngo, "Planning the emergency collision avoidance strategy based on personal zones for safe human-machine interaction in smart cyber-physical system," *Complexity,* vol. 2022, no. 1, pp. 1–21, Mar. 2022, doi: 10.1155/2022/2992379.
16. D. K. Misra, P. Kumar, and M. Khari, "Achieving the security in cyber world through blockchain: a taxonomy," in *Proceedings of the 17th INDIACom; 2023 10th International Conference on Computing for Sustainable Global Development, INDIACom 2023*, 2023, pp. 1552–1557.
17. V. Varadarajan, P. Kommers, and V. Piuri, "Preface of special issue on advanced techniques and emerging trends in smart cyber–physical systems," *Futur. Gener. Comput. Syst.*, vol. 135, pp. 299–302, 2022, doi: 10.1016/j.future.2022.05.011.
18. S. N. Narayanan, A. Joshi, and R. Bose, "ABATe: automatic behavioral abstraction technique to detect anomalies in smart cyber-physical systems," *IEEE Trans. Dependable Secur. Comput.,* vol. 19, no. 3, pp. 1673–1686, May 2022, doi: 10.1109/TDSC.2020.3034331.
19. Supriya, "Manet security breaches: threat to a secure communication platform," *Int. J. AdHoc Netw. Syst.,* vol. 2, no. 2, pp. 45–51, 2012, doi: 10.5121/ijans.2012.2205.
20. M. Chen, Z. Li, P. Chen, W. Liu, and A. Liu, "A novel differential dynamic gradient descent optimization algorithm for resource allocation and offloading in the COMEC system," *Int. J. Intell. Syst.*, vol. 37, no. 11, pp. 8365–8386, Jun. 2022, doi: 10.1002/int.22943.
21. L. Horváth, "Developing strategies in system level model of smart cyber physical system," *Acta Polytech. Hungarica*, vol. 18, no. 5, pp. 55–76, 2021, doi: 10.12700/APH.18.5.2021.5.5.
22. A. Lohachab, S. Garg, B. H. Kang, and M. B. Amin, "Performance evaluation of hyperledger fabric-enabled framework for pervasive peer-to-peer energy trading in smart cyber–physical systems," *Futur. Gener. Comput. Syst.*, vol. 118, pp. 392–416, 2021, doi: 10.1016/j.future.2021.01.023.
23. B. B. Gupta and T. Akhtar, "Analysing smart power grid against different cyber attacks on SCADA system," *Int. J. Innov. Comput. Appl.,* vol. 12, no. 4, p. 195, 2021, doi: 10.1504/ijica.2021.10038618.
24. M. Molinara, A. Bria, S. De Vito, and C. Marrocco, "Artificial intelligence for distributed smart systems," *Pattern Recognit. Lett.*, vol. 142, pp. 48–50, 2021, doi: 10.1016/j.patrec.2020.12.006.
25. M. Hamlich, L. Bellatreche, A. Mondal, and C. Ordonez, "Special issue on smart applications and data analysis for smart cyber–physical systems," *J. Syst. Archit.*, vol. 121, p. 102305, 2021, doi: 10.1016/j.sysarc.2021.102305.
26. R. Verma, "Smart city healthcare cyber physical system: characteristics, technologies and challenges," *Wirel. Pers. Commun.*, vol. 122, no. 2, pp. 1413–1433, Aug. 2022, doi: 10.1007/s11277-021-08955-6.

27. T. Akhtar and B. B. Gupta, "Analysing smart power grid against different cyber attacks on SCADA system," *Int. J. Innov. Comput. Appl.*, vol. 12, no. 4, pp. 195–205, 2021, doi: 10.1504/IJICA.2021.116656.
28. M. Khari, M. Kumar, S. Vij, P. Pandey, and Vaishali, "Internet of Things: Proposed security aspects for digitizing the world," in *Proceedings of the 10th INDIACom; 2016 3rd International Conference on Computing for Sustainable Global Development, INDIACom 2016*, 2016, pp. 2165–2170.
29. M. Singh, G. S. Aujla, and R. S. Bali, "A deep learning-based blockchain mechanism for secure internet of drones environment," *IEEE Trans. Intell. Transp. Syst.*, vol. 22, no. 7, pp. 4404–4413, 2021, doi: 10.1109/TITS.2020.2997469.
30. S. Tiku, P. Kale, and S. Pasricha, "QuickLoc: Adaptive deep-learning for fast indoor localization with mobile devices," *ACM Trans. Cyber-Physical Syst.*, vol. 5, no. 4, pp. 1–30, Sep. 2021, doi: 10.1145/3461342.
31. D. Weyns, J. Andersson, M. Caporuscio, F. Flammini, A. Kerren, and W. Löwe, "A research agenda for smarter cyber-physical systems," *J. Integr. Des. Process Sci.*, vol. 25, no. 2, pp. 1–21, Aug. 2021, doi: 10.3233/jid-210010.
32. P. Thakur and V. Kumar Sehgal, "Emerging architecture for heterogeneous smart cyber-physical systems for industry 5.0," *Comput. Ind. Eng.*, vol. 162, p. 107750, 2021, doi: 10.1016/j.cie.2021.107750.
33. L. F. Rivera, M. Jiménez, G. Tamura, N. M. Villegas, and H. A. Müller, "Designing run-time evolution for dependable and resilient cyber-physical systems using digital twins," *J. Integr. Des. Process Sci.*, vol. 25, no. 2, pp. 1–32, Oct. 2021, doi: 10.3233/jid-210014.
34. F. C. Delicato, A. Al-Anbuky, and K. I. K. Wang, "Editorial: smart cyber–physical systems: toward pervasive intelligence systems," *Futur. Gener. Comput. Syst.*, vol. 107, pp. 1134–1139, 2020, doi: 10.1016/j.future.2019.06.031.
35. J. Tavčar, J. Duhovnik, and I. Horváth, "From validation of medical devices towards validation of adaptive cyber-physical systems," *J. Integr. Des. Process Sci.*, vol. 23, no. 1, pp. 37–59, Jan. 2020, doi: 10.3233/JID190008.
36. S. Von Solms and L. A. Futcher, "Adaption of a secure software development methodology for secure engineering design," *IEEE Access*, vol. 8, pp. 125630–125637, 2020, doi: 10.1109/ACCESS.2020.3007355.
37. K. Kaippilly Radhakrishnan, H. D. Chinh, M. Gupta, S. K. Panda, and C. J. Spanos, "Context-aware plug-load identification toward enhanced energy efficiency in the built environment," *IEEE Trans. Ind. Appl.*, vol. 56, no. 6, pp. 6781–6791, 2020, doi: 10.1109/TIA.2020.3016621.
38. L. ben Othmane, D. Jacobson, and E. Weippl, "Editorial: special issue on security and privacy in smart cyber-physical systems," *Comput. Secur.*, vol. 88, p. 101612, 2020, doi: 10.1016/j.cose.2019.101612.
39. Z. Cai and X. Zheng, "A private and efficient mechanism for data uploading in smart cyber-physical systems," *IEEE Trans. Netw. Sci. Eng.*, vol. 7, no. 2, pp. 766–775, Apr. 2020, doi: 10.1109/TNSE.2018.2830307.
40. P. P. Khargonekar and M. Sampath, "A framework for ethics in cyber-physical-human systems," *IFAC-PapersOnLine,* vol. 53, no. 2, pp. 17008–17015, 2020, doi: 10.1016/j.ifacol.2020.12.1251.
41. M. Shafique et al., "Robust machine learning systems: challenges, current trends, perspectives, and the road ahead," *IEEE Des. Test*, vol. 37, no. 2, pp. 30–57, 2020, doi: 10.1109/MDAT.2020.2971217.

42. S. N. Narayanan, A. Joshi, and R. Bose, "ABATe: automatic behavioral abstraction technique to detect anomalies in smart cyber-physical systems," *IEEE Trans. Dependable Secur. Comput.*, vol. 19, no. 3, pp. 1673–1686, Oct. 2022, doi: 10.1109/TDSC.2020.3034331.
43. D. Gupta, S. Bhatt, M. Gupta, and A. S. Tosun, "Future smart connected communities to fight COVID-19 outbreak," *Internet of Things (Netherlands)*, vol. 13, p. 100342, Dec. 2021, doi: 10.1016/j.iot.2020.100342.
44. M. Khari, G. Shrivastava, S. Gupta, and R. Gupta, "Role of cyber security in today's scenario," in *Cyber Security and Threats: Concepts, Methodologies, Tools, and Applications*, IGI Global, 2018, pp. 1–15. doi: 10.4018/978-1-5225-5634-3.ch001.
45. S. N. Narayanan, A. Joshi and R. Bose, "ABATe: Automatic Behavioral Abstraction Technique to Detect Anomalies in Smart Cyber-Physical Systems," in *IEEE Transactions on Dependable and Secure Computing*, vol. 19, no. 3, pp. 1673-1686, 1 May-June 2022, doi: 10.1109/TDSC.2020.3034331
46. S. Tepjit, I. Horváth, and Z. Rusák, "The state of framework development for implementing reasoning mechanisms in smart cyber-physical systems: A literature review," *J. Comput. Des. Eng.*, vol. 6, no. 4, pp. 527–541, Apr. 2019, doi: 10.1016/j.jcde.2019.04.002.
47. L. Ben Othmane, D. Jacobson, and E. Weippl, "Special issue on security and privacy in smart cyber-physical systems," *Comput. Secur.*, vol. 87, p. 101611, 2019, doi: 10.1016/j.cose.2019.101611.
48. X. Zhu, G. Srivastava, and R. M. Parizi, "An efficient encryption algorithm for the security of sensitive private information in cyber-physical systems," *Electron*, vol. 8, no. 11, p. 1220, Oct. 2019, doi: 10.3390/electronics8111220.
49. M. Búr, G. Szilágyi, A. Vörös, and D. Varró, "Distributed graph queries over models@run.time for runtime monitoring of cyber-physical systems," *Int. J. Softw. Tools Technol. Transf.*, vol. 22, no. 1, pp. 79–102, Sep. 2020, doi: 10.1007/s10009-019-00531-5.
50. M. M. Rana and R. Bo, "IoT-based cyber-physical communication architecture: Challenges and research directions," *IET Cyber-Physical Syst. Theory Appl.*, vol. 5, no. 1, pp. 25–30, Sep. 2020, doi: 10.1049/iet-cps.2019.0028.
51. L. H. Melnyk, "Anatomy of phase transitions in economic systems: Enterprise, region, macroeconomics," *Mech. an Econ. Regul.*, no. 1, pp. 6–32, 2019, doi: 10.21272/mer.2019.83.01.
52. Kenrick, Nicole, Lucia Svabova, and Elvira Nica (2019). "Real-Time Health-related Data, Wearable Medical Sensor Devices, and Smart Cyber-Physical Systems," American Journal of Medical Research 6(2): 25–30. doi:10.22381/AJMR6220193
53. Y. Chen, T. Du, C. Jiang, and S. Sun, "Indoor location method of interference source based on deep learning of spectrum fingerprint features in smart cyber-physical systems," *Eurasip J. Wirel. Commun. Netw.*, vol. 2019, no. 1, pp. 1–12, Feb. 2019, doi: 10.1186/s13638-019-1363-y.
54. L. Ma, Y. Huo, C. Hu, and W. Li, "Security and privacy for smart cyber-physical systems," *Secur. Commun. Networks*, vol. 2019, pp. 1–2, Jan. 2019, doi: 10.1155/2019/7045862.
55. T. Ma, S. Ali, T. Yue, and M. Elaasar, "Testing self-healing cyber-physical systems under uncertainty: a fragility-oriented approach," *Softw. Qual. J.*, vol. 27, no. 2, pp. 615–649, Mar. 2019, doi: 10.1007/s11219-018-9437-3.

56. T. Bures, D. Weyns, B. Schmerl, and J. Fitzgerald, "Software engineering for smart cyber-physical systems," *ACM SIGSOFT Softw. Eng. Notes*, vol. 43, no. 4, pp. 54–54, Jan. 2019, doi: 10.1145/3282517.3282542.
57. F. Liang, C. Qian, W. G. Hatcher, and W. Yu, "Search engine for the internet of things: lessons from web search, vision, and opportunities," *IEEE Access*, vol. 7, pp. 104673–104691, 2019, doi: 10.1109/ACCESS.2019.2931659.
58. M. A. P. Chamikara, P. Bertok, D. Liu, S. Camtepe, and I. Khalil, "An efficient and scalable privacy preserving algorithm for big data and data streams," *Comput. Secur.*, vol. 87, p. 101570, 2019, doi: 10.1016/j.cose.2019.101570.
59. E. Ilie-Zudor, D. Preuveneers, A. Ekárt, and E. Hohwieler, "Smart cyber-physical System applications in production and logistics: special issue editorial," *Int. J. Comput. Integr. Manuf.*, vol. 32, no. 4–5, pp. 327–328, May 2019, doi: 10.1080/0951192X.2019.1616152.
60. A. Ahadov, E. S. Asgarov, and I. El-Thalji, "A summary of adapting Industry 4.0 vision into engineering education in Azerbaijan," *IOP Conf. Ser. Mater. Sci. Eng.*, vol. 700, no. 1, pp. 012063--, Nov. 2019, doi: 10.1088/1757-899X/700/1/012063.
61. K. Bou Chaaya, M. Barhamgi, R. Chbeir, P. Arnould, and D. Benslimane, "Context-aware system for dynamic privacy risk inference: application to smart IoT environments," *Futur. Gener. Comput. Syst.*, vol. 101, pp. 1096–1111, 2019, doi: 10.1016/j.future.2019.07.011.
62. T. Bures, D. Weyns, B. Schmerl, and J. Fitzgerald, "Software engineering for smart cyber-physical systems," *ACM SIGSOFT Softw. Eng. Notes*, vol. 43, no. 4, pp. 54–54, Jan. 2019, doi: 10.1145/3282517.3282542.
63. J. Tavcar and I. Horvath, "A review of the principles of designing smart cyber-physical systems for run-time adaptation: Learned lessons and open issues," *IEEE Trans. Syst. Man, Cybern. Syst.*, vol. 49, no. 1, pp. 145–158, 2019, doi: 10.1109/TSMC.2018.2814539.
64. T. Bures et al., "Software Engineering for Smart Cyber-Physical Systems (SEsCPS 2018) - workshop report," *ACM SIGSOFT Softw. Eng. Notes*, vol. 44, no. 4, pp. 11–13, Dec. 2019, doi: 10.1145/3364452.3364465.
65. K. W. Cosgrave, "The smart cyber-physical systems of sustainable industry 4.0: Innovation-driven manufacturing technologies, creative cognitive computing, and advanced robotics," *J. Self-Governance Manag. Econ.*, vol. 7, no. 3, pp. 7–13, 2019, doi: 10.22381/JSME7320191.
66. M. V. R. Kumar, N. Bhalaji, and S. Singh, "An augmented approach for pseudo-free groups in smart cyber-physical system," *Cluster Comput.*, vol. 22, no. 1, pp. 673–692, Mar. 2019, doi: 10.1007/s10586-018-2353-2.
67. D.-N. Le, R. Kumar, B. K. Mishra, M. Khari, and J. M. Chatterjee, "Cyber Security in Parallel and Distributed Computing," *Cyber Security in Parallel and Distributed Computing*. Scrivener Publishing LLC: Vietnam. Hal, 2019. doi: 10.1002/9781119488330.
68. M. Barhamgi, C. Perera, C. Ghedira, and D. Benslimane, "User-centric privacy engineering for the internet of things," *IEEE Cloud Comput.*, vol. 5, no. 5, pp. 47–57, 2018, doi: 10.1109/MCC.2018.053711666.
69. J. Al-Jaroodi and N. Mohamed, "PsCPS: A distributed platform for cloud and fog integrated smart cyber-physical systems," *IEEE Access*, vol. 6, pp. 41432–41449, 2018, doi: 10.1109/ACCESS.2018.2856509.

70. I. Horváth, "A computational framework for procedural abduction done by smart cyber-physical systems," *Designs*, vol. 3, no. 1, pp. 1–27, Dec. 2019, doi: 10.3390/designs3010001.
71. S. Abrol, A. Mehmani, M. Kerman, C. J. Meinrenken, and P. J. Culligan, "Data-Enabled Building Energy Savings (D-E BES)," *Proc. IEEE,* vol. 106, no. 4, pp. 661–679, 2018, doi: 10.1109/JPROC.2018.2791405.
72. T. Bures, D. Weyns, B. Schmer, and J. Fitzgerald, "Software engineering for smart cyber-physical systems," *ACM SIGSOFT Softw. Eng. Notes*, vol. 43, no. 4, pp. 42–44, Dec. 2018, doi: 10.1145/3282517.3302401.
73. A. Panwar, V. Bhatnagar, M. Khari, A. W. Salehi, and G. Gupta, "A Blockchain Framework to Secure Personal Health Record (PHR) in IBM cloud-based data lake," *Comput. Intell. Neurosci.*, vol. 2022, pp. 1–14, 2022, doi: 10.1155/2022/3045107.
74. A. Panwar, M. Khari, S. Misra, and U. Sugandh, "Blockchain in agriculture to ensure trust, effectiveness, and traceability from farm fields to groceries," *Futur. Internet*, vol. 15, no. 12, p. 404, Dec. 2023, doi: 10.3390/fi15120404.
75. A. Panwar, Shyla, and V. Bhatnagar, "Blockchain-based web 4.0: Decentralized web for decentralized cloud computing," in *Cloud IoT: Concepts, Paradigms, and Applications*, 1st ed., Jitendra Kumar Verma, Deepak Kumar Saxena, Vicente González-Prida Díaz, Vira Shendryk, Ed. CRC Press, 2022, pp. 219–233, doi: 10.1201/9781003155577-19.

Chapter 15

Enhancing operational efficiency in cloud computing through computational intelligence for environmental sustainability

Nidhi Singh, Prabha Arya, Sapna Dadwal, Shitiz Upreti, Ramesh Chandra Poonia, and Rituraj Jain

15.1 INTRODUCTION

There have been recent discussions in various forums on the impact of heavy computations on greenhouse gases [1], and many scholars have adopted generative artificial intelligence solutions, which have become essential for organizations. This usage requires large-scale data centers that not only consume significant amounts of water but also emit pollutants that affect the environment. One of the primary concerns now is how to mitigate their negative impact without hindering the adoption of new developments in the field of today's computational-aware world. Hence, there is considerable interest in exploring the interplay between computational intelligence (CI) and sustainability in cloud operations. This study examined how these three domains interact to enhance operational efficiency. As the demand for cloud computing continues to rise, optimizing cloud processes for sustainability is both a strategic necessity and ethical obligation for society [2]. This chapter introduces the concept of cloud sustainability and how key advancements in smart intelligent solutions address the current issues. Before delving into these specifics, it is essential to define the key concepts that are frequently encountered throughout this chapter. Computational intelligence refers to computational methods that emulate human capabilities using tools, such as machine learning, adaptive learning neural networks, and evolutionary computation. Sustainability, in contrast, emphasizes meeting present needs in such a way that it does not exploit existing nonrenewable resources. The cloud refers to the collection of various systems at one place for the delivery of computing services over the internet, encompassing data storage, computationally intensive tasks, and virtual computing resources to meet various needs of society at large. The optimization of cloud operations is linked to broader sustainability goals. With the adoption of green technologies, clouds have the potential to optimize resource utilization [3] and centralize, thereby reducing energy consumption. By leveraging computational intelligence and optimization, organizations can make informed decisions that align cloud practices with sustainability objectives. However, this connection presents several challenges. Current research indicates gaps

DOI: 10.1201/9781003569251-15

in the understanding of how computational intelligence can effectively operate within the constraints of sustainable cloud practices, and various authors hold different views on this matter. This study examines the integration of computational intelligence technologies into cloud-based systems to enhance operational efficiency with a focus on environmental sustainability. By employing advanced intelligence in sensors kept at data centers, dynamic solutions can be created to improve resource management [4] and reduce the environmental impact of the cloud infrastructure. Unlike traditional methods that focus solely on hardware optimization, the application of computational intelligence facilitates real-time decision-making and adaptive strategies that can significantly decrease energy consumption.

15.2 BACKGROUND

The definition of "cloud" given by the National Institute of Standards and Technology (NIST) is the most widely accepted. The cloud is defined as having three services and four deployment models. The first service is *Infrastructure as a Service (IaaS)*, which is the provisioning of processing time, storage, networks, and other fundamental computing resources made available by cloud service providers for use by customers [5, 6]. These resources can be used to deploy and run arbitrary software, including operating systems and applications. The service provider is responsible for managing the underlying cloud infrastructure [5, 6]. The second service is *Platform as a Service (PaaS)*, which provides the capability for cloud service users to develop and host applications [5, 6]. The customer does not manage or control the underlying infrastructure that includes the network, servers, operating systems, and software for application development, which are directly managed by the service provider [5]. The third type of service is *Software as a Service (SaaS)*, which is the delivery of software applications over the Internet by the service provider [5, 6]. Customers are not concerned with the regular maintenance, upgrading, and management of the underlying computing infrastructure used to host applications [5, 6]. Applications are accessible from various client devices through web browsers. According to NIST, there are four types of clouds: private, public, hybrid, and community. A *public cloud* is a cloud in which services are available to the general public or a large group of companies; resources are shared. It is owned by a third-party selling cloud services. This is common in today's world, and many organizations are leveraging services from Google, Microsoft, Amazon, and other large vendors. The second type of cloud system is a *private cloud*, which is a cloud that operates solely for a single organization. Private cloud infrastructure may be set up on or off premises and managed either internally or by third-party service providers. Private clouds offer the highest level of security and control but are expensive. The third type is a *hybrid cloud*, wherein the cloud uses a combination of two or more clouds (public, private, or community). Clouds can be individually managed by multiple cloud service providers but are bound by proprietary technology to enable data and application portability. Azure from Microsoft provides the services in this space. Last, but not widely used, is the *community cloud*, whose infrastructure is shared by several organizations and supports a specific community that has shared concerns. The community cloud may be deployed on or off premises and managed by organizations collectively or by a third-party cloud service provider [5].

Furthermore, the utilization of these clouds and leveraging their services are done via various drivers of the cloud, such as pay-as-you-go and multitenancy. In the *pay-as-you-go* model, the cloud reduces the significant upfront capital expenditure needed to purchase and maintain IT infrastructure such as hardware and software systems. Pricing in the cloud is performed on a usage-based or subscription-based model that converts fixed costs into variable costs. The advantage of being cost-efficient makes the cloud an attractive option, particularly for small and medium enterprises (SMEs) [7]. According to one study, cloud-based services generate higher cost differentials for smaller companies compared to large enterprises. The savings in using cloud services have helped various organizations invest in their core areas. In *multitenancy*, cloud resources run in multitenancy mode, which means that multiple users can access the infrastructure simultaneously from different organizations. Multitenancy also allows cloud service providers to take significant cost advantages and better profitability, allowing them to make more R&D investments. The benefits are passed on to cloud users, making the cloud solution a cost-effective option. The current trend is to identify various sustainable initiatives that can help reduce the carbon footprint and be environmentally friendly [7].

Cloud services are available on demand, which allows customers to upscale/downscale capacity without incurring any capital expenditure [7]. Service providers of the cloud are able to juggle between servers and datacenters to shift resource utilization as per the demand patterns and user-specific service-level agreements (SLAs). This provides agility and flexibility to users. In addition, service providers can implement smart predictive sensors to detect and correct anomalies in cloud infrastructure and balance the load from one system to another. Using cloud-based applications is much easier than building new applications and data centers. Moreover, the entire management of the resources rests with the cloud service providers themselves. The user interfaces of cloud solutions are also very user-friendly compared with the interfaces of traditional software. The most important feature of the cloud is that users can use computing capabilities as and when they require them without any human intervention from the service provider of the cloud. The user would not have to worry about commissioning/decommissioning the computing infrastructure as per utilization/requirements [7].

15.3 LITERATURE REVIEW

In recent years, computational intelligence technologies and cloud operation systems have gained traction in the market and present a promising avenue for increasing operational efficiency while reducing energy consumption and promoting environmental impacts. Currently, working toward investigating the intersection of computational intelligence, cloud operations, and sustainability, the study also examines the challenges in the implementation of intelligent systems and issues related to scalability, high initial cost, data privacy, technology complexities, and lack of framework. Most studies have focused on the application of machine learning algorithms and neural networks to optimize resource allocation and energy consumption in cloud data centers. Researchers have also explored the implementation of fuzzy logic systems and algorithms to improve load balancing and workload prediction in cloud environments. These computer intelligence-driven approaches help reduce energy consumption, minimize carbon footprints, and enhance overall system performance, thereby contributing to more sustainable cloud computing practices.

Recently, the development of artificial intelligence, machine learning, neural networks, and fuzzy logic has expanded the possibilities for energy management. The application of these techniques has allowed many organizations and industries to develop advanced systems that can predict and manage energy consumption in real-time at various stages. For example, some authors have shown that cloud storage systems have implemented machine-learning algorithms to optimize energy consumption during storage [8]. Resource allocation and energy efficiency have also been addressed using simulation tools in computing systems [9]. The merging of computational intelligence with cloud environments offers a promising direction for sustainable and eco-friendly practices. With the growing popularity of cloud services, optimizing their operation for eco-sustainability has become a primary concern [10]. This literature review highlights the conjunction of computational intelligence, cloud operations, and sustainability to clarify the structure, theoretical frameworks, and empirical evidence required to create energy-efficient solutions. Applying contemporary techniques in cloud operations alongside sustainable practices can aid organizations in reducing their ecological footprint while increasing operational efficacy [11].

Learning algorithms, neural networks, and even evolutionary computation fall under machine learning; as such, they can significantly optimize cloud operations, which is incredibly useful. These methods allow for the development of advanced systems capable of real-time forecasting and management of energy consumption at different levels. AI systems such as Green Cloud have developed a model that incorporates machine-learning algorithms that consider emissions and renewable energy to optimize data storage in the cloud. This model improves the energy efficiency of cloud storage. This planning builds on previous models regarding the optimization of energy resources, energy management, and dynamic decision-making using computational intelligence in the cloud environment [12]. Other systems, such as CloudSim, have integrated simulation tools to evaluate the resource allocation and energy efficiency within computing systems. All of these frameworks can be considered building blocks and can be built on each other. Another relevant topic is the cloud federation, which is the amalgamation of separately owned clouds. Such a federation can lead to a significant reduction in carbon emissions and other environmental impacts with the help of intelligent control of virtual machine operations [13].

Artificial neural networks (ANNs) and other machine learning methods have become essential components for the effective management and balance of cloud ecosystems. Currently, artificial intelligence models are nowadays used to forecast energy consumption, control load balancing, automate energy-aware routing, and distribute resources within data centers. One example is the application of deep learning using graph convolutional networks (GCNs) that augment cloud management by enhancing data processing and task scheduling [14]. Moreover, advanced artificial intelligence techniques have been leveraged to further catalyze effectiveness initiatives. These strategies focus on developing sustainable ecosystems for data centers by implementing a minimal footprint of efficient resource management at the software level. Such strategies enhance the alignment of cloud computing with green energy and significantly decrease the carbon footprint. Currently, artificial intelligence-aware application utilization in these solutions can ensure optimal energy usage without diminishing the system performance.

One of the core focuses and visible solutions for data centers is the optimization of energy consumption, thereby advancing sustainability in cloud services. Large-scale data centers have traditionally been considered some of the most energy-hungry systems and are known to have a very large energy footprint. Some scholars have also analyzed energy-saving techniques at the software level, ranging from virtualization to application-level management. Their survey underlined the important contributions that software technologies can make toward achieving green data centers by reducing power consumption and lessening the negative effects of cloud services on the environment.

More advanced cloud computing features are currently being developed along with possible problems in sustaining support for these services. Shifting paradigms of cloud computing is the next big thing. Several studies have explored that the promise of cloud operation lies in the flexibility, scalability, and cost benefits made available through the "as-a-service" paradigm [15]. For example, a blockchain-based decentralized federated learning framework was developed by Wang and Tsai [16], which enables secure private data sharing on cloud networks and reduces computation-intensive tasks in a sustainable manner. Regardless, most of the solutions are still in their developmental stages, and achieving full sustainability throughout cloud environments also faces other obstacles. Issues such as data security, data privacy, and the identification of the application of environmentally friendly technologies pose prominent barriers of their own. Some scholars [17, 18] have clearly described these issues and recommended that more advanced policies and architectures are required to help achieve sustainability objectives [19].

The incorporation of computational intelligence into cloud computing resolves the issues related to sustainability and energy efficiency. The examined frameworks and empirical research indicate that AI-powered models, energy-efficient data management processes, and optimization techniques serve an important function in making cloud ecosystems more sustainable. Despite the issues still being faced in terms of technology integration and regulatory frameworks, the sustained development of AI, green technologies, and green computing has led to promising prospects for sustainable cloud computing. Nevertheless, with the advent of the cloud, organizations are once again at the crossroads of technology. Early adopter organizations of the cloud can provide an opportunity to transform their business models and gain a competitive edge [6]. While cost reduction is one of the benefits, several other benefits accrue to organizations [6]. Several studies have stated that organizations can concentrate on their core competencies while leaving the task of cloud infrastructure to cloud service providers. Furthermore, through the adoption of the cloud, organizations are expected to become more nimbler [6], focus on their core competencies, and utilize the cloud for their needs rather than investing in setting up dedicated infrastructure. The cloud helps them quickly provide resources for business opportunities, as they can obtain insights from the data, which could otherwise be lost. Rather than taking and investing in licensing the analytical tools, today, organizations are able to use the tools provided by cloud service providers, which are able to negotiate better prices from vendors and pass the benefits to the organizations utilizing their services. The cloud also enables organizations to build virtual and "open" business processes, enabling its various stakeholders, including customers, business partners, suppliers, etc., to connect and do business more seamlessly.

The integration of computational intelligence (CI) into cloud storage systems has been widely explored for optimizing resource allocation, load balancing, and energy consumption. However, while several studies have highlighted the potential of CI in enhancing cloud efficiency, there is a notable gap in understanding how these technologies can specifically drive sustainability within cloud ecosystems. Research on green cloud computing [20] and energy-efficient data centers [21] has emphasized the role of CI in reducing energy consumption. However, there is insufficient empirical analysis of the actual environmental benefits of eco-friendly cloud storage solutions. Moreover, the existing literature often fails to address barriers to the integration of sustainable technologies in cloud operations, particularly from a practical implementation perspective. Research has also underexplored the connection between sustainable technologies and the effectiveness of CI frameworks in real-world scenarios. The following key research questions were explored:

1. How can computational intelligence technologies be integrated into cloud storage systems to improve sustainability and operational efficiency?
2. What environmental benefits do eco-friendly cloud storage solutions offer in comparison with traditional data center models, particularly in terms of energy consumption and carbon footprint reduction?
3. What are the key challenges in adopting sustainable technologies in cloud storage, and how can computational intelligence frameworks address these barriers to achieve greater sustainability in the industry?

15.4 RESEARCH METHODOLOGY

This study employed a qualitative research methodology [22] to explore the integration of computational intelligence and sustainability practices in cloud operations. The qualitative research approach provides an in-depth understanding of the practical challenges [23] and the success of implementing sustainable solutions in real-world cloud environments. By focusing on sustainable cloud solution providers, this study examines how a company uses computational intelligence technologies, such as machine learning and artificial intelligence, to reduce the environmental impact of cloud operation systems and optimize energy consumption.

15.4.1 Data collection

Data for this study were collected through a combination of semistructured interviews, existing document analysis, and further observational studies in line with the case study methodology. Semistructured interviews were conducted with key personnel within the organization, including technical leaders and cloud operations managers. These interviews provided rich qualitative insights into the companies' sustainability strategies, involvement of computational intelligence with artificial intelligence integration, and current energy optimization practices as being done in the respective organizations. The interviews were structured around the implementation of green cloud computing, use of intelligence in energy optimization, and overall sustainability goals. In addition to interviews, some internal reports, a few project documentation reports, and other reports from the organization were reviewed jointly to understand and validate the landscape and discussions with their leaders.

15.4.2 Research design

This study follows a qualitative approach, allowing for a comprehensive exploration of sustainability practices from multiple perspectives within the organization. The main focus of this study is to examine the integration of computational intelligence and sustainability across different operational levels, including corporate strategy, product development, and data center management. This study also considers how these perspectives interact with and influence one another in practice.

15.4.3 Data analysis

Using NVivo 12 software, we synthesized a thematic analysis by checking and performing systematic coding of interview transcripts, documents, and observational data. The data were initially coded according to their relationships with the corporate, product, or operational aspects. This classification allowed for the identification of key themes across different levels of the organization. After the initial coding, the data were further evaluated and analyzed to identify specific sustainability themes related to energy consumption, artificial intelligence integration, and green cloud computing practices. These themes were further explored in more detail to understand how they were implemented within the context of the organization's operations. NVivo also helped to identify the frequency of key recurring themes, which provided insights into the areas of sustainability that were more central to the organization's strategy.

15.4.4 Participant selection

A purposive sampling strategy was employed to select interview participants within the organization. The sample included technical leaders, cloud operations managers, and others who helped provide their insights and viewpoints into the implementation of computational intelligence, their sustainability initiatives, and day-to-day challenges and successes in addressing and optimizing the cloud for energy efficiency. Fourteen semistructured interviews were conducted with key stakeholders across the corporate headquarters and operational teams. Each interview lasted between 45 and 90 minutes, allowing for detailed discussions on the company's sustainability strategies. The participants were assured of confidentiality, and informed consent was obtained from all individuals involved in the study.

15.4.5 Semistructured interview data

Semistructured interviews, a qualitative research method, were used here as they combine predetermined questions with flexibility to explore and gain further insights into topics in greater depth based on the knowledge of the personnel whose interviews are being conducted. This approach allows researchers to gather rich and detailed data, while maintaining a consistent structure across multiple interviews. The resulting semistructured interviews provided valuable insights into participants' experiences, perspectives, and opinions on the topic. This helps obtain contextual information that may not be captured by more rigid data collection methods. Researchers analyze the data that are used as input to various qualitative techniques, such as thematic analysis and content analysis, to further identify patterns, themes, and relationships within the responses provided.

1. Interview with technical leader

   Theme: Use of artificial intelligence in energy optimization
   - Detailed definition: The integration of artificial intelligence into clouds is crucial for energy efficiency. Relevant genetic and other machine learning and adaptive algorithms are employed to optimize resource usage and reduce the environmental impact of data centers, thereby minimizing energy consumption without compromising performance.
   - Supporting quotes
   - "We rely heavily on machine learning algorithms and artificial intelligence to predict the most energy-efficient routes for data storage and retrieval. This allows us to reduce the load on certain parts of our infrastructure during peak times, leading to significant savings in energy consumption."
   - "Our systems are continually learning and adapting to the patterns of our cloud services, which enables us to fine-tune energy use without impacting performance."
   - "With current ability to predict server load and dynamically adjust resources, we have managed to cut our energy consumption by 20% in the last year."
2. Interview with sustainability officer

   Theme: Sustainability practices in cloud operations
   - Detailed definition: Sustainability practices in organizational cloud operations include the use of renewable energy-efficient sources, efficient cooling systems, and energy-efficient hardware. The sustainability officer highlighted the importance of integrating environmental objectives into the core of the cloud service offerings.
   - Supporting quotes
   - "Our data centers were powered 100% by renewable energy. This is a crucial part of our commitment to reducing carbon footprint and promoting sustainable practices."
   - "We've made a substantial investment in the development of green infrastructure, like advanced cooling systems that has helped in

minimizing the energy consumption while maintaining optimal temperature conditions for our servers to run without any issue."
- "We are constantly evaluating new technologies to reduce our carbon footprint even further. Whether it's in the form of energy-efficient processors or using relevant intelligent sensors to manage power, sustainability is at the forefront of our decisions."

3. Interview with cloud operations manager
   Theme: Efficiency of cloud infrastructure
   - Detailed definition: Cloud operation managers at a company are responsible for overseeing the energy efficiency of cloud systems. They ensured that the infrastructure was dynamically optimized to maintain high performance while minimizing energy use.
   - Supporting quotes
   - "Energy efficiency is about maintaining a balance between cost-effectiveness and environmental responsibility with keeping the purpose of cloud setup in mind. By using computational intelligence, we have been able to optimize our cloud solutions on real-time basis."
   - "The integration of smart and context aware algorithms in our cloud infrastructure allows us to dynamically shift workloads to areas of the data center that are less energy-intensive at any given moment."
   - "We have worked closely with our computational teams to develop systems that autonomously detect and make real-time decisions to balance our cloud needs with our sustainability targets."

15.5 FINDINGS

Through the analysis conducted using NVivo, various key themes emerged, highlighting the integration of sustainability and computational intelligence. The first theme, "optimization of energy efficiency through computational intelligence," emphasizes the critical role of learning algorithms and smart infrastructure in enhancing energy efficiency [24]. Subthemes include energy load balancing, where automated agents predict server loads and shift workloads to less energy-intensive areas, and real-time decision-making systems, enabling autonomous optimization of energy use without compromising the performance of the cloud [25]. Predictive resource management is another important subtheme in which artificial and machine learning tools are leveraged to forecast energy needs and efficiently allocate resources. The integration of intelligence into cloud infrastructure has a directional relationship with minimizing energy consumption, thus demonstrating the potential of computational intelligence in enhancing both sustainability and efficiency.

The second theme, "sustainability practices in cloud operations," captures the various initiatives geared toward integrating sustainability objectives with cloud operations. Recently, the world has been talking about sustainability and ensuring that future generations are not let down because of the exploitation of Earth's resources. Hence, it is imperative for organizations to improve efficiency by using collaborative solutions that include smart, computationally aware systems that can self-heal and utilize energy efficiently without any intervention. The key subthemes identified were the utilization and adoption of renewable energy, wherein data centers are powered with the support of renewable energy, which is environmentally friendly. Green infrastructure investments, such as those using advanced cooling systems and energy-efficient processors, are also critical for reducing energy consumption [26]. The theme also explores sustainability-focused decision-making, where environmental goals play a key role in influencing the adoption of new technologies with the aim of reducing energy consumption and reducing the cost of day-to-day operations of cloud services. Ultimately, the goal of these initiatives is to minimize the carbon footprint of the data center and reduce operational costs [27]. This can be achieved by continuously evaluating and incorporating sustainable practices and implementing the changes to the systems. The findings show that organizations deeply committed to sustainability focus on integrating and using renewable energy for running cloud centers, adopting green infrastructure for efficiency, and smart intelligent systems to make appropriate decisions.

The third theme, "challenges in adopting sustainable technologies," outlines the hurdles organizations face in transitioning to and adopting sustainable practices. One of the biggest barriers is the high upfront investment required to enable it, which can deter organizations from making necessary changes [28]. Technological integration is another subtheme, where organizations struggle to incorporate those technologies with existing systems owing to the modifications required and the risk of impacting the current working cloud infrastructure. Another challenge is balancing sustainability with performance, as generally, sustainable solutions might impact or degrade the performance that organizations would like to avoid in the fear of losing business and reputation, which inhibits prioritization and obtaining approval from the business. Thus, it is imperative that organizations ensure that their performance levels are not compromised while pursuing environmental goals [29], as they might have a more harmful impact on their going concern. Additionally, stakeholder alignment and support become crucial to ensure that all levels of the organization are committed and that leaders buy in to initiating change management for such sustainability initiatives. The findings indicate that despite these challenges, intelligent aware systems can play an essential role in helping organizations achieve greater sustainability, while minimizing the impact of existing solutions on existing performance.

For the fourth theme, "environmental benefits of eco-friendly cloud solutions," the data highlight the substantial advantages of adopting sustainable cloud solutions [30]. One of the key findings was the subtheme of energy efficiency gains. There have been new developments in cooling systems and new processors that produce less heat. The adoption of these technologies and the replacement of old systems can contribute to long-term environmental gains. Furthermore, the theme touches on long-term resource endurance as new systems are designed to keep in mind the utilization of minimal water and energy [31]. The findings reveal that such approaches and practices offer tangible benefits in reducing energy consumption and carbon emissions, with long-term positive environmental impacts.

The fifth and final theme, "future of sustainable cloud technologies," explores current emerging trends such as smart-aware intelligent systems for power management that optimize power consumption [32] and look for the right utilization and energy-efficient processors instead of guaranteeing processors of yesteryears, which are becoming seriously considered in calculations of long-term benefits. The adoption of sustainability practices within the cloud, such as reusing and repurposing [30, 32] the existing intelligent system, is also gaining traction across the industry. Finally, the scalability of sustainable solutions is a critical subtheme, as organizations look for ways to implement them [29]. The findings suggest that the future of clouds will increasingly depend on the scalability and integration of computationally smart sustainable technologies, particularly those that enable better resource management and minimize environmental impact on a global scale. The aforementioned themes and subthemes provide a comprehensive high-level overview of the current and future states, showcasing the integration of computationally aware intelligence technologies and the significant environmental benefits they offer. The findings suggest that although challenges exist, the integration of such smart systems, along with a focus on sustainable practices, will prove a transformative solution in cloud-based environments.

15.6 IMPLICATIONS

This study is limited by current technology adoption in enabling sustainable solutions for data center cloud operations. Because changes in such infrastructure require heavy investments, new technology adoption is slow. We believe that advanced and upcoming computational intelligence technologies with their subsequent integration into cloud infrastructure systems will become increasingly essential for achieving operational efficiency and sustainability. Future studies should explore the scalability of such solutions across diverse cloud environments, including large-scale global infrastructure and smaller regional data centers. Additionally, the continued development of efficient hardware with a reduced thermal footprint, renewable energy utilization, and sustainable economic practices will play a pivotal role in shaping the future of sustainable cloud solutions. It is also important to investigate how businesses can overcome the financial and technological barriers to adopting these technologies, with an emphasis on developing cost-effective models that align with the environmental goals. Furthermore, as consumer and regulatory pressures increase, future research should assess how these advancements in computational intelligent systems can help drive sustainability, with an emphasis on corporate responsibility, reduced carbon footprint, and meeting regulatory standards. Technological advancements and financial limitations indirectly contribute to a decrease in organizations running their own infrastructure or cloud-based solutions. Moreover, stringent regulations such as (Restriction of the Use of Certain Hazardous Substances (RoHS) in Electrical and Electronic Equipment Regulations and the Waste Electrical and Electronic Equipment Directive (WEEE) in Europe also force suppliers as well as organizations in other parts of the world to reengineer their processes and move to greener IT solutions in order to remain competitive.

15.7 LIMITATIONS

A key limitation of this study is that it primarily focuses on the integration of computational intelligence in cloud computing and examines sustainability practices within cloud systems from a qualitative perspective, based on interviews with a limited number of industry professionals. Although some valuable insights into the current trends and challenges can be captured, they are limited to the type of professionals involved in providing the inputs and are not representative of all. Additionally, the study does not cover potential variations in the adoption and utilization of sustainable practices and smart technologies across different industries, data center sizes, and geographies. This may affect generalization, and more studies should be conducted. Furthermore, the long-term impacts of implementing these technologies, particularly in terms of economic feasibility and scalability, have not been fully explored. Additionally, cloud vendors choose a center for operations that is cost-effective and can support their operations (availability of bandwidth, power, etc.). However, these locations may not be suitable from a customer's perspective due to security and regulatory compliance restrictions. For example, projects in Iran may not like to use US-based hosting environments because the US has access to all such projects under its Patriot Act, which could lead to a breach of privacy and concern. There are other aspects of taxation wherein service utilization or consumption can be taxed at the source and endpoints. In addition, some governments provide incentives for setting up data centers by providing relaxations, which also play a key role in the executive decisions of setting up cloud service operations. These points generally influence the setup of data centers and investment in computing resources, which is not very straightforward and is dependent on changing the IT landscape. Finally, the subjective nature of interview-based data has a good chance of introducing bias, and it is recommended that quantitative studies and interview pools be created to obtain clearer insights into this area.

15.8 DISCUSSION

The data analysis findings highlight that leveraging smart technologies plays a significant role in improving the operational efficiency of cloud systems. The increasing usage of data brings its own overhead, as the current environment is such that delayed decisions can take one out from the business. Hence, businesses consume data in gigabytes (GBs) and terabytes (TBs). Ensuring that no loss is observed, cloud providers have to invest in technologies and infrastructure, which is becoming a headache for them as it requires investment to add more infrastructure, energy, and space. Such high consumption forces business leaders, especially in cloud center domains, to look at sustainable practices and utilize intelligent adaptive systems that can work efficiently with less consumption. By utilizing adaptive algorithms and smart energy-efficient infrastructure [4], cloud providers cannot only monitor but also optimize energy usage by their data centers and infrastructure [15] and balance workloads dynamically [19], leading to significant energy consumption reductions [23]. These technologies, along with sustainability practices, can contribute to substantial reductions in carbon emissions and overall environmental impact. However, challenges such as high initial upfront investments, complexity of management and integration, government policies and laws, and environmental concerns remain prominent and key concerns. Despite these barriers, these technological adoptions provide critical support by enabling real-time optimization [30] and autonomous decision-making that ensures that cloud systems can become more energy efficient with minimal impact on service performance. Looking ahead, the adoption of emerging green technologies and sustainability practices, coupled with increased collaboration with external partners who are able to bring fresh insights, will drive further progress in sustainable cloud-efficient solutions. The overall discussion suggests that integrating and adopting computational intelligence with sustainability not only addresses the reduction in the carbon footprint but also enhances the long-term viability and efficiency of cloud operational efficiencies. The in-depth interviews provided good insights into the impact of government incentives and country-specific policies, along with macro- and microeconomic considerations, technology-aware resources, and relevant infrastructure and site availability, which play a major role in deciding the setup and operation of data centers. Thus, there are many variable parameters that play out in decision making and take a long-term view in sustainable operations that we were able to just scratch the surface. There is a general awareness of global warming and ecological considerations, but it is also dependent on proactive support from all levels and needs buy-in from key stakeholders. Organizations have understood the importance of using sustainable practices and environmentally friendly solutions, as they are not only humane to do but also provide the long-term benefit of the life expectancy of computational systems and hence are looking more to adopt computational intelligence-based solutions in their cloud operations.

15.9 CONCLUSION

In conclusion, the use of computational intelligence technologies in cloud operations presents a transformative solution to enhance both sustainability and operational efficiency. The findings highlight how adaptive algorithms and smart infrastructure can optimize energy usage, reduce carbon emissions, and improve resource utilization in cloud operations. The long-term benefits of investment in sustainable practices far outweigh the upfront high investment, integration, and revamping of existing systems. As the industry moves toward more sustainable practices, the adoption of renewable energy and energy-efficient hardware will further drive environmental benefits in the long term. Ultimately, this research underscores the importance of leveraging adaptive smart-aware algorithms and infrastructure to achieve long-term sustainability goals while minimizing the impact on existing performance in cloud operations, suggesting a promising future for eco-friendly cloud-enabled systems.

REFERENCES

1. Yaman, C. (2024). A review on the process of greenhouse gas inventory preparation and proposed mitigation measures for reducing carbon footprint. *Gases*, *4*(1), 18–40. https://doi.org/10.3390/gases4010002
2. Shyamala, K., & Sunitha Rani, T. (2015). Analysis of efficient resource-allocation mechanisms in cloud computing. *Indian Journal of Science and Technology*, *8*(9), 814. https://doi.org/10.17485/ijst/2015/v8i9/50180
3. Ghobaei-Arani, M., Rahmani, A. M., Rahmanian, Ali. A., & Souri, A. (2018). A moth-flame optimization algorithm for web service composition in cloud computing: Simulation and verification. *Software: Practice and Experience*, *48*(10), 1865–1892. https://doi.org/10.1002/spe.2598
4. Liu, Z., Wierman, A., Wang, Z., Gmach, D., Chen, Y., Hyser, C., Marwah, M., & Bash, C. (2012). Renewable and cooling-aware workload management for sustainable data centers. *ACM SIGMETRICS Performance Evaluation Review*, *40*(1), 175–186. https://doi.org/10.1145/2318857.2254779

5. Simmon, E. (2018). Evaluation of cloud computing services based on NIST SP 800-145. *NIST Special Publication*, *500*(322), 500-322. (Accessed April 27, 2025)
6. Abualkibash, M., & Elleithy, K. (2012). Cloud computing: The future of IT industry. *International Journal of Distributed and Parallel Systems*, *3*(4), 1. (Accessed April 27, 2025).
7. Stanton, B., Theofanos, M., & Joshi, K. P. (2015). Framework for cloud usability. In *Human Aspects of Information Security, Privacy, and Trust: Third International Conference, HAS 2015, Held as Part of HCI International 2015, Los Angeles, CA, USA, August 2–7, 2015. Proceedings 3* (pp. 664–671). Springer International Publishing. (Accessed April 27, 2025)
8. Xiong, A.-P., & Xu, C.-X. (2014). Energy efficient multiresource allocation of virtual machine based on PSO in cloud data center. *Mathematical Problems in Engineering*, *2014*(1), 1–8. https://doi.org/10.1155/2014/816518
9. Chou, L.-D., Chen, H.-F., Chang, Y.-J., Chao, H.-C., & Tseng, F.-H. (2018). DPRA: Dynamic power-saving resource allocation for cloud data center using particle swarm optimization. *IEEE Systems Journal*, *12*(2), 1554–1565. https://doi.org/10.1109/jsyst.2016.2596299
10. Buyya, R., Arroba, P., & Ilager, S. (2023). Energy-efficiency and sustainability in new generation cloud computing: A vision and directions for integrated management of data centre resources and workloads. *Software: Practice and Experience*, *54*(1), 24–38. https://doi.org/10.1002/spe.3248
11. Mahadasa, R., & Surarapu, P. (2016). Toward green clouds: Sustainable practices and energy-efficient solutions in cloud computing. *Asia Pacific Journal of Energy and Environment*, *3*(2), 83–88. https://doi.org/10.18034/apjee.v3i2.713
12. Gupta, A., Sharma, A. K., Sharma, V. P., Vats, P., Jain, D., & Singh, P. (2022). *A Sustainable Green Approach to the Virtualized Environment in Cloud Computing* (pp. 751–760). Springer Nature, Singapore. https://doi.org/10.1007/978-981-16-9967-2_71
13. Garg, S. K., Yeo, C. S., & Buyya, R. (2011). *Green Cloud Framework for Improving Carbon Efficiency of Clouds* (Vol. 6852, pp. 491–502). Springer, Berlin, Heidelberg. https://doi.org/10.1007/978-3-642-23400-2_45
14. Khan, A. R. (2024). Dynamic load balancing in cloud computing: Optimized RL-based clustering with multi-objective optimized task scheduling. *Processes*, *12*(3), 519. https://doi.org/10.3390/pr12030519
15. Chen, X., Lu, S., Fu, X., Li, W., & Zhou, Z. (2018). Efficient resource allocation for on-demand mobile-edge cloud computing. *IEEE Transactions on Vehicular Technology*, *67*(9), 8769–8780. https://doi.org/10.1109/tvt.2018.2846232
16. Wang, R., & Tsai, W.-T. (2022). Asynchronous federated learning system based on permissioned blockchains. *Sensors*, 22(4), 1672. https://doi.org/10.3390/s22041672
17. Mohammed, M. A., Ahmed, M. A., & Hacimahmud, A. V. (2023). Data-driven sustainability: Leveraging big data and machine learning to build a greener future. *Babylonian Journal of Artificial Intelligence*, *2023*, 17–23. https://doi.org/10.58496/bjai/2023/005

18. Uzondu, N., & Lele, D. (2024). Multifaceted impact of renewable energy on achieving global climate targets: Technological innovations, policy frameworks, and international collaborations. *International Journal of Applied Research in Social Sciences*, *6*(7), 1520–1537. https://doi.org/10.51594/ijarss.v6i7.1338
19. Liu, J., Pahl-Wostl, C., Hull, V., Chung, M. G., Li, S., Godfray, H. C. J., Sun, J., Hoff, H., Gleick, P., Xu, Z., & Tilman, D. (2018). Nexus approaches to global sustainable development. *Nature Sustainability*, *1*(9), 466–476. https://doi.org/10.1038/s41893-018-0135-8
20. Lim, W. M. (2024). What is qualitative research? An overview and guidelines. *Australasian Marketing Journal*, *17,* 17–23. https://doi.org/10.1177/14413582241264619
21. Kyngäs, H. (2019). *Qualitative Research and Content Analysis* (pp. 3–11). Springer. https://doi.org/10.1007/978-3-030-30199-6_1
22. Venkateswaran, N., Boopathi, S., Ayyannan, M., Chavan, S. M., Vidhya, K., & Sekar, K. (2023). *A Study on Smart Energy Management Framework Using Cloud Computing* (pp. 189–212). IGI Global. https://doi.org/10.4018/978-1-6684-8634-4.ch009
23. Mohammed, M. A., & Ţăpuş, N. (2023). A novel approach of reducing energy consumption by utilizing big data analysis in mobile cloud computing. *Mesopotamian Journal of Big Data*, *2023*, 110–117. https://doi.org/10.58496/mjbd/2023/015
24. Mahadasa, R., & Surarapu, P. (2016). Toward green clouds: Sustainable practices and energy-efficient solutions in cloud computing. *Asia Pacific Journal of Energy and Environment*, *3*(2), 83–88. https://doi.org/10.18034/apjee.v3i2.713
25. Centofanti, C., Santos, J., Gudepu, V., & Kondepu, K. (2024). Impact of power consumption in containerized clouds: A comprehensive analysis of open-source power measurement tools. *Computer Networks*, *245*, 110371. https://doi.org/10.1016/j.comnet.2024.110371
26. Lv, Y. (2023). Transitioning to sustainable energy: opportunities, challenges, and the potential of blockchain technology. *Frontiers in Energy Research*, *11*. https://doi.org/10.3389/fenrg.2023.1258044
27. Jayaraman, K., Jayashree, S., & Dorasamy, M. (2023). The effects of green innovations in organizations: Influence of stakeholders. *Sustainability*, *15*(2), 1133. https://doi.org/10.3390/su15021133
28. Patel, Y. S., Mehrotra, N., & Soner, S. (2015). *Green Cloud Computing: A Review on Green IT Areas for Cloud Computing Environment. 1*, 327–332. https://doi.org/10.1109/ablaze.2015.7155006
29. Ryan, A., Nomura, K., Abe, O., Tilbury, D., & Blaze Corcoran, P. (2010). Sustainability in higher education in the Asia-Pacific: Developments, challenges, and prospects. *International Journal of Sustainability in Higher Education*, *11*(2), 106–119. https://doi.org/10.1108/14676371011031838
30. Albreem, M. A., Mohd Yasin, M. N., Alsharif, M. H., Jusoh, M., & Sheikh, A. M. (2021). Green Internet of Things (GIoT): Applications, practices, awareness, and challenges. *IEEE Access*, *9*, 38833–38858. https://doi.org/10.1109/access.2021.3061697

31. Rahmani, H., Carvalho, N. B., Costanzo, A., Masotti, D., Alimenti, F., Kettle, J., Palazzi, V., Mezzanotte, P., Shetty, D., Ghasempour, Y., Grosinger, J., Roselli, L., Correia, R., Wagih, M., & Vital, D. (2023). Next-generation IoT devices: Sustainable eco-friendly manufacturing, energy harvesting, and wireless connectivity. *IEEE Journal of Microwaves*, *3*(1), 237–255. https://doi.org/10.1109/jmw.2022.3228683
32. Ahmad, S., Sharma, V., & Mishra, S. (2023). *Green Computing for Sustainable Future Technologies and Its Applications* (pp. 241–256). Emerald. https://doi.org/10.1108/978-1-80455-562-020231016

Chapter 16

The evolution of cyber-physical systems

From embedded computing to smart automation

Rupanshi Agarwal, Akash Sanghi, Gaurav Agarwal, Kushal Johari, Kamal Upreti, and Ashish Kumar Sharma

16.1 INTRODUCTION

Cyber-physical systems (CPS) are a novel paradigm that integrates computer skills with physical processes to enhance control, monitoring, and optimization in many different domains. CPS involve a fusion of physical infrastructure (sensors, actuators, control systems) and embedded computing (software, algorithms, real-time data), and are central to the CPS concept. CPS involves the dynamic interactivity of digitization and composing a cohesive system to deliberate, make decisions, and control autonomously in the real world. Historically, CPS is a recognizable evolution from industrial automation and traditional embedded systems, both of which would be reinvented, incorporating today's cutting-edge computation, real-time communication, and adaptive control [1].

Definition: Cyber-physical systems (CPS) are complex, engineered systems that allow for real-time monitoring, interconnected control, and decision-making in distributed and heterogeneous environments autonomously. There are physical processes that are tightly linked and entangled, with computer algorithms and networked digital technologies embedded within structures of CPS.

In other words, cyber-physical systems are integrated systems consisting of computational processes (aka algorithms) and physical processes. CPS allows for the real-time monitoring, control, and decision-making of physical processes across interconnected networks.

CPS comprises the integration of the physical processes and computational intelligence. CPS uses sensors, actuators, embedded systems, and advanced algorithms to monitor and control a physical phenomenon in real time [2, 3]. CPS provides an architecture for creating such smart, autonomous, and more efficient behaviors in a variety of industries, such as manufacturing, energy use, transportation, healthcare, and agriculture [3]. Figure 16.1 illustrates the architecture of a basic CPS, and the core components and interactions of a CPS.

DOI: 10.1201/9781003569251-16

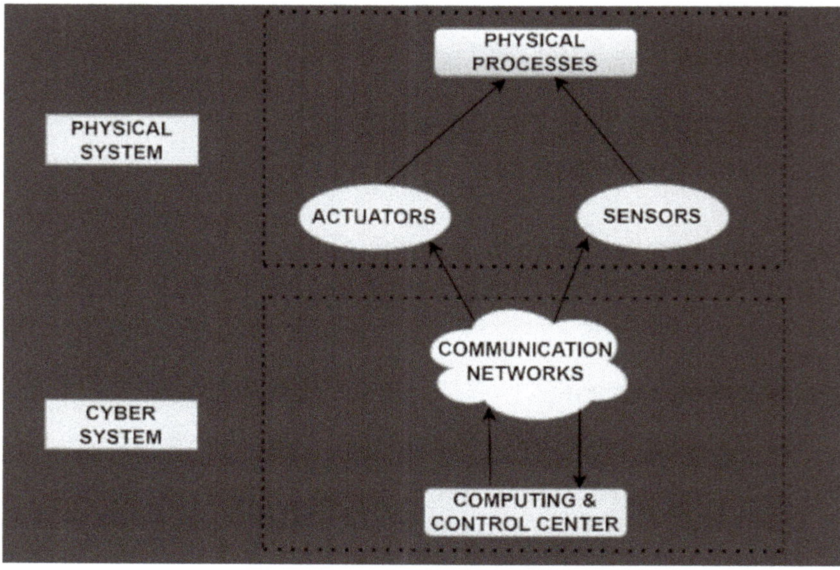

Figure 16.1 Fundamental components of CPS

Safety is the distinctive feature of CPS. CPS gathers key environmental data by means of real-time monitoring that facilitates decision-making processes that achieve optimal efficiency while reducing the risk of failures [4, 5]. Automation is one of the other important characteristics associated with CPS. In CPS applications, such as autonomous vehicles and smart manufacturing production lines, this means automation of complex processes previously managed by humans (e.g., driving, production). In addition, CPS technologies incorporate advanced computational models, such as artificial intelligence (AI) and machine learning (ML), and drive intelligent decision-making that allow them to adapt to changes and continuously improve performance [6, 7]. CPS has the potential to transform many sectors because of the improved combination of production, security, and innovation. Some of these sectors are:

- Healthcare: CPS provides real-time patient monitoring for the healthcare sector through wearable technologies and smart implants, enabling continuous care and timely intervention. CPS robot-assisted surgery enables greater accuracy and optimal patient outcomes [2].
- Transportation: CPS powers smart traffic systems and autonomous cars in the transportation sector, enhancing safety, streamlining traffic, and lowering human error [6]. The creation of smart cities depends on these systems.
- Manufacturing: Industry 4.0 in manufacturing is made possible by CPS, which paves the way for smart factories wherein machinery can

self-monitor, change parameters, and optimize production processes in real time. CPS-powered predictive maintenance reduces downtime and increases operational effectiveness [8].

- Energy and infrastructure: Real-time data guarantees efficient energy distribution and consumption in energy and infrastructure, and CPS propels the development of smart grids and energy systems [8, 9]. Incorporating renewable energy sources is also made possible by CPS, guaranteeing steady and sustainable energy consumption.
- Agriculture: By using automated equipment and real-time animal and crop monitoring, CPS helps precision farming in agriculture, which increases yields and makes better use of available resources.

These capabilities support new paradigms of human–machine collaboration, wherein CPS collaborate with humans to enhance decision-making and enhance overall system performance, in addition to enabling automation and optimization. CPS enables the possibility for previously unthinkable levels of automation, safety, and operational efficiency by fusing the digital and physical worlds.

It is important to acknowledge the revolutionary effect that CPS has already had on various industries and the enormous promise it offers for the future as we delve deeper into the elements, architecture, and applications of CPS in the sections that follow. CPS is positioned as a key technology propelling the next generation of industrial and societal innovation because of its integration of automation, intelligent decision-making, and real-time monitoring [6, 10].

16.2 CORE COMPONENTS OF CYBER-PHYSICAL SYSTEMS

Cyber-physical systems are made of many different key elements; thus, it is easy to carefully integrate computer processes with physical mechanisms. By enabling this real-time interaction of these parts, the physical environment can be more easily monitored, regulated, and/or optimized. The four major components of a CPS are communication networks, processors, actuators, and sensors. By integrating these elements, the crucial feedback loop that empowers CPS to adaptively choose the right path, dynamically adjust to changing scenarios, and balance the efficiency and reliability of the whole system is achieved. In this section, we discuss details about the technical specifics that comprise each of these basic components and implications among them.

16.2.1 Sensors

Since the physical context is to be sensed and monitored online, sensors are the foundational elements for CPS. They convert physical constants into signals readable by the computer system, like movement, pressure, temperature, and chemical concentration. For the system to understand the state of the physical environment and make intelligent decisions, these signals are of this type. We can consider sensors to be functions that transduce physical processes to measurable digital data:

$$S(p(t)) = d(t)$$

where

S is the sensor function.

$p(t)$ is the physical parameter being monitored over time.

$d(t)$ is denoted by the matching digital data that is entered into the computer systems.

16.2.1.1 Type of sensors

- Temperature sensors are used to detect the change of temperature and display to be watched or adjusted in an industrial system, health system, and environment control.
- Pressure sensors are used to measure the fluid or gas pressure and detect abnormal conditions in aircraft, industrial machinery, and medical devices [2–3, 5].
- Motion sensors are used to detect movement and modify operations in robotics, autonomous vehicles, and security systems.
- Chemical sensors are widely employed in healthcare and environmental monitoring to identify the presence of chemical molecules, such as gas or glucose levels [11].

16.2.1.2 Sensor technology

Sensors use a variety of physical concepts to identify environmental changes.

- Pressure sensors that use piezoelectric sensors work by using the piezoelectric effect to produce electrical signals when they are under mechanical stress.
- Used in light sensors, photodiodes change light into voltage or current according to how much light is shining on the sensor [6].
- Thermocouple devices measure temperature by generating an electrical voltage from temperature variations at junctions [6].

Within CPS, the relationship between actuators and sensors creates a key feedback loop. The system can continuously adjust to its surroundings thanks to this loop, which also guarantees that physical processes run as efficiently as possible. Take an autonomous car, for instance. Actuators use the data from sensors that collect information about the vehicle's environment—such as its speed and distance from objects—to modify the vehicle's steering, braking, and acceleration.

16.2.2 Actuators

Actuators carry out the physical operations necessary to modify or regulate a system in response to decisions made by the computing core, while sensors collect data. Actuators are devices that translate electrical inputs into mechanical movements, flow control, or sound production [2, 11, 12].

The model for this procedure is

$$A(c(t)) = p(t)$$

where

- A is the actuator function.
- $c(t)$ represents the control signal.
- $p(t)$ is the physical state.

16.2.2.1 Type of actuators

Electric motors are actuators that transform electrical energy into mechanical motion. They are commonly seen in robotics, industrial machinery, and driverless autos.

- Hydraulic actuators generate large forces and are powered by pressurized fluids in large equipment such as airplane landing gear and excavators [13].
- Actuators are devices that compress the air for motion with a high frequency and are mostly employed in production automation systems.

16.2.2.2 Actuator mechanisms

There is a range of energy sources used to generate motion from actuators.

Electromagnetic actuation: Motors and solenoids provide electromagnetic energy that creates a magnetic field to move the mechanical component.

Hydraulic and pneumatic actuators use fluid or air pressure to move linearly or rotate.

16.2.3 Processors

The CPS computational core has processors used to sense, control, and choose in response to sensor data [3, 9]. Embedded processors in CPS are often designed to perform in real time executing control algorithms to make sure the system performs as defined. The performance and stability of the system are guaranteed by these processors that make intricate decision-making algorithms. The system state can be characterized by means of a discrete-time model.

$$x[k+1] = f(x[k], u[k], w[k])$$

where

The state of the system is given by $x[k]$, at time step k. $u[k]$ is the control input at the same time step and $w[k]$ is the external disturbances on the system at the same time step.
f is the function that describes the system states evolution.

CPS currently employs the control techniques like model predictive control (MPC) and proportional–integral–derivative (PID) control by which fine-grained and real-time adjustments are made in physical processes. Thus, such algorithms should consider many constraints such as time, energy, and storage in particular, which are important applications in healthcare and industrial automation [9–10].

16.2.4 Communication networks

CPS uses a communication network to handle data transfer between the sensors, processors, and actuators of a system and coordinate the distribution of members of a system. CPS is therefore made possible in a variety of settings ranging from small-scale systems in intelligent factories to large-scale systems in smart grids and intelligent transportation systems [13, 14]. The communication network latency (L), bandwidth (B), and reliability (R) are key factors for CPS performance. Following is a possible method to express the delay in transmission between components:

$$L = d/v + p$$

where

- d represents the physical distance between nodes.
- v represents the transmission speed.
- p is the processing delay in network nodes.

In wireless CPS applications, the selection of communication technology—such as Wi-Fi, Bluetooth, Zigbee, or 5G—is crucial for low-latency and high-reliability data transfer. Time-sensitive networking (TSN), for example, is widely used in industrial CPS since it provides the necessary real-time guarantees for mission-critical applications.

16.2.5 Real-time systems

One of the most prominent characteristics of CPS is its ability to operate as a real-time system. This means that all computing operations, such as data analysis, decision-making, and control actions, must be completed by a tight deadline. Failure to meet these temporal limitations, especially in safety-critical applications such as medical devices or autonomous driving, may result in unstable systems or catastrophic failures [11, 15]. In CPS, the description of real-time systems is often done as hybrid systems, that is, continuous processes and discrete control actions. Such systems can be described through differential equations for discrete control and continuous dynamics, respectively.

$$x(t) = f(x(t), u\ (t)), x[k + 1] = g(x[k], u[k])$$

where

$x[k + 1]$ is the discrete state of the system at the next time step.
$f(x(t))$ is the mechanism used to drive the system's continuous evolution.
$x(t)$ represents the behavior of a physical system in continuous time.
The $g(x[k], u[k])$ is the update function for the discrete control.

In order to ensure timely completion of CPS operations, scheduling techniques such as earliest deadline first (EDF) and rate monotonic scheduling (RMS) are used. In real-time systems, such failures may occur due to delayed deadlines, and these algorithms are particularly critical.

16.3 ARCHITECTURE OF CPS

CPS is a collective of communication technology, computational decision-making, and physical processes that enable systems to have autonomous control, monitoring, and functioning in an optimized way. For the seamless integration of the physical components with the ones that are digital, the architecture of a CPS is critical. Strong communication and changes in the architecture based on dynamically changing conditions can produce an efficient architecture for accomplishing real-time data processing. This section looks at the basic structure of CPS, highlighting its key components. From the physical components to the decision-making and network levels, Figure 16.2 illustrates the hierarchical architecture of a CPS, highlighting the functions and interactions of each layer. A CPS's fundamental architecture is made up of several layers, each of which carries out unique but related tasks to provide real-time control and feedback. There are three main layers to the architecture: physical, cyber, and communication.

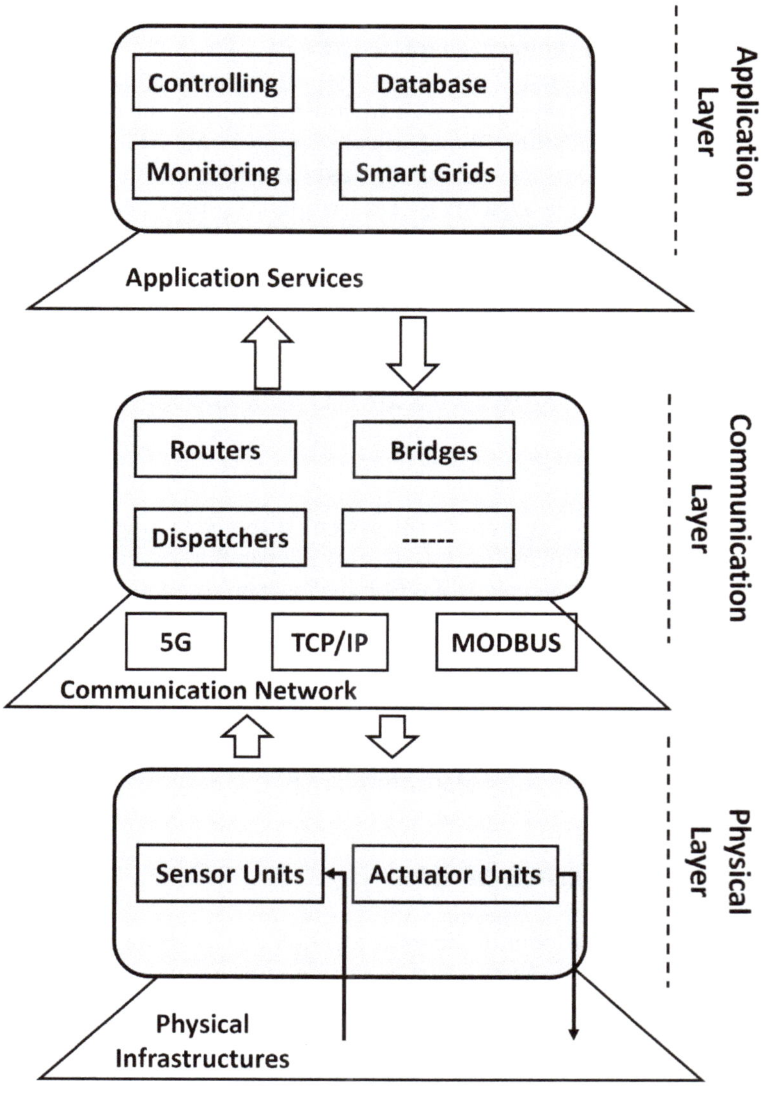

Figure 16.2 Architecture of CPS

16.4 PHYSICAL LAYER (PERCEPTION LAYER)

The perception layer is the initial layer in the CPS architecture [6]. Everything that will understand the physical occurrences, convert them into electrical signals, and then turn those signals into information is stored in this layer. Aggregators, actuators, sensors, transducers, cameras, radio-frequency identification (RFID) tags, lasers, and any other intelligent equipment of the so-called "factory floor" are some examples of the equipment that make up this initial layer [4, 9]. The objective of this layer is to gather real-time process data for the purpose of physical system planning, monitoring, and control. Because of these elements, this layer is sometimes referred to as the "sensing layer" and the "recognition layer" in the literature. This layer is an essential component of any CPS since it is in charge of data gathering and direct environmental interaction.

Actuators and sensors that communicate directly with the physical environment make up the physical layer. Real-time environmental data, such as temperature, motion, pressure, or chemical concentrations, are gathered by sensors, which then translate the data into digital signals. Actuators, on the other hand, take the control signals produced by the computing unit and translate them into actual physical actions, such as varying the speed of a vehicle, opening a valve, or adjusting a robotic arm [15, 16].

16.4.1 Sensors

The physical layer's sensors are in charge of continuously gathering environmental data.

Depending on the particular use of the CPS, they may include temperature sensors, pressure sensors, motion detectors, cameras, chemical sensors, and more. The state of the physical process under observation and management is reflected in the data they collect. As an illustration:

- Sensors in autonomous cars record information about the speed, distance to objects, and state of the road.
- Sensors track power flow and energy use in smart networks.

16.4.2 Actuators

Actuators take on the opposite function, translating the control signals produced by the cyber layer's computational algorithms into actual physical actions. Actuators regulate a wide range of equipment and mechanisms, including brakes, valves, robotic arms, and motors [12].

As an illustration:

- Actuators in smart manufacturing modify the speed or precise alignment of machinery by utilizing data feedback.
- Actuators may regulate the release of medication in automated drug delivery systems used in healthcare CPS.

16.4.3 Feedback loop integration

The closed-loop feedback mechanism of CPS, which permits ongoing communication between the physical and cyber levels, is reliant on sensors and actuators. But here, the hardware specifications are not as important as the data flow and control logic [17, 18]. For data flow, sensors gather environmental data, which is transmitted to the cyber layer's processing units for examination. The actuators receive the control decisions made by these units and modify the physical system accordingly. The system may function independently and react flexibly to changes in real time thanks to this feedback loop [17]. Example: In autonomous cars, sensors identify impediments and the state of the road, and the cyber layer uses this information to decide whether to brake and steer. Actuators then modify the vehicle's motion in accordance with these determinations.

16.5 CYBER LAYER (COMPUTATIONAL LAYER)

A CPS's brain is located in the cyber layer. Data processing units, control algorithms, and embedded computers make up the layer. These components analyze sensor data, make decisions, and provide control signals for the actuators. This layer manages all computations, optimization, and real-time control.

Embedded processors on the cyber layer perform real-time data processing. This data goes through algorithms that analyze sensor data and provide control signals to these processors. These processors are typically implemented in systems that are close to the physical environment [17, 18] to allow swift, real-time responses.

For example, in industrial automation, robotic arms are governed by processors based on sensor input that ensure precision for operations like material handling and assembly. Autonomous car processors make decisions about steering and braking based on sensor input from cameras, LiDAR, and other sources [13].

Control algorithms:

- The cyber layer implements control algorithms to ensure that the physical processes run as planned, maintain system stability, and optimize performance. These algorithms could be as basic as PID controllers, which are a type of classical control, or as sophisticated as a machine learning-based control or model predictive control (MPC) [19].
- A PID controller calculates the error between a process variable and the desired setpoint by measuring the variable. After that, it adjusts the system's behavior to reduce the mistake [19].
- MPC uses a dynamic model of the system to forecast future states and optimize control inputs across a finite horizon, modifying control actions in real-time to minimize a cost function [2, 18].
- A machine learning-based control system can learn the best control strategies based on past data and get better over time thanks to machine learning models like reinforcement learning.

Machine learning models may be taught to identify anomalies in sensor data patterns that point to possible system failures in order to perform predictive maintenance and fault detection.

The cyber layer is also in charge of storing historical data, which is useful for predictive analysis and long-term optimization. Large datasets can be processed and stored locally at the edge (in embedded devices or local servers) or in the cloud (described in the communication layer), depending on the complexity of the system.

16.6 COMMUNICATION LAYER (NETWORK LAYER)

Data exchange between the cyber layer (processors and control algorithms), physical layer (sensors and actuators), and any external systems (e.g., cloud platforms or remote monitoring systems) is made possible by the communication layer. For real-time system operation, effective communication is essential, particularly in distributed CPS where components may be spread geographically [16, 17, 20].

16.6.1 Wired and wireless communication networks

Depending on the speed, latency, and reliability needs of the application, communication in CPS can take place over wired or wireless networks [11].

- Wired communication: Common in industrial CPS, wired connections are favored due to their high data speed and dependability (e.g., Ethernet, fiber optics).
- Wireless communication: Because of their adaptability and simplicity of installation, wireless technologies (such as 5G, Wi-Fi, Zigbee, and Bluetooth) are preferred in applications such as smart grids and autonomous cars.

A number of things influence the network choice:

- The time it takes for data to travel from sensors to processors and back to actuators is known as latency.
- Bandwidth: The network's ability to manage large amounts of data.
- Dependability: Ensuring error-free data transmission, especially in systems that are vital to missions.

16.6.2 Edge computing

Sending all data to the cloud for processing in time-sensitive CPS applications causes delays that could impair system performance. Edge computing handles data nearer to its source, or at the "edge" of the network, in order to lessen this (inside local devices or nearby servers). Edge computing dramatically lowers latency and speeds up response times for applications requiring instantaneous decision-making [19–21].

For instance:

- Edge computing allows real-time sensor data processing (such as obstacle detection) in autonomous cars without depending on cloud-based servers.
- Local controllers in smart factories process machine data to quickly detect and fix performance problems.

16.6.3 Cloud integration

Cloud computing offers centralized, massive data storage, analysis, and coordination, whereas edge computing manages immediate data processing. Tasks requiring substantial computational resources, such as the following, are performed on the cloud:

- Machine learning model training: Training of machine learning models is handled by cloud servers using enormous datasets, which are then incorporated into real-time systems through the use of edge devices.
- Coordination across the entire system: In smart grids, for instance, cloud systems optimize electricity distribution by using information about global energy usage.

To put it briefly, the structure of CPS is made up of discrete yet linked layers that allow the digital and physical domains to interact seamlessly. The physical layer acts as an interface between the environment and the system, using actuators to carry out control operations and sensors to gather data. The cyber layer (which is the layer that decides and controls) processes this data in some way with sophisticated algorithms and comes up with control signals. These components need to rely on the communication layer for timely and reliable data flow to each other and to external systems, for example, cloud platforms, in such a way as to allow distributed, scalable, and intelligent operations [20, 21]. CPS can autonomously perform challenging tasks under dynamic changes in their environment due to the layered design at heart [7, 8].

16.7 APPLICATIONS OF CPS

Some industries have been revolutionized with such automation, intelligent decision-making, and real-time monitoring by CPS. CPS can "optimize" complicated systems that operate with the least amount of human, exhibit real-time response to change, and sense change and react in dynamic situations. The fusion of computational intelligence and physical processes has made it possible to apply it to applications such as smart cities, healthcare, and industrial automation. The following details out different domains where CPS is applied.

16.7.1 CPS in smart cities

By making it possible to simultaneously deploy sensors across the network, integrate them with networks, and corresponding data analytics, CPS can drive improvement of municipal services in real time, enhance quality of life, improve environmental sustainability, and foster sustainable urban development [21, 22].

16.7.1.1 Traffic management

Intelligent traffic management is one of the most well-known applications of CPS in smart cities. A traffic signal system (CPS) is a system of cameras, connected vehicles, and traffic sensors that are always monitoring and evaluating traffic conditions. Real-time updates offered by the system can further reduce gridlock and can also shorten travel times.

CPS-based systems continuously monitor traffic flow in cities such as Singapore and Amsterdam, and immediately move traffic along in order to avoid congestion. These systems have the capability to give emergency vehicles priority, which means they will reach their destination quickly.

16.7.1.2 Smart grids

Advanced power networks, known as smart grids, emphasize energy distribution optimization in real time as well as integrating such energy sources as renewable energy. Their operation depends on CPS. Sensors and meters are placed throughout the grid to collect data on energy generation and energy consumption. This data then travels through the cyber layer to be processed and changed so as to prevent overloads, reduce energy waste, and maintain supply and demand equilibrium. CPS also makes it possible to incorporate renewable energy sources into the system by predicting energy production based on environmental data. It paves the way to sustainable energy sources and ensures grid stability.

16.7.1.3 Waste and water management

CPS is also used in smart cities to regulate trash as well as water disposal. Trash cans with built-in sensors can be used to signal that they are full and thus help streamline waste collection routes. Similarly, utility sensors installed in water distribution networks can detect leaks or unusual usage patterns, and utilities can put a stop to water wastage and take any appropriate remedial actions.

16.7.2 Healthcare in CPS

The healthcare sector has greatly benefited from the application of CPS, particularly in the areas of patient monitoring, robotic surgery, and diagnostics. CPS has revolutionized healthcare delivery, making it more precise, timely, and customized.

16.7.2.1 Real-time patient monitoring

One of the most important applications of CPS in the medical industry is real-time patient monitoring. Wearables with sensors continuously collect data on blood pressure, heart rate, body temperature, and glucose levels. For data processing, anomaly detection, and, if necessary, alerting medical specialists, these devices connect to a centralized system. Real-time monitoring, for example, allows doctors to remotely check on the health of patients with long-term conditions like diabetes or heart problems and to intervene early if deterioration is observed. In telemedicine, CPS-based remote monitoring has been shown to be extremely beneficial, enabling ongoing care outside of the hospital environment.

16.7.2.2 Robotic surgery

With the creation of robotic surgical systems like the Da Vinci Surgical System, CPS has completely transformed the surgical sector. Surgeons can now execute minimally invasive treatments with more control and precision owing to these devices. Actuators accurately regulate the movement of surgical instruments in response to input from the surgeon, while sensors integrated in the instruments offer real-time data on tissue resistance and pressure [22–24].

By ensuring a quick, accurate, and dependable feedback loop between the surgeon's motions and the robotic equipment, CPS lowers the possibility of human mistakes while also enhancing patient outcomes. Furthermore, competent surgeons can operate CPS-powered surgical robots remotely to perform treatments on patients in different geographic areas.

16.7.2.3 Personalized medicines

CPS is used in the field of personalized medicine to analyze real-time data and customize treatments for specific patients [25]. To ensure the best possible therapeutic results, intelligent drug delivery systems, for instance, can modify a patient's prescription dosage in response to changes in their condition. These devices have biosensors that identify particular biomarkers in the blood and modify medicine delivery to reduce adverse effects and increase effectiveness.

16.7.3 CPS in industrial automation

The foundation of Industry 4.0, or the fourth industrial revolution, is CPS. In these factories, automated and optimized production processes are achieved through the use of linked systems. Real-time control, predictive maintenance, and flexibility are made possible in manufacturing with CPS, which boosts productivity and decreases downtime.

16.7.3.1 Smart factories

CPS enables real-time manufacturing process monitoring and control in smart factories. Machine sensors gather information on operation, energy usage, and wear and tear. The cyber layer receives this data and uses algorithms to maximize output, cut down on energy use, and minimize waste [22, 25]. For instance, in the production of automobiles, CPS continuously monitors the assembly line, identifying inefficiencies and modifying equipment to guarantee seamless operations. To minimize downtime, the system notifies operators or initiates automated modifications if a machine is not operating at optimal efficiency or needs maintenance [18].

16.7.3.2 Predictive maintenance

The most important advantages of CPS for the industrial field are predictive maintenance. CPS allows condition-based maintenance in case of continuous monitoring of the state of the equipment instead of relying on normal maintenance schedules. Machine learning algorithms examine sensor data [26] in order to identify trends that will indicate whether a machine is wearing, damaged, or about to fail.

This tactic also allows manufacturers to save money from unexpected problems and expensive repairs. For instance, in aerospace, the predictive maintenance systems track engine vibrations, temperature, and pressure. It facilitates airlines to plan repairs well before a problem occurs, which enhances safety and lowers expenses.

16.7.3.3 Human–robot collaboration

Another role that CPS plays is to enable human–robot cooperation in an industrial environment. Cobots are collaborative robots that assist human operators in doing tasks of accuracy, dexterity, or endurance. The robots' sensors monitor their environment so that they can interact with people in a way that won't make them vulnerable.

Cobots help carry out work like soldering, putting small pieces together, and checking the components for defects in electronics production fields. It is ensured by CPS that cobots are able to adapt to changes in the environment and respond fast to human commands [18, 25].

16.7.4 CPS in autonomous transportation systems

CPS is needed to build driverless cars and intelligent transportation network. Based on combination of sensors, live data processing and algorithms, CPS is used to support autonomous driving, V2V (vehicle-to-vehicle) communication, as well as enhancing road safety.

16.7.4.1 Autonomous vehicles

Arguably, the most known use of CPS is in autonomous vehicles (AVs). In fact, these cars depend on a large number of sensors including GPS, radar, cameras, and LiDAR to sense their environment. Real-time processing of acquired data allows for the vehicle to maneuver, collision avoid, make speed, lane change, and route optimization decisions [26]. The CPS architecture of AV makes sure that all the sensor data is examined in milliseconds, which then allows the car to respond quickly to changing traffic situations. Using CPS, Tesla's autonomous driving system continuously collects information from sensors and based on that continuously makes judgments about how to drive in real time.

16.7.4.2 Intelligent traffic systems

CPS is also essential for automated traffic systems that replace individual cars with an automated traffic system that regulates the flow of vehicle traffic in real time to ease congestion and promote safety. In smart cities, the data of vehicle speed, traffic volume, and road condition is collected by the sensors such as those incorporated in traffic lights, public transportation systems, and roads. The door also opens for processing this data to dynamically change or modify traffic light conditions, route traffic to less congested areas, and even help aid the travel of public transportation vehicles. With regard to Barcelona and Los Angeles, problems like traffic congestion and relying too heavily on public transportation can be solved by CPS-based traffic solutions.

The extensive list of CPS applications in Table 16.1 gives a better idea of the diverse range of uses for CPS. It provides examples, use cases, and a range of sectors beyond the previously offered content.

Table 16.1 Key applications of CPS

Industry	*Application*	*Description*	*Example*
Healthcare	Remote surgery	Remote surgery is made possible by CPS, which gives surgeons exact, real-time control over robotic instruments.	The Da Vinci Surgical System makes minimally invasive, remote surgery possible.
Telecommunications	Smart networks	Real-time network resource optimization is managed by CPS, which also handles traffic management, bandwidth adjustments, and failure prediction.	5G networks: Real-time network optimization and dynamic resource allocation.
Logistics and supply chain	Autonomous delivery systems	Drones and automated cars carry products from warehouses to clients, cutting expenses and delivery times.	Starship Technologies: Package and food delivery by self-governing robots.
Public safety	Emergency response systems	In order to coordinate immediate reactions to emergencies like fires, medical emergencies, or natural catastrophes, CPS incorporates sensors and communication technologies.	Real-time data integration for quicker emergency response is known as Smart911.
Education	Smart classrooms	Smart equipment with sensors keep an eye on the lighting, temperature, and student engagement in the classroom.	ClassDojo: Classroom administration and real-time student behavior monitoring.
Entertainment	Immersive virtual reality	Virtual reality (VR) systems powered by CPS generate immersive experiences for entertainment, teaching, and gaming in response to real-time physical inputs.	CPS-driven virtual reality headsets for interactive gaming are called Oculus Rift.
Aviation	Aircraft health monitoring systems (AHMS)	Real-time condition monitoring of airplane parts by CPS helps to anticipate problems and optimize maintenance schedules.	Real-time engine health monitoring is provided by Boeing's AHMS.

16.8 INTEGRATION OF CPS WITH OTHER TECHNOLOGIES

The way industries and infrastructures function has been completely transformed by the convergence of CPS with cutting-edge technologies like blockchain, 5G, AI, ML, and IoT [19]. By enhancing CPS capabilities, these integrations help systems operate with greater autonomy, intelligence, and efficiency. When these technologies are combined, they generate synergistic systems that can manage enormous amounts of data, predict outcomes, enhance security, and ensure real-time performance across a variety of domains. The following sections detail how CPS integrates with important new technologies.

16.8.1 Integration with IoT

IoT is a network of networked devices that connect and share information online. CPS integration with IoT facilitates a seamless connection between digital control systems and physical assets, hence enhancing real-time awareness, control, and decision-making. IoT, along with its distributed architecture, enables real-world data collection, followed by data processing, and hence the use of data to serve as a performance improvement tool for a CPS [19, 20]. IoT-CPS is the ongoing collection of data via sensors in physical objects that transmit this data to the computing elements involved in the CPS. In this way, system responsiveness and efficiency are improved through the ability to monitor and control physical processes in real time. Distributed and decentralized control provides the scalability for CPS since the increasing number of devices that are connected to the system through IoT [24] increases. For example, smart grids use IoT-enabled sensors to collect data as to how much energy is being used in real time. They use this data to redistribute energy, balance the loads, and recognize the renewable energy sources.

16.8.2 Integration with AI and ML

Consequently, AI and ML lead to CPS getting far more intelligent, autonomous, and flexible. Large-scale sensor data processing using AI, ML algorithms to do anomaly detection, predictive maintenance, and real-time decision-making in CPS is facilitated. A particular benefit of supervised learning models to CPS is that they allow systems to optimize operations using historical data and predict equipment failures. Among other things, unsupervised learning speeds up defect diagnosis without requiring labeled data by spotting anomalies, revealing latent patterns in system behavior. Further, reinforcement learning serves to reinforce this even through autonomous systems such as self-driving cars or smart manufacturing, being self-learning by receiving feedback from the environment and improving in this way. Furthermore, Explainable AI (XAI) is integrated into CPS for the critical sectors such as autonomous cars and healthcare, making sure that AI decisions are transparent, understandable, and in compliance with the safety regulations [25]. For the decentralized AI integration with CPS, in particular those possessing autonomous fleets and the healthcare, federated learning (FL) is key due to bandwidth and privacy constraints.

By enabling scattered nodes in CPS to collaboratively train the global models with the protection of privacy and data security, FL shares raw data. For large-scale distributed systems like fleets of driverless cars or smart hospitals where local changes take place and global coordination is necessary, this is extremely useful, as they constitute multiagent reinforcement learning (MARL), algorithms that optimally cooperate without a central controller, whereas CPS agents learn from each other and their environment. In the smart city and robotic swarm scenario where agents have to function together to achieve the best result in a dynamically changing environment, this decentralization is an improvement over CPS scalability and endurance. Further, employing AI in the form of digital twins further enhances CPS by allowing proactive maintenance, predictive modeling, and system optimization via creation of virtual copies of the actual devices in real time. CPS's fault tolerance and self-healing features through its AI-based features make systems automatically self-correct mistakes, thereby making systems more reliable and reducing the downtime in critical infrastructure like smart grids and telecom networks [1, 4].

16.8.3 Integration with 5G network

Massive device connectivity, ultralow latency, and high-speed data that is enabled through the incorporation of 5G networks into CPS leads to revolutionary change. These features are crucial for real-time applications in which data is to be analyzed and decisions to be made in real time, such as industrial automation, smart cities, and driverless automobiles. CPS benefits from 5G's ultrareliability low latency communication (URLLC) feature, which enables CPS to finish time-sensitive tasks whenever possible without having any such disruptions. Here, for instance, 5G allows for real-time communication between vehicles and the road infrastructure in the case of autonomous driving to enable the fastest decision-making to minimize collisions, optimize routes, or manage traffic flow. The real-time connectivity to CPS is very good for remote robotic surgery and remote industrial processes. Moreover, it improves system resilience, whereby it can automatically adjust to changing conditions without human intervention [24, 25].

One of the impressive attributes of 5G in CPS is its ability to provide a massive machine-type communication (mMTC) capacity for seamless connection of the huge installation of IoT equipment. This is especially attractive for smart cities, which have to transfer data produced by hundreds of sensors and devices toward the CPS core for real-time analysis and control.

In a sense, 5G can integrate the distributed data from various IoT devices and optimize energy distribution across the entire city by balancing the use of energy in real time. In addition, 5G enables remote control and teleoperation of CPS for the quick and efficient use of equipment, drones, or cars from a distance. Productivity and safety in logistics, mining, and agriculture sectors increase with it. By integrating 5G networks with CPS, systems turn into a new world with unprecedented levels of scalability, speed, and reliability that enable a large set of real-time, mission-critical applications [25, 26].

16.8.4 Integration with blockchain technology

A safe, decentralized foundation for documenting exchanges of data and transactions amongst CPS components is offered by blockchain technology. Blockchain technology improves data security, accountability, and transparency when it is connected with CPS. This increases CPS's resistance to data manipulation and cyberattacks. Blockchain guarantees that data is auditable and tamper-proof, enabling CPS to safely transfer data between sensors, actuators, and external systems. Due to the distributed ledger's recording of every data transaction, it is very difficult for malicious actors to erase or alter data without being discovered [26].

As an illustration, blockchain-integrated CPS in the healthcare industry guarantees the safe transmission and storage of sensitive patient data obtained by wearable technology, resulting in an unchangeable record of health data that can be shared with various healthcare providers. When CPS is coupled with smart contracts, which are blockchain-based self-executing contracts, it becomes possible to automate specific tasks based on predetermined criteria. For example, when a specific event occurs, such as a change in energy use in response to price signals or environmental conditions, a smart contract might be utilized to trigger a CPS action [27]. For example, supply chain management automates product tracking via blockchain-enabled CPS. When goods move from one place to another, smart contracts immediately update the blockchain with shipment information, ensuring real-time visibility and traceability of commodities.

16.8.5 Integration with edge computing and cloud computing

System performance and scalability are enhanced by the distribution of computational workloads made feasible by the integration of CPS with edge and cloud computing. Cloud computing offers centralized processing capability for large-scale data analysis and long-term storage, while edge computing guarantees low-latency processing close to the data source. By processing data locally at the edge of the network, edge computing reduces the delay associated with cloud-based processing. Because sensor data may be processed quickly instead of being sent to a distant cloud server, this allows for faster administration and decision-making in CPS. For example, edge-integrated, CPS real-time analysis of machine sensor data in smart manufacturing reduces downtime and allows for quick adjustments to manufacturing processes [7, 10].

Cloud computing offers the processing capacity required for large-scale data analysis, machine learning model training, and distributed system administration, while edge computing manages real-time operations. When CPS is cloud-integrated, it can delegate complicated processing jobs to the cloud, including training AI models or optimizing entire systems. For example, CPS in the form of precision agriculture uses IoT sensors to gather crop health data, weather, and soil moisture data. To maximize the crop yield, this data is worked out in the cloud using AI algorithms to forecast the best time to plant and harvest crops [18].

16.9 CHALLENGES IN CPS

CPS, though promising, has a lot of challenges even as it aims to transform industries and increase automation. The reason these problems exist is that there is complexity involved with the physical processes combined with real-time operation of such systems, as well as security and scalability issues in large-scale implementations. To make CPS function reliably, securely, and effectively across a variety of industries, these challenges must be met.

16.9.1 Security and privacy

Security in CPS is one of its most important problems. With CPS managing critical infrastructure like power grids, autonomous vehicles, or healthcare systems, any breach of the system or any cyberattack may have disastrous impacts. The attack vectors on CPS are:

- Violation of data integrity: Malicious parties can modify or change data as it's being sent from sensors to the cyber layer to make decisions that are erroneous and cause system failures.
- Denial of Service (DoS) attacks: These types of attacks attempt to stop CPS components from being able to communicate with one another, thereby delaying or even stopping some essential operations. Such systems, namely, time-sensitive systems such as industrial robots or self-driving cars, are particularly well-suited for this kind of approach.

Attacks, including control hijacking, allow attackers to somehow take control of actuators and adjust physical processes, with destabilizing effects like equipment failure or auto accidents to risky medical procedures. One of the crucial issues in healthcare CPS is privacy. As in healthcare CPS, sensitive patient data is being collected from wearable technology or monitoring systems. The challenge is to keep this data away from an attacker while allowing quick usability in the form of making decisions in real time. Some of the solutions include real-time intrusion detection systems, a decentralized security framework based on blockchain technology, and robust encryption algorithms.

16.9.2 Real-time operation and latency

CPS usually works in settings where instantaneous choices are required to regulate physical processes. Delays in decision-making can have disastrous effects in applications such as industrial automation, robotic surgery, and autonomous cars. One of the biggest challenges in CPS is ensuring low-latency functioning, especially as these systems expand to larger and more complicated contexts.

- Communication latency: Near real-time data from sensors needs to be sent to processors for analysis and conversion into control actions. In this process, network latency can cause the system to respond more slowly, which could have hazardous or less-than-ideal effects. Although 5G networks and other solutions are being used to lower this latency, it is still difficult to maintain dependable low-latency communication in dynamic contexts.
- Computational delays: The computational complexity of implementing AI/ML algorithms, especially those requiring deep learning or reinforcement learning, might slow down decision-making. Edge computing, which allows data processing closer to the sensors, is one way to mitigate this. Maintaining minimal latency while distributing the computing load between the edge and cloud can be difficult, though.

16.9.3 Scalability and complexity

Complexity and scalability problems in CPS can lead to some significant issues as the CPS becomes more widely spread. Strong coordination is required in smart transportation systems, smart cities, and industrial automation between hundreds of networked devices each having its own sensors, actuators, and processors. Due to the size of these large-scale systems, it becomes difficult to manage and control such systems with reliable and real-time responses.

- Overload: With thousands of sensors continuing to generate data, it is very much possible that the data will overload CPS. Organizing data and effective screening methods are required to ensure only the proper data is used in decision-making.

- Complexity: For example, as the number of devices grows, the complexity increases, and there needs to be more coordination between a myriad of components and systems in distributed CPS, such as car networks or smart factories. One of the problems with scaling CPS is that synchronization needs to be preserved, conflicts in decision-making need to be avoided, and consistency in the communication between systems must be ensured.
- Multiple heterogeneous supply chains: This includes microwave hardware devices and heterogeneous software platforms from different sellers. While the creation of standardized protocols and architectures is a work in progress, it is necessary to ensure all systems and devices in a CPS network work together perfectly.

As more cyber-physical systems are deployed, it will be necessary to create scalable architectures and strong coordination mechanisms so that the systems can be depended upon and efficient.

16.9.4 Fault tolerance and reliability

Due to the potentially serious harm, financial losses, or other consequences that a single system malfunction could bring, CPS are widely used in critical situations such as high stakes, where a single system's failure can lead to unsafe outcomes. Consequently, and especially in such applications as energy management, autonomous mobility, and healthcare, it is important to achieve fault tolerance and assure system dependability.

- Dependability of CPS: For the dependability of CPS, errors or malfunctions have to be detected timely. Actuators, sensors, or the communication network must also have self-monitoring so that systems may identify if they are not operating correctly. By integrating AI-based anomaly detection systems into CPS, these errors from the process can be anticipated and identified before they get worse; but it is not easy to give guarantees on the precision and reliability of the detection systems themselves.
- Self-healing systems: This implies that CPS should be designed to heal itself from errors without human intervention. For it to have complete self-healing, sophisticated algorithms need to be developed that can move the system to dynamically isolate and repair malfunctioning components or even communication routes.

There is a multitude of obstacles that CPS must overcome by combining physical and digital processes in real time. There are various critical issues that need continuous research and development in order to make them secure and keep up with real-time speed, able to maintain a similar level of scalability and interoperability, and achieving fault tolerance and power economy. A solution to these issues will therefore be necessary to fully exploit CPS across industries and to ensure that these systems will operate safely, dependably, and effectively in a great variety of dynamic environments.

16.10 UPCOMING DEVELOPMENTS AND STUDIES IN CPS

CPS is very rapidly developing owing to the technological breakthroughs that improve its intelligence, scalability, and autonomy. As CPS becomes seamlessly integrated with advanced technologies such as AI, ML, and next-generation communication networks, it is expected to transform a wide range of businesses. However, the future of CPS also raises serious ethical, legal, and cybersecurity issues. Then this section explores the avenues of the emerging trends and research in the area of CPS.

16.10.1 Advancements in AI and machine learning

The main important area of development in CPS is to integrate deep learning and reinforcement learning. They will help systems learn from their past experiences and come up with sophisticated decisions in unknown environments. For example, in the future, deep reinforcement learning may be used by future autonomous vehicle systems to continuously improve their technique of navigation around obstructions, with minimum fuel consumption, and adapt to changing traffic patterns. Additionally, federated learning is expected to be important for CPS, especially when privacy is of utmost importance. For example, federated learning ensures the privacy compliance required in the healthcare industry by allowing hospitals to train common machine learning models without disclosing confidential patient data, and yet enjoy the benefits of the collectiveness of intelligence. The motivation of this work is to improve the effectiveness of federated learning algorithms so that learning can take place faster with less bandwidth and computer power.

Finally, sectors including manufacturing, energy, and smart cities will see an increase in the use of AI-driven digital twins. With the use of digital twins, CPS is able to model, forecast, and enhance the performance of physical systems by building virtual versions of them in real time. Across all industries, these systems' capacity to self-correct and optimize physical processes in response to AI forecasts will result in more sustainable and effective operations.

16.10.2 5G and Beyond

Beyond 5G, the emergence of 6G networks holds even more potential for CPS. The goals of 6G technology research are to facilitate larger-scale device connectivity, achieve better bandwidth, and even lower latency—possibly in the microsecond range. Real-time holographic communication and immersive applications are anticipated to be made possible by 6G, pushing the limits of CPS in sectors like remote healthcare and smart cities.

16.10.3 Quantum computing in CPS

Though current quantum hardware is still in development, the practical application of quantum computing in CPS remains a long-term aim. The best way to incorporate quantum computing into current CPS frameworks and go from classical to quantum-based systems without affecting operations is still being researched.

16.11 CONCLUSION

Advances in AI, next-generation communication networks (such 5G and beyond), quantum computing, and the development of ethical and regulatory frameworks are all expected to have an impact on the future of CPS. By making it possible for systems to be more intelligent, adaptive, and autonomous, AI and ML will continue to push the frontiers of CPS. More sophisticated CPS applications will be supported by the emergence of 5G and eventual 6G networks, which will offer quicker, more dependable communication over enormous networks of devices. The promise of quantum computing is to solve computational problems that are currently insurmountable for classical systems, while ethical and legal issues will guarantee the responsible and secure application of CPS. To fully realize the potential of CPS and tackle the forthcoming problems, it will be imperative to do continuous research in these domains as CPS becomes increasingly interwoven with essential infrastructure and everyday life.

REFERENCES

1. Khazraei, A., Hallyburton, S., Gao, Q., Wang, Y., & Pajic, M. (2022). Learning-based vulnerability analysis of cyber-physical systems. *2022 ACM/IEEE 13th International Conference on Cyber-Physical Systems (ICCPS)*, Milano, Italy, pp. 259–269. https://doi.org/10.1109/ICCPS54341.2022.00030.
2. Oks, S. J., Jalowski, M., Lechner, M. *et al.* (2022). Cyber-physical systems in the context of Industry 4.0: a review, categorization and outlook. *Information Systems Frontiers*, vol. 1. https://doi.org/10.1007/s10796-022-10252-x
3. Zhang, K., Shi, Y., Karnouskos, S., Sauter, T., Fang, H., & Colombo, A. W. (Jan 2023). Advancements in industrial cyber-physical systems: an overview and perspectives. *IEEE Transactions on Industrial Informatics*, vol. 19, no. 1, pp. 716–729. https://doi.org/10.1109/TII.2022.3199481.

4. Duo, W., Zhou, M., & Abusorrah, A. (May 2022). A survey of cyber attacks on cyber physical systems: recent advances and challenges. *IEEE/CAA Journal of Automatica Sinica*, 9(5), 784–800. https://doi.org/10.1109/JAS.2022.105548.
5. Cook, M., Paterson, C., Marnerides, A. K., & Pezaros, D. (2022). Anomaly diagnosis in cyber-physical systems. *ICC 2022 - IEEE International Conference on Communications*, Seoul, Korea, pp. 5445–5450. https://doi.org/10.1109/ICC45855.2022.9838968.
6. Nickerson, R. C., Varshney, U., & Muntermann, J. (2013). A method for taxonomy development and its application in information systems. *European Journal of Information Systems*, 22(3), 336–359. https://doi.org/10.1057/ejis.2012.26
7. Gulzar, B., Ahmad Sofi, S., & Sholla, S. (2024). Cognitive transformation in personal IoT: pioneering intelligent automation. *Cyber-Physical Systems*, vol. 1, 1–58. https://doi.org/10.1080/23335777.2024.2411736
8. Rotibi, A. O., Saxena, N., Burnap, P., & Read, C. (2024). System-level operational cyber risks identification in industrial control systems. *Cyber-Physical Systems*, vol. 1, 1–32. https://doi.org/10.1080/23335777.2024.2373388
9. Liu, L., Ye, Z., Zhang, D., & Li, Y. (2024). Automated parking trajectory planning: a hybrid curve search and nonlinear optimisation approach. *Cyber-Physical Systems*, vol. 1, 1–28. https://doi.org/10.1080/23335777.2024.2385910
10. Kim, S., Park, K. -J., & Lu, C. (2022). A survey on network security for cyber–physical systems: from threats to resilient design. *IEEE Communications Surveys & Tutorials*, 24(3), 1534–1573. https://doi.org/10.1109/COMST.2022.3187531.
11. Dafflon, B., Moalla, N., & Ouzrout, Y. (2021). The challenges, approaches, and used techniques of CPS for manufacturing in Industry 4.0: a literature review. *The International Journal of Advanced Manufacturing Technology*, 113(7–8), 2395–2412. https://doi.org/10.1007/s00170-020-06572-4.
12. Gil, M., Albert, M., Fons, J., & Pelechano, V. (2019). Designing human-in-the-loop autonomous cyber-physical systems. *International Journal of Human-Computer Studies*, 130, 21–39. https://doi.org/10.1016/j.ijhcs.2019.04.006.
13. Haque, S. A., Aziz, S. M., & Rahman, M. (2014). Review of cyber-physical system in healthcare. *International Journal of Distributed Sensor Networks*, 2014, 1–20. https://doi.org/10.1155/2014/217415.
14. Lasi, H., Fettke, P., Kemper, H. G., Feld, T., & Hoffmann, M. (2014). Industry 4.0. *Business & Information Systems Engineering*, 6(4), 239–242. https://doi.org/10.1007/s12599-014-0334-4.
15. Trunzer, E., Wullenweber, A., & Vogel-Heuser, B. (2020). Graphical modeling notation for data collection and analysis architectures in cyber-physical systems of systems. *Journal of Industrial Information Integration*, 19, 100155–100165. https://doi.org/10.1016/j.jii.2020.100155.
16. Zhao, S., Ge, M., & Xu, Z. (2022). A survey on cyber-physical systems: An overview of architecture, design, and challenges. *IEEE Access*, 10, 45682–45703. https://doi.org/10.1109/ACCESS.2022.3164598

17. Yang, X., Zhou, Y., & Zhu, C. (2021). Toward resilient cyber-physical systems: A comprehensive survey on the critical infrastructure protection. *IEEE Internet of Things Journal*, 8(9), 7243–7258. https://doi.org/10.1109/JIOT.2020.3043339
18. Gunes, V., Peter, S., Givargis, T., & Vahid, F. (2020). A survey on concepts, applications, and challenges in cyber-physical systems. *KSII Transactions on Internet and Information Systems*, 14(2), 1628–1658. https://doi.org/10.3837/tiis.2020.02.019
19. Lin, W., Yang, Q., & Wang, Z. (2023). Edge computing-enabled cyber-physical systems: Architecture, challenges, and opportunities. *Journal of Systems Architecture*, 141, 102481. https://doi.org/10.1016/j.sysarc.2023.102481
20. Nunes, F. R., Dantas, M. A., & Schmitt, A. J. (2023). Cyber-physical systems in smart agriculture: A comprehensive review of technologies, challenges, and future directions. *Computers and Electronics in Agriculture*, 205, 107527. https://doi.org/10.1016/j.compag.2023.107527
21. Chen, X., Liu, M., Li, K., & Zhang, L. (2022). A review of security issues in cyber-physical systems: An industrial control system perspective. *Computers & Security*, 121, 102847. https://doi.org/10.1016/j.cose.2022.102847
22. Zhu, T., Yu, F. R., Wang, Y., Ning, B., & Tang, T. (2020). Big data analytics in intelligent transportation systems: A review. *IEEE Transactions on Intelligent Transportation Systems*, 21(8), 3838–3856. https://doi.org/10.1109/TITS.2019.2948287
23. Das, M., & Morris, T. H. (2021). A survey on cyber-physical system security vulnerabilities, attacks, and countermeasures. *IEEE Internet of Things Journal*, 8(8), 5952–5969. https://doi.org/10.1109/JIOT.2021.3059719
24. Zhao, W., Jiang, C., Gao, H., Yang, S., & Luo, X. (2021). Blockchain-enabled cyber–physical systems: A review. *IEEE Internet of Things Journal*, 8(6), 4023–4034. https://doi.org/10.1109/JIOT.2020.3014864.
25. K. D. Singh, P. Singh, R. Chhabra, G. Kaur, A. Bansal and V. Tripathi, "Cyber-physical systems for smart city applications: A comparative study," *2023 International Conference on Advancement in Computation & Computer Technologies (InCACCT)*, Gharuan, India, 2023, pp. 871-876, https://doi.org/10.1109/InCACCT57535.2023.10141719.
26. Chen, Hong. (2017). Applications of cyber-physical system: A literature review. *Journal of Industrial Integration and Management,* 2, 1750012. https://doi.org/10.1142/S2424862217500129.
27. Khazraei, A., Hallyburton, S., Gao, Q., Wang, Y., & Pajic, M. (2022). Learning-based vulnerability analysis of cyber-physical systems. *2022 ACM/IEEE 13th International Conference on Cyber-Physical Systems (ICCPS),* Milano, Italy, pp. 259–269. https://doi.org/10.1109/ICCPS54341.2022.00030.

Index

For Product Safety Concerns and Information please contact our EU
representative GPSR@taylorandfrancis.com
Taylor & Francis Verlag GmbH, Kaufingerstraße 24, 80331 München, Germany

www.ingramcontent.com/pod-product-compliance
Ingram Content Group UK Ltd.
Pitfield, Milton Keynes, MK11 3LW, UK
UKHW022215190926
475702UK00003B/372
* 9 7 8 1 0 3 2 9 4 1 6 5 3 *